ADVANCED TECHNOLOGIES APPLIED TO TRAINING DESIGN

DEFENSE RESEARCH SERIES

ADVANCED TECHNOLOGIES APPLIED TO TRAINING DESIGN

Edited by

Robert J. Seidel

United States Army Research Institute
Alexandria, Virginia

and

Paul R. Chatelier

United States Navy (Ret.)
Springfield, Virginia

SPRINGER SCIENCE + BUSINESS MEDIA, LLC

Library of Congress Cataloging in Publication Data

Advanced technologies applied to training design / edited by Robert J. Seidel and Paul R. Chatelier.
p. cm. — (Defense research series; v. 4)
Includes bibliographical references and index.
ISBN 978-0-306-44308-4 ISBN 978-1-4615-3014-5 (eBook)
DOI 10.1007/978-1-4615-3014-5
1. Military education — Methodology. 2. Educational technology, I. Seidel, Robert J., date. II. Chatelier, Paul R. III. Series.
U405.A38 1993 92-26766
355.5′0285 — dc20 CIP

Proceedings of a workshop on Advanced Technologies Applied to Training Design, sponsored by the NATO Defense Research Group, Panel 8, Research Study Group 16, held October 22–24, 1991, in Venice, Italy

ISBN 978-0-306-44308-4

Originally published by Plenum Press, New York in 1993

FOREWORD

This collection of papers is the result of a workshop sponsored by NATO's Defense Research Group Panel 8 in the Fall of 1991. The workshop is the second of a series, the first of which was held in the Spring of 1985. As you study these papers, recall that this workshop occurred during the time that many changes were occurring in Eastern Europe and world wide. The need to identify training technologies for maintaining a capable and ready force during times of decreases in military force structure was, and is currently, our challenge. The opportunities for these technologies to provide a service and opportunity for nonmilitary usage is our future. Therefore this workshop maintained its focus on technology and application, regardless of the user. These and other statements made herein are personal and reflect the opinions of the author(s) and in no way represent the official position or policy of our individual governments.

PREFACE

The truly international contributions to this book reinforced our belief that training technology must be collaborative and data widely shared to strengthen our future.

We want to thank the authors of these papers for their abilities to see beyond the near horizon. Their contributions, and the support of the organizations that sponsored their work is greatly appreciated. We also gratefully recognize the contributions of all who attended the workshop. We appreciate their active participation for without this dialogue we would have been unable to guide the Research Study Group into its next phase of work. The stimulating discussions were quite useful in making this book one that we hope will contribute to both the practitioner and user alike. We hope that these contributions will be but one of many that the participants will be able to provide the international training community.

We would also like to acknowledge the professional and organizational support provided by Dr. Milton Katz, Director of the U.S.Army Research Institute Field Office in London, England, Dr. Paul Noja, and his GAJON Institute of Technology, Genoa, Italy.

Special thanks are due Admiral Barachio and his staff at the Arsenale di Venezia. Without their assistance and the use of their historic facility, we would not have been able to have such a successful workshop

R. Seidel
P. Chatelier

CONTENTS

INTRODUCTION: ADVANCED TECHNOLOGIES APPLIED TO TRAINING DESIGN

R. Seidel[1] and P. Chatelier[2]

[1]Army Research Institute
[2]United States Navy (Ret.)

The purpose of NATO workshops are "to establish active exchange of information among member nations and to stimulate the defense applications of research." In 1985, NATO's Defense Research Group sponsored its first workshop on computer based training. During the next 6 years, there have been so many changes in technology, national priorities, and training needs that in 1991, a second workshop was held. This workshop covered advanced training technologies and the products that they could provide to the alliance. The workshop focused on the following 6 technological areas plus existing data base structures that could provide both researchers and practitioners with ready sources of information. The following six technological categories comprise the major sections of the book and are to serve as the initial entry points for the eventual data base:

1. AUTHORING SYSTEMS
2. MODELS
3. EMBEDDED TRAINING & SIMULATION
4. ADVANCED HARDWARE TECHNOLOGY
5. USE OF COGNITIVE APPROACHES
6. EXPERT SYSTEMS

Authoring systems was chosen because it is a very high cost driver for perhaps the most important factor of effective education and training. Most of this cost is driven by the amount of human involvement in traditional training development. This, when multiplied by the variations of procedures and the personal touches added to each course design, prevents the use of efficient authoring techniques. Computer aided techniques offer opportunities to overcome this problem.

The next three sections all relate to what some are calling the technological revolution of education and training. Each of these three categories has experienced exponential growth in capabilities due to the changes in computer and display technologies. Models, while a very imprecise and often misunderstood term, offer great opportunities to reduce error from variations in approach, offer standardized formats, and provide interoperability of training and education. Embedded training and simulation technologies now offer the operators an alternative to having to buy expensive dedicated training equipment. Such embedded training capabilities also offer the user the chance to choose if they want to use the real equipment with real signals or simulated signals. These decisions are decided on the basis of safety, environmental impact, and cost. Advanced hardware technology is often the limiting factor in the application of advanced training and education software. It was critical that researchers focus on what delivery system capabilities were anticipated to be available during the next decade. Without such planning, it is all too common to design a training capability that fails to meet the advances and constraints of the next generation delivery systems.

Advanced Technologies Applied to Training Design, Edited by
R.J. Seidel and P.R. Chatelier, Plenum Press, New York, 1993

The use of cognitive approaches and expert systems were the two subjects most solidly a part of academia and basic science. The papers and discussions on cognitive approaches helped the scientists to understand some of the newer theoretical underpinnings to improve and to further design improvements using better authoring, embedded training, and models. In addition, this section lays the foundation for discussions in the last section on expert systems, especially on neural networks.

The section on expert systems focuses more on application and use than on theory since the root of all expert system technology is pure computer science and its theory was considered well outside the scope of this workshop which was to provide guidance for improving training design.

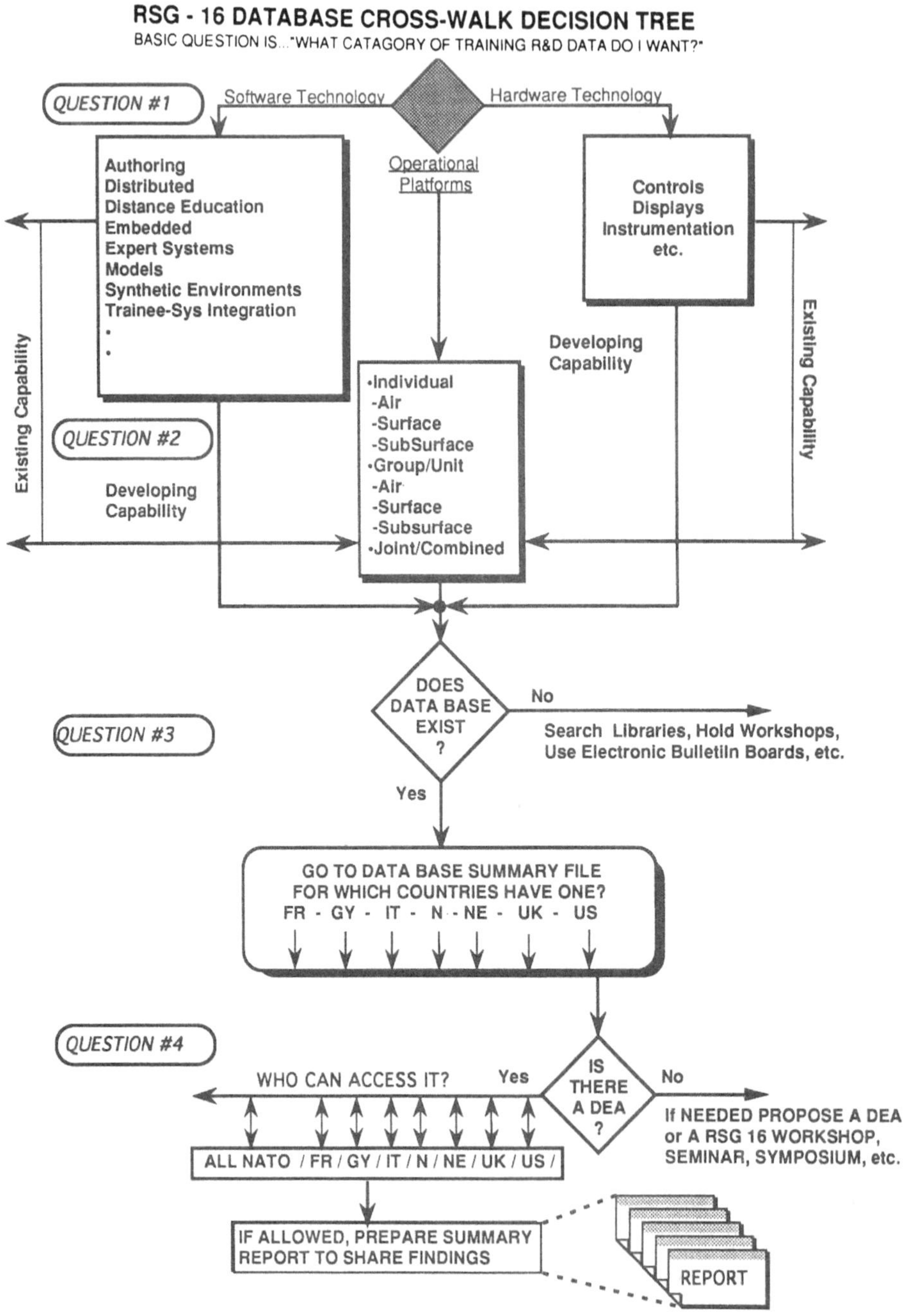

Figure 1

However, all of these most stimulating chapters are short lived in eventual usefulness if the data and information are not continually available. There is so much valuable knowledge available throughout the alliance that a usable system for recording, and retrieving needed information is a major requirement. Each country has or is developing its own repository of education and training data. Each of these repositories have their own constraints regarding access and their own unique data fields and categories which are a part of their culture.

It is not the intent to try to standardize on any one data base type, format, etc. Rather one of the objectives of our effort is to develop a data inquiry system that would be a tool for enabling multi-cultural, database, cross-walks to direct those in search of data to the existing repositories. The value of such a product is that an architecture and a format are provided so that a NATO user, whether trainer or system designer, would have available the latest information to make his job more effective and efficient. This also provides the NATO researcher the opportunity to update any agreed upon data base by adding new technologies or new findings. Therefore, it is not time-linked as are most summaries of literature gathered at this point.

In order to fully appreciate the practical value of the data base architecture we are developing, we provide a hypothetical tour for the reader. The accompanying flow chart (Figure 1) illustrates the process. Note that the questions in the text are referenced in the chart as well.

If you were a user of the Advanced Technology data base the following scenario describes the ways in which you might traverse its structure. We make two assumptions concerning your needs and then we follow with possible questions and answers to assist you in understanding how to use the data base.

ASSUMPTION #1: That someone wants to find out if there are training systems related data/information available for their use.

ASSUMPTION #2: The person(s) in search of the data/information has or can obtain the appropriate "need to know."

Question 1.0: AM I INTERESTED IN SOFTWARE TECHNOLOGY, HARDWARE TECHNOLOGY OR AM I NOT SURE SINCE I HAVE TO WORK ON A TRAINING PROGRAM RELATED TO A TYPE OF OPERATIONAL MILITARY PLATFORM?

Discussion: If I am interested in hardware or software technology I then have to determine what type or category of data I am interested in or might I use a data dictionary with key words.

After I have chosen the appropriate entry point, I must also decide if I am interested in what exists now (existing capability) or if I want to know what is under development. Of course the R&D person might desire to know both. At any rate this leads us to Question 2.0.

Question 2.0: AM I INTERESTED IN WHAT CAPABILITY EXISTS AND/OR THAT WHICH IS UNDER DEVELOPMENT?

Discussion: If I am interested in what exists and I was rather confident that I knew what to look for, I would have to go immediately to the next question (Q 3.0) concerning the existence of a data base. However, if I was not sure what to use as a search category or what to ask, I might go to the third list of search categories which are based on operational platforms.

For example, I might have been asked to provide an embedded training capability for a tank. I might look at "Jane" listing of training equipment and find that the Navy has embedded training on its destroyers so I decide to study what they have done and determine if the technology is transferable to the Army. If the answer is yes, I go to question 3.0. If it is not I change my search technique to look at developing capability and then go to question 3.0 but with a different "set of parameters and expectations

This same strategy would be true if I was interested in hardware such as a thin film EL display or some non-intrusive instrumentation for use on a training range, etc.

A point of difficulty might be to determine if it is hardware or software. For example, voice recognition systems. Some people might think of it as software (the algorithms) and others the computer equipment (box with algorithms, displays, etc.). In these cases, I might want to search both technologies unless I was clearly focused on a specific issue or need.

Question 3.0: DOES A DATA BASE EXIST THAT WOULD HELP ME ANSWER THIS QUESTION?

Discussion: If the answer is YES, then I would have to be prepared to look consider such looking for pointers to the data base. In some cases the pointer might be located in a service, so I would have to look to the Army, Navy, Air Force. (e.g., the Royal Air Force (RAF) has a good simulator data base). Perhaps I have to look at equipment data base in general (the United States Training and Performance Data Center (TPDC) has a good training equipment data base). Or I might look at R&D for an appropriate data base. From any of these entry points, I would eventually find myself having to focus on the country or countries that own the data bases. This would then lead me to Question 4.0.

If the answer is NO then I would have to decide how important these data are to me and to other members of the alliance. If they are very important and there is a consensus or majority of the RSG 16 members agree, then a Workshop, Seminar, or Symposium should be proposed to focus on the issues related to these data and perhaps the risk or effectiveness of the solution.

Question 4.0: DOES A DATA EXCHANGE AGREEMENT EXIST BETWEEN THE COUNTRY ASKING THE QUESTION AND THE COUNTRY/COUNTRIES HAVING THE DATA BASE?

Discussion: If the answer is YES, then arrange for either bilateral or multilateral access to the data. If it is only Bilateral, then appropriate security and data sharing rules must be followed and RSG 16 might not be able to either provide any assistance or prepare a summary report showing who had what type of data. It is hoped that the existence of such a data base is not too controlled so that at least the members of the alliance can know who had such a data base and if they need access can work with their government to obtain the appropriate clearance to get the data. If it is an open or Multilateral access data base then RSG 16 should consider preparing at least a summary report about the data base and provide as much information as possible to the US's Defense Technical Data Center in Alexandria, VA. Since this data base is accessible to all members of the alliance, it would serve as an excellent hub for training related data.

If the answer is NO, then we would follow the same procedures listed in Question 3.0. I would have to decide how important these data are to me and to other members of the alliance. If they are very important and there is a consensus or majority of the RSG 16

DATA BASE SURVEY SHEET

SERVICE

- Army
- Navy
- Air Forc
- Marine

OPERATING PLATFORM

- Aircraft
- Ship
- Submarine
- Ground Vehicle
- Space System
- Ground Weapons
- Missiles

TYPE REPORT

- Technologies
- Critical Data
- References
- R&D
- T&E

FOR SELECTING TECHNOLOGIES

HARDWARE

- CAI
- I.V.D.
- DVI
- Simulators
- Workstations
- Virtual Displays

APPLICATIONS

- Propriatory Systems
- Authoring Systems
- Multi-Media Systems
- Models
- Embedded Training
- Cognitive Models
- Expert Systems
- Constructive Simulations
- (War Games)
- Logistics
- Others

SOFTWARE (MODELS)

- Human Interaction Design
- (Head, Time, Space, etc.)
- Human Performance
- (Individual, Group, Team, HQ)
- Curriculum Development
- Semi Automated Force

FOR SELECTING CRITICAL DATA

STATUS

- R&D
- Operational
- Multi-uses
- Multi-users

EVALUATION

- Effectiveness
- Time Reduction
- Costs
- Cost Effectiveness
- Career Retention

ENTRY LEVEL

- No Experience
- Apprentice
- Expert

Figure 2

RSG 16 DIRECTORY of DATABASES

CURRENT DATE [ ] PC Type ☐ Mac ☐ DOS

DATA BASE TITLE

COUNTRY OF ORIGIN

TYPE DATA BASE

- ☐ E'Mail
- ☐ Hard Copy
- ☐ Soft Copy
- ☐ Optical
- ☐ Classified

KEY WORDS

POINT(S) OF CONTACT:

DATA BASE CATEGORIES

- ☐ Advanced Hardware
- ☐ Authoring System(s)
- ☐ Cognitive Processes
- ☐ Distance Learning
- ☐ Embedded Training
- ☐ Expert System(s)
- ☐ Models
- ☐ Synthetic Environments
- ☐ Team/Group/Unit Training
- ☐ Individual
- ☐ Simulators/Training Devices
- ☐ Reserve Training
- ☐ Army
- ☐ Navy
- ☐ Air Force

BRIEF DESCRIPTION OF DATA BASE :

Figure 3

members, then a Workshop, Seminar, or Symposium should be proposed to focus on the issues related to these data and perhaps the application or risk or effectiveness of the solution.

The proposed elements underlying this crosswalk could be organized in a hierarchical structure as shown in Figure 2, data base guide. As one might perform on an on line search, he would use a mouse to click at the highest level on identification of Service, Operating Platform, and Type of Report desired. At the next level of the hierarchy the user would be able to link to the above by clicking on the Technologies; HARDWARE, APPLICATIONS, and SOFTWARE (including MODELS). Finally, the user would click on the CRITICAL DATA ELEMENTS of interest linked to his previous choices in the hierarchy This hierarchical structure is ideally what we would desire for our RSG 16 data base. However, the extent to which this is obtainable would depend on the ability to standardize it on a relational data base program. The outcome of such a search process could be represented as shown in Figure 3, the RSG 16 Directory of Data Bases.

The last chapter in this volume provides a discussion of the advantages of an integrated data base architecture. It also gives some detailed descriptions of U.S. data bases which might be accessed by a NATO inquiry system which could also act as a catalog of data bases.

It is our intent to provide the readers with both the latest in advanced technologies that the NATO Defense Research community has been able to bring together and provide a quick reference source to both articles, processes and points of contact for you. The editors are anxious to hear your constructive comments regarding this book.

SECTION I
ADVANCED AUTHORING TOOLS

The goal of this section is to point out how advances in authoring can reduce the cost of creating training materials whether they be for simulators, embedded training, CBI, or a combination of these delivered with the latest in hardware technology.

EVALUATION OF CBT AUTHORING SYSTEMS: NEED A DATA BASE

Robert J. Seidel and Ok-choon Park

U.S. Army Research Institute for the
Behavioral and Social Sciences
Alexandria, Virginia, USA

ABSTRACT

A computer-based training (CBT) authoring system is a special kind of software tool-kit for training developers with no programming skills and limited instructional design experience to use in developing CBT materials (i.e.,courseware). Many authoring systems have taken into account the needs of computer-naive users so that the learning requirements for using the systems is supposed to be minimal. However, the selection of the best authoring system for a given purpose requires a systematic evaluation because many systems have been developed to serve basically the same purpose with different procedural approaches, functional features and power. The purposes of this paper are: (a) to discuss evaluation methods for selecting an authoring system; (b) to examine important evaluation criteria; (c) to propose a set of evaluation procedures; (d) to discuss the need for developing a CBT authoring tool data base and provide initial inputs for the development of the data base.

INTRODUCTION

Most of today's computer users do not have the time to develop programming skills. They need to select and use the most appropriate software tools for their information applications. Most of recently developed software tools have taken into account the needs of computer-naive users so that the learning requirements for using the tools is supposed to be minimal. However, the selection of the best software tool for the given purpose requires a systematic evaluation because many software tools have been developed to serve basically the same purpose with different procedural approaches, functional features and power.

A computer-based training (CBT) authoring system is a special kind of software tool-kit for training developers with no programming skills and limited instructional design experience to use in developing CBT materials (i.e., courseware). A complete authoring system should provide aids for the training developer through the entire process of a system approach to training (SAT) development, including front-end or training need analysis, analysis of training program requirements, design and development, formative evaluation, delivery/implementation, and summative and maintenance evaluation. However, it is difficult in a single system to build all the aids necessary for the required activities in every stage of the process because quite different types of tasks are required for the different stages. If all aids are built in, the system would become too complex and sophisticated for most training developers to learn and use. Therefore, it is recommended to develop an independent tool for each of the stages and to integrate them in the operating hardware system so that they can interface with each other.

Advanced Technologies Applied to Training Design, Edited by
R.J. Seidel and P.R. Chatelier, Plenum Press, New York, 1993

In a broad sense, an authoring system includes the various types of automated tools for supporting activities required in the different stages of SAT. However, the definition of an authoring system in a narrow sense is usually limited to a specific type of automated tools directly used for developing (i.e., programming) courseware. The evaluation criteria and evaluation procedures discussed in this paper are to select a CBT authoring system in the narrow definition.

The selection of a good authoring system is one of the most critical factors for the success of computer-based training (CBT) development (Avner, Smith & Tenczar, 1984; Fairweather & O'Neal, 1984). To provide systematic information for the selection of an authoring system, a number of evaluation studies have been conducted (e.g., Gillingham, et al., 1986; MacKnight & Balagopalan, 1988-89; cf., Journal of computer-Based Instruction, a special issue on authoring tools, 1984; Young, 1984). However, most of these studies have reviewed the authoring systems as a form of software product rather than as a form of user-oriented tool. Although some studies comparatively analyzed the important functional features of authoring languages (e.g., Gillingham, et al., 1986) or systems (e.g., MacKnight & Balagopalan, 1988-89), the systems evaluated were selected based on the availability of the systems and the evaluator's convenience. Thus, only a few systems (three or four in most cases) were included in the evaluation and the evaluation procedures were not systematic.

The purpose of this paper is: (a) to discuss evaluation methods for selecting an authoring system; (b) to examine important evaluation criteria; (c) to propose a set of evaluation procedures; (d) to discuss the need for developing a CBT authoring tool data base and provide initial inputs for the development of the data base.

EVALUATION METHODS OF CBT AUTHORING SYSTEMS

Although several methodological approaches to the evaluation of CBT authoring systems have been proposed, most evaluations have been conducted with a checklist method or a benchmark method.

Locatis and Carr (1985) proposed a checklist for selecting an authoring system. However, this checklist has four basic problems. First, questions in the checklist are too abstract and ambiguous. For example, the checklist asks "Does the system provide adequate levels of authoring flexibility?". In this question, it is not clear what "authoring flexibility" means. It could mean a flexibility to develop various formats of courseware materials (e.g., tutorial, games, simulations, etc.), or to adopt various instructional principles and strategies, or to interface with other software development tools, or to edit and modify already developed materials, or to adapt to various levels of user skills and experience. Many questions in the checklist have similar problems and are difficult to use as comparative criteria for selecting a system among many alternatives. Second, the checklist ignored some important factors that should be evaluated (e.g., instructional design aids, instructional strategy selection aids, etc.). Third, the checklist asks questions mostly about a single dimension of the authoring feature in a dichotomous form (i.e., is a given function present or not?). Other dimensions are ignored such as the ease of using the specific function, level of the functional sophistication, etc. Fourth, most parts of the evaluation are subjective judgements because no comparative criterion is available and evaluator's personal preference influences the judgment.

Another approach to conducting a comparative evaluation of authoring systems is to develop benchmark lessons using the systems (Hillelsohn, 1984, Hunter, 1987). An advantage of the benchmark approach is the ability to obtain a variety of empirical evaluation measures of the systems (e.g., development time, difficulty to learn and use the system, functional capability to handle specific features, instructional design helps, the quality of the developed materials, etc.). However, the benchmark approach also has several problems: (a) It is generally not feasible to expand the benchmark to include every possible characteristic required to develop all different types of courseware materials. Most lessons to be developed using an authoring system are likely to require only limited system features because most benchmark materials developed for evaluation are segmented portions of a lesson and the function for integrating them into a complete lesson structure is frequently ignored. (b) Evaluation results are influenced by the similarity and difference between the evaluator's prior experience (including instructional design orientation) and the instructional development procedures built into the system. (c) The evaluator's experience in one system influences his/her use and evaluation of other systems. Although some other methods exist like key-stroke measures (Locatis & Carr, 1988) and expert review, most evaluations of CBT authoring systems have been conducted with a checklist or benchmark method. However, the system features and characteristics assessed in the checklists and in the benchmark lessons are varied among the evaluators mainly because no systematic guidelines to develop the checklist items or benchmark lessons has been available.

CRITERIA FOR SELECTING CBT AUTHORING SYSTEMS

Evaluation criteria for selecting an authoring system should be determined on the basis of unique needs for the given situation (see "Evaluation Procedures"). However, there are some essential requirements that an authoring system should have and be examined in the evaluation process. They are: (a) built-in instructional design aids; (b) built-in development (programming) aids; (c) expandability of the system capability; (d) user friendliness. In addition, (e) cost/ affordability is an important factor.

Instructional Design Aids

The process of courseware development consists of two basic phases: designing and developing (i.e., programming). The design specifications of courseware to be developed are usually provided in a document form. Thus, most authoring systems mainly provide programming aids to develop the courseware as specified in the design document. However, the design document prepared in advance is usually not specific enough to develop the detailed characteristics of the courseware. In addition, the document itself needs frequent revisions during the development process. Therefore, instructional design aids is one of the primary requirements for an authoring system. Important instructional design aids include the following:

1. Aids for Laying Out the Courseware Structure. A typical aid for the first stage of courseware development is to provide a framework in which representations (e.g., titles, icons, etc.) of the courseware components (e.g., displays such as text and graphics, and interaction procedures) can be arranged into an organized structure. The framework should be flexible enough to support the following requirements:

- design single or combined types of CBI formats (e.g., tutorial, drill and practice, games, simulations, modeling, etc.) for various domain contents;
- design any size of courseware by allowing the organization of the representations into unlimited levels of a hierarchical structure;
- reorganize the courseware structure from the rearrangements of the representations in the design framework;
- develop or revise the courseware structure and components at any time.

2. Aids for Selecting Instructional Strategies. This aid allows the author to select various types of instructional strategies from the built-in options. Representative examples include types of questions, feedback, sequence, etc. (for a taxonomy of computer-based instructional strategies, refer to Park, 1983).

3. Provision of Templates for Developing Specific Lesson Components. This aid provides a pre-formatted screen form which the author uses to develop a specific courseware component by filling in the blanks.

4. Aids for Selecting Specific Data Collection Options. This aid allows the author to select from the built-in options various types of data to be collected during the courseware implementation or to define new types of data.

Development (Programming) Aids

An authoring system should provide users without programming skills aids to develop various types of courseware components. Even if programming skills are available, it should be more efficient to use the aids. The courseware components can be divided into two main categories: (a) presentation displays and (b) student-computer interaction specifications.

1. Aids for Developing Presentation Displays. An authoring system should have the following utilities to develop various types of presentation displays without requiring programming skills:

- A text editor
- A graphic (including animation) editor
- A sound generator/editor
- A video production aid

2. Aids for Developing Interaction Procedures. These aids allow the author to specify the student-computer interaction procedures, including the sequence of the presentation displays. The following is the list of specific aids that the author may need to define the interaction procedures:

- An aid for selecting question types and required response modes
- An aid for specifying the number of questions to present
- An aid for specifying the presentation order of questions
- An aid for specifying the replacement of question items
- An aid for specifying number of trials allowed for the student to answer to a given question
- An aid for limiting the student's response time
- An aid for specifying types of feedback
- An aid for specifying the time to give the feedback
- An aid for specifying branching control options (learner control versus program control)
- An aid for selecting student input devices

3. Aids for developing data collection and management procedures. These aids provide built-in facilities to specify what kinds of data should be collected during the instruction and how the data should be used and managed. The data are collected for two general purposes: instructional diagnosis and training management decisions. Important data collection and management aids include the following:

- An aid for collecting student input data
- An aid for collecting student on-task time data on various levels of instruction
- An aid for defining and generating new data

Expandability of the System Capability

It is not feasible to build all the features and functions required in courseware development and delivery into a single authoring system because of the technical complexity to build such a system and the difficulty involved in learning and using it. A solution to this problem is to extend the system's functional capability by allowing the system to have flexible interfaces with other software tools. The specialized software tools are usually more powerful and convenient for handling the special functions than an authoring system designed for many different functions. Also, it is important that the authoring system has a capability to extend the hardware system's inherent capability by allowing the operating system to interface with other hardware systems.

1. Capability to Interface with Other Software. Important types of software that an authoring system should be able to interface with include the following:

- Other development tools (e.g., word processor, graphic editor)

- One or more programming languages (e.g., BASIC, PASCAL, C)
- Built-in utilities in the operating system (e.g., calculator, clock)
- Utility software programs (e.g., spreadsheet, HyperCard)
- Training management system (e.g., test developer, computer managed instructional system)

2. Capability to Interface with Other Hardware Systems. An authoring system can extend the hardware system's inherent capability by having the following capabilities:

- Interface with selected hardware peripherals (e.g., interactive video disc, printer)
- Interoperability among independent hardware systems
- On-line communication and network capability
- Use of variety of input devices

User Friendliness

Learning and using an authoring system require significant effort and time. Therefore, degree of user friendliness is an important factor in evaluating an authoring system. Criteria for evaluating user friendliness include the following:

- Flexible changes of design and development modes
- Ease of development and revision procedures
- Provision of built-in development templates
- Ease of integrating the developed materials: Overlays and lesson integration
- Adaptation to author skill levels
- Development in the student mode
- Help features built into the system
- An adequate manual

Cost/Affordability

The cost factor needs to be examined in terms of the affordability and cost-benefit of the system. Important considerations for examining the cost factor include the following:

- Hardware requirements (main system and configuration requirements)
- System price
- Contract terms
- Training expenses

Most of the above evaluation criteria are explained in detail in Park and Seidel's paper (1989).

Park and Seidel have proposed to use the above criteria in the development of an evaluation checklist and in the selection of courseware characteristics to be developed in a benchmark evaluation. However, many other variables unique to the given training development situation should be considered in the evaluation. Important variables to be considered include the training and experience of the system users (i.e., training developers), general characteristics of the subject domains, instructional goals and objectives to be achieved with the courseware, general types and formats of courseware to be developed, existing hardware and software, and the expected period and frequency for using the system.

PROCEDURES FOR EVALUATING AUTHORING SYSTEMS

Although a comprehensive set of criteria for evaluating authoring systems is provided as listed in the previous section, the mere application of the criteria to the development of the checklist or to the specification of courseware characteristics for a benchmark method would not produce desirable results per se, because each of the method falls short by itself. Thus, we propose to combine the two approaches, checklist and benchmark methods. A simple combination or simultaneous use of the two methods does not automatically eliminate the problems inherent in the methods. Functional features to be included in the evaluation checklist and benchmark lesson-development should be selected carefully not only to minimize the limitations of each method, but also to complement the advantages of the other method as much as possible.

The specific evaluation procedures for selecting an authoring system should be determined after examining important situational variables. The following is a general guideline for developing a comparative evaluation procedure that combines a checklist method and a benchmark method in a way that their strong features complement each other.

First, list all the important situational variables for using an authoring system and describe the current status of the variables.

Example: Intended Users (Training or Instructional Developers)

> Knowledge in instructional theories and strategies - *high (e.g., graduate degree in Instructional Systems)*
> Experience in instructional development - *average 4 years*
> Experience in the use of authoring systems - *None (e.g., no experience in any authoring system)*
> Computer programming skills - *little (e.g., a few short programings with BASIC)*

Second, list five main evaluation criteria and their subcriteria presented in the earlier section of this paper, and give a weight to each of the subcriteria on the basis of the analysis results of the important situational variables as presented in the first stage (for example, very important - 5; important - 4; somewhat important - 3; little important - 2; not important - 1).

Example: Instructional Design Aids

> Aids for laying out courseware structure - *3 (The user is familiar with instructional design).*
> Aids for selecting instructional strategies - *3 (The user has in-depth knowledge of instruction and learning.)*
> Provision of development templates - *5 (The user does not have programming skills.)*

Third, check whether the given authoring system to be compared has each of the subcriteria listed in a checklist.

Example: Instructional Design Aids

> Aids for laying out courseware structure - *No*
> Aids for selecting instructional strategies - *No*
> Provision of development templates - *Yes*

Fourth, assess the functional capability for performing each of the sub-criteria from the analyses of user manuals and/or from user surveys, and provide justifications for the assessment.

Example: Graphic Editor

> Aid for developing an animation - *limited to very simple ones (Allows one to develop only squares, lines, circles)*
>
> Ease of using the animation development aid - *difficult to use (must use programming language-like commands; need to specify every location to move the object by pixel numbers)*

Fifth, analyze the evaluation results obtained in the above stages, and select the one or two best system(s) on the basis of the evaluation results, cost factors listed in the evaluation criteria, and other important situational variables like the frequency and expected period of the use.

Sixth, if it is necessary to empirically validate the evaluation results, conduct a benchmark evaluation as follows:

(a) Design a lesson (including detailed storyboards) which represents the characteristics of most courseware materials that need to be developed in the given situation.

(b) Train one or two persons to use the selected authoring system. The representative user selected for a benchmark evaluation should not have much experience in the system since information concerning the naive user's needs to learn the system should be collected. Through observation, prepared questionnaire, and/or interview, evaluate the difficulty level and required time to learn the system. This result should be interpreted in comparison with information provided by the vendor.

(c) Have the trained persons develop the lesson designed in (a). Through observation, prepared questionnaire, and/or interview, evaluate the difficulty level and required time to develop the lesson. The evaluation data should be collected for every important segment or object in the lesson.

Seventh, after checking the consistency between the checklist evaluation results (step one through five) and the benchmarking results obtained in step six, select a system from those being considered. If the checklist and benchmark evaluation results are not consistent (particularly if the benchmarking results do not support the checklist results), the outcomes should be reexamined to determine whether they resulted from a weakness in the specific function of the system or the developer's inability. This can be reexamined by having another person develop the segment of the lesson following the same procedures.

To review, the final selection decision should be made not only on the basis of the evaluation results of the system's functional capability and ease of learning and use, but also on the basis of cost factors and other important situational variables. The evaluation criteria and procedures also should be applied selectively or modified according to the unique needs of the given situation.

NEED FOR AN AUTHORING SYSTEM DATA BASE

According to our estimate, about two hundred authoring systems are currently available in the market. Among these systems, prices, functional capabilities and required hardware are widely different. Thus, the initial decision for selecting an authoring system is to choose the systems to be evaluated. Because of the time, skills and expenses required to evaluate authoring systems, the minimum number of systems should be selected for the evaluation. For the initial selection of the systems to be evaluated, the prospective user must rely on descriptive information mainly provided by vendors, partially available evaluation information, and recommendations from experts and some experienced users.

In order to provide some basic information which can be used for selecting the evaluation candidate systems, a number of authoring system directories (e.g., T.H.E. Journal, 1988; Weingarten Publications, 1988) have been published. This kind of directory information would be helpful if it is restructured and stored in a data base so that the user could selectively access to the information according to his/her needs.

The second type of information that can be stored in the data base is evaluation results. Although several evaluation studies (Gillingham, et. al., 1987; MacKnight & Balagopalan, 1988-89, Journal of Computer Based Instruction special issue on authoring systems, 1984) have been published, the reported information is partial and subjective. Also, every CBT development situation is unique. Therefore, the final selection of an authoring system should be made through a systematic evaluation in which unique situational requirements are adequately considered. However, existing evaluation information may be useful for selecting the initial pool of candidate systems and for developing evaluation procedures and instruments. Thus, the evaluation information, including results and procedures, can be stored systematically in a data base.

Development of a data base for storing descriptive information and evaluation results of many authoring systems requires many decisions concerning the data-base structure and procedures for accessing and retrieving the information.

The descriptive information which is mostly provided by the vendors could be organized in the following structure:

- The tool and vendor information

  1) Name of the tool
  2) Vendor's name, address, and telephone and fax number
  3) Price

- The type of authoring tools

  1) Authoring systems
     a) Main-frame computer based
     b) Mini-computer based
     c) Micro-computer based
        (a) IBM pc and compatible
        (b) Apple family
        (c) Macintosh
  2) Authoring languages
  3) Other authoring aids
     a) Computer-managed instructional system
     b) Instructional design aids
     c) Knowledge acquisition aids
     d) Test generator

The CBT Guide (Weingarten Publications, 1988), a paper-based directory of 80 different authoring systems and languages, includes various descriptive types of information. The Guide first categorized authoring systems into four main groups according to the types of base hardware systems:

- Micro-based systems
- Apple-based systems
- Mini-based systems
- Mainframe-based systems.

Authoring languages and other authoring tools are added as separate groups. Each authoring system in the Guide has basic descriptive information such as:

- The company which markets the system
- Authoring mechanism (e.g., menu-driven, icon-based, authoring language access, etc.)
- Required hardware for authoring and delivering
- Recommended training methods (e.g., manual, class, etc.)
- Maintenance supports
- Cost

Also, each system has descriptive information about some important functional capabilities such as:

- Basic method and procedure for response analysis
- Options for student control
- Basic programming tools (e.g., built-in editors, external editors, screen capture, etc.)
- Basic method of graphic development (e.g., bit-mapped, character, etc.)
- Animation
- Use of interactive videodisc

This kind of information needs to be considered when developing a computerized data base for authoring systems, although the structure of a computer data base should be different from that of a paper-based directory.

Evaluation information can be categorized into different types of evaluation studies (e.g., checklist, product review, benchmark, no evaluation, etc.). It may include not only measured ratings of important functions, but also deficiencies addressed in the evaluation process. Before developing the data base, all of the necessary decisions (including the base hardware system to develop the data base, types and formats of information to be included, procedures for accessing, sorting and retrieving necessary information) should be specified in a design document.

Finally, procedures for maintaining the data base, including modification of the existing information and addition of new data, should be specified. The distribution responsibility and the maintenance effort also should be considered.

An initial effort to develop a CBT authoring tool data base has begun in the U. S. A. at the National Library of Medicine (NLM) (Lo, Locatis, Ullmer, Car, Banvard, Le, and Williamson, 1991). At present time, the data base consists of about 90 authoring systems and the information included is mainly vendor-provided system descriptions. Authoring systems in the data base are organized in an alphabetical order and they can be reviewed in the sequential order. However, the data base has a menu-based search capability and the systems can be selectively reviewed by choosing a value for each of the given menu parameters: (a) supporting hardware; (b) graphic capability; (c) record keeping capability; (c) capability to develop different types of instructional programs; and (d) cost. The sample description of an authoring system in the NLM's data base is attached.

In order to avoid the overlapping effort to develop the CBT authoring systems data base, the NLM and the U.S. Army Research Institute (ARI) are currently negotiating the procedures and responsibilities to upgrade and maintain the NLM's data base. We are planning to upgrade the data base by incorporating information presented in the CBT Directions (Weingarten Publications, 1988; Raybold, 1990), T.H.E. Journal's Directory of Authoring Software (1988, 1990), Instructional Delivery Systems' Directory of Authoring Systems (1991), information supplied by vendors, and other descriptive and evaluative information published in journals and magazines (e.g., Dean, 1987a, 1987b, 1988). Also, Twenty nine authoring systems from the NATO member countries have been provided by France for the candidate additions to the data base. We look forward to the possibility of making this data base more dynamic and comprehensive with opportunities to share updated versions with all interested users.

REFERENCES

ADCIS. A special issue on authoring systems. Journal of Computer-Based Instruction, 11, 67-102, 1984.

Avner, A., Smith, S., & Tenczer, P. CBI authoring tools: Effects on productivity and quality. Journal of Computer-Based Instruction, 11, 85-89, 1984.

Dean, C. Authoring systems for CBT - Part 1. Training Technology, 12-16, 1987.

Dean, C. Authoring systems for CBT - Part 2. Training Technology, 16-20, 1987.

Dean, C. Authoring systems for CBT - Part 3. Training Technology, 11-14, 1988

Gillingham, M., Murphy, P., Cresci., K., Klevenow, S., Sims-Tucker, B., Slade, D., & Wizer, D. A evaluation of computer courseware authoring tools and a corresponding assessment instrument for use by instructors. Educational Technology, 26, 7-18, 1986.

Hillelsohn, M. J. Benchmarking authoring systems. Journal of Computer-Based Instruction, 11, 95-97, 1984.

Instructional Delivery Systems. Directory of Authoring Systems. Instructional Delivery Systems, 5, 16-23, 1991.

Locatis, C., Ullmer, E., Car, V., Banvard, R., Le, Q., & Williamson, M. AuthorBase: A database of authoring systems software. Proceedings of the 15th Annual Symposium on Computer Applications in Medical Care, Los Alamitos, California: IEEE Computer Society Press, 1991.

Locatis, C., & Carr, V. Selecting authoring systems. Journal of Computer-Based Instruction, 12, 28-31, 1985.

MacKnight, C. B., & Balagopalan, S. Authoring systems: Some instructional implications. Journal of Educational Technology Systems, 17, 123-134, 1988-89.

Park, O., & Seidel. J. R. Evaluation criteria for selecting a CBI authoring system. T.H.E. Journal, 17, 61-68, 1989.

Park, O. Instructional Strategies: A Taxonomy. Minneapolis, MN.: Control Data Corporation, 1983.

Raybold, B. Choosing the right hypertext product for performance supports, CBT Directions, 3, 13-20. Boston, M.A.: Weingarten Publications, 1990.

T.H.E. Journal. Directory of authoring software. T.H.E. Journal, 15, 19-24, 1988.

T.H.E. Journal. Authoring Packages Continued to Mature. T.H.E. Journal, 17, 10-18, 1990.

Weingarten Publications. CBT Guide: A directory of authoring systems and courseware compiled by the editors of CBT Directions. Boston, M.A.: Weingarten Publications, 1988.

Young, J. L. The case for using authoring systems to develop courseware. Educational Technology, 24, 26-28, 1984.

Attachment:

Sample Description Of An Authoring System In The Data Base Of The National Library Of Medicine

Date: September 18, 1990

Title: Hyper Card 2.0

Source: Claris Corporation
P.O. Box 526
Santa Clara, CA 95014
(408) 727-8227
FAX: (408) 987-7105

Packaging: Manual and Five 3.5 inch DS/DD disks

Computer Requirements: Macintosh Plus or higher
1 M RAM, 2 M RAM if used with Multi-finder
4.0M Hard disk space
System 6.0 or higher

System Options: Macintosh compatible monochrome monitor
Videodisc -- (Optional)
Input devices -- Keyboard, mouse

Instruction: Hypermedia

General Description:

Menus are used to create and assign properties to multiple windows, cards, graphics, and other objects. Cards, buttons, and other objects can be linked. Students select buttons and objects to answer questions and brows information. There are pull-down menus to go to next, previous, first, last, and other cards and windows. The Hypertalk programming language can be used to provide additional functions or existing XCMD"s written in Pascal or C can be used or authored to incorporate effects such as color. Older HyperCard stacks (version 1.x) can run under 2.0 but must be converted to be edited. 2.0 stacks and converted stacks do not run under earlier HyperCard versions. 2.0 has variable card sizes, multiple windows, different style text in fields, hottest (selectable-editable test that links other information) and more Hypertalk commands. Although color graphics cannot be authored, those crated with other programs can be called up and linked (without using xcmds) and text fields in cards can be "marked" to do boolean searches. Drivers supplied by the Apple Programmer Development Association or Voyager can be used to control videodisc.

ATR trainer: INTELLIGENT CBT FOR THE REST OF US?

Dik Gregory[1] and Paul French[2]

[1]Dent Lee Witte plc
Riverside 3
Hester Road
London SW11 4AN
United Kingdom

[2]Advanced Training Research
Kingston Business School
Kingston Hill
Kingston upon Thames, Surrey
United Kingdom

ABSTRACT

The last decade has seen a great deal of research on Intelligent Tutoring Systems (ITS), with very few (if any) commercially viable solutions. There are those who have been tempted to apply the more mature expert systems technology to learning, having noticed that expert system users tend automatically to learn from such systems.

The problem with relying on straightforward expert systems technology for learning concerns the blindness of such systems to student progress. Although the process of reasoning towards some solution is rendered accessible in expert systems, essentially, they do not need to care whether users understand this process. This means that to use them effectively, learners need to have learned how to learn in order to create and support their own learning objectives.

ATR trainer is a rule-based expert system shell with student modelling and tutorial facilities. The system runs in a 640K MS-DOS environment and is commercially available. Significantly, its logic constitutes an example description language that enables the system to generate novel, learner-centred examples and test material on the fly. The system is particularly interesting for its ability to function in both a tutorial and an expert system mode, using the same knowledge structure.

This paper examines the evolution of ATR trainer and outlines future developments in the light of commercial experience with Version 1.

ORIGINS

Examples of Intelligent Teaching Systems (ITS) began emerging from AI labs in the 1970's. Carbonell's SCHOLAR (1970), Clancey's GUIDON (1979), Goldstein's GENETIC GRAPH (1979), Brown et al's SOPHIE (1980), Sleeman's MAL-RULES (1982) and O'Shea's 5 RING MODEL (1984) are all well documented, intensively researched systems that emerged from this era.

While these systems were designed essentially to explore different techniques and metaphors for intelligent CBT, the fact remains that, two decades later, conventional CBT remains largely unchallenged in the marketplace.

One potential exception to this — ATR trainer — is explored in this paper. (See also French, 1987, 1990; Gregory, 1990; and Wreyford, 1990). ATR trainer (henceforward, *trainer*) is a PC-based shell for authoring and delivering intelligent CBT. It runs in a standard 640K MS-DOS environment equipped with a VGA screen. It has been developed over the last four years by a group at Kingston Business School in the UK and funded jointly by the Department of Trade and Industry and a consortium of a dozen or so British commercial organisations.

Trainer's Prolog-based engine is an implementation of expert systems technology, requiring subject matter authors to represent knowledge as executable high level rules.

The temptation to borrow expert systems technology for application to learning is very great. Many have noticed that expert system users tend automatically to learn from such systems. Expert systems technology is arguably more mature than ITS, and systems are relatively easy to construct using modern high level shells. Since the expert knowledge encoded in the system is, by definition, explicitly accessible by users, such systems seem able to provide useful learning environments where questions can be asked and 'what-if' games can be played. This latter facility is important because it means that expert systems are able to provide the modelling environment that is so important to an ITS.

There is, however, a central problem with relying on straightforward expert systems technology for learning. Although the process of reasoning towards some solution is rendered accessible in expert systems, essentially, they do not need to *care* whether users understand this process. This means that expert systems are limited in their usefulness as learning devices to those students for whom learning is already an explicit skill, and who can generate for themselves the questions they need to ask.

KEY FEATURES

There are two key features of *trainer* that allow it to function as much more than an application of expert systems technology:

- The use of its logic as an *example description language*
- The provision for *student modelling*

Before examining these features, it is important to understand the way in which the author can use *trainer*'s logic to represent a subject matter.

ATR trainer's logic

The subject matter that the author creates takes the form of a set of learning topics, each of which is described by one or more rules which manipulate concepts.

The basic rule syntax is as follows:

Concept-1 **is** Value 1 **if** Concept-2 **is** Value 2	**Rule Syntax**

For example:

animal **is** mammal **if** baby-food **is** milk	**Animals:Rule 1**

In this case, "animal" is the topic to be taught. It is the subject of a rule in which the "if" conditional must be examined in order to determine if the conclusion (ie that animal is a mammal) is true.

It is important to understand that the learner is never confronted with an actual rule in *trainer* — even though rules are written at this very high level. Instead, the system presents learners with a screen of text and graphic information that the author creates to describe the *meaning* of the rule. The concept of "baby-food" is effectively a primitive unless the author turns it into a topic by creating a rule (or rules) for it. A primitive in *trainer* is, therefore, simply a concept which the author decides needs no further distinction. Similarly, a topic is a concept for which a rule has been created to distinguish between two or more values.

Topics can be created from concepts even if no rule is authored to distinguish them from other concepts. The author is free to create expository material for delivery via a descriptive overview screen, rather than as a description of a rule. In this case, of course, the distinctions that the author creates are not available to the rule base, and cannot form part of the subject matter's executable knowledge.

Let us allow our author to create a few more rules:

baby-food **is** milk **if** food-system **is** mammary gland	**Animals:Rule 2**

animal **is** bird **if**
    birth-process **is** hatching **and**
    body-covering **is** feathers
	Animals:Rule 3

animal **is** mammal **if** homeostatic-system **is-not** cold-blooded	**Animals:Rule 4**
work-value **is** high **if** animal **one-of** (horse mule donkey sheepdog)	**Animals:Rule 5**

The relationships between these 5 rules can be seen better in *Figure 1*, which presents them in the form of a topic map. In *Figure 1*, solid lines connect topics; dotted lines point to concepts which have not yet been turned into topics via the creation of rules to distinguish them further.

Consider the rule for "baby-food" (Animals:Rule 2). In adding this rule, the author effectively creates a sub-topic from the perspective of the topic "animal". A learner could choose to move "deeper" in the structure to learn about babyfood, having first accessed information about the topic "animal".

Note that the relationship between topics and sub-topics in *trainer* has little to do with (absolute) hierarchical level, but rather is an expression of *conceptual distance*. Further, this conceptual distance is itself relative and a direct reflection of the way in which the author conceives the topic structure and defines the teaching objectives. In *Figure 1* the addition of a rule for work-value (Animals:Rule 5) not only establishes a new topic, but also renders "animal" a sub-topic (relative to work-value).

By creating topics from concepts via the specification of (groups of) rules, the author produces a mesh-like rule structure containing many potential pathways. The actual pathway taken by a learner is only decided by that learner when s/he chooses which topic to access first, and from there, which to move on to next.

Animals:Rules 3, 4 and 5 demonstrate a variety of connectors (and combinations of connectors) that are allowed by *trainer*'s rule syntax (ie in addition to the basic "is"). Other allowable suffix connectors include greater-than; less-than; includes; excludes; includes/excludes-all-of; includes/excludes-some-of. These connectors allow concepts to take values that are either numerical or text strings.

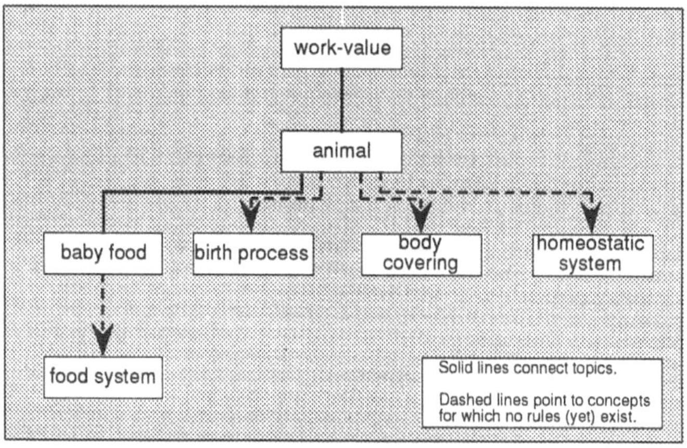

Figure 1. Topic map for the "Animals" domain

Example Description Language

The conventional expert system view is that a rule base uses an *inference process* based on information supplied by a user. For example, a user may be trying to determine which Macintosh program to employ for creating a certain type of drawing.

Using the following rules...

program-type **is** object-oriented **if** drawing type **is** flow diagram	**Macintosh:Rule 1**
representation **is** postscript **if** the output device **is** Linotronic	**Macintosh:Rule 2**
product **is** Adobe Illustrator **if** program-type **is** object-oriented **and** representation **is** postscript	**Macintosh:Rule 3**

the system might create the following inference structure to reach a conclusion:

Inference process	
the drawing type is flow diagram **so** conclude that the program type is object-oriented	**Stage 1**
the output device is Linotronic **so** conclude that the representation is postscript	**Stage 2**
the program type is object-oriented and the representation is postscript **so** conclude that the product is Adobe Illustrator	**Stage 3**

(Here, the user (or other rules) would be needed to provide information about the values of "drawing type" and "output device".)

However, the logic in *trainer* is purely descriptive and has no implied method of use. The inference process is only one way of using the rule structure. An alternative is to use the logic as an *example description language*. This gives *trainer* a generative capacity which serves as the basis of its distinctive ability to construct illustrative examples and tests for the student on the fly. For example, the following *trainer* rule will be true if the drawing type is "painting":

program-type **is** bitmap **if** drawing-type **is** painting	**Macintosh:Rule 4**

If the student requests an example of this rule, *trainer* simply generates the value for drawing type as "painting", presents this as a supposition to the student and retrieves explanatory materials associated with program types to demonstrate the link.

If *trainer* decides to test this rule, it once again assigns "painting" as the value of drawing type, presents this information to the student and asks what the program type should be. When the student responds, *trainer* assesses the answer by executing the rule and comparing its "expert" result with the student's. For tests of compound rules which need to call other rules to provide other input values to the rule under test, this quickly becomes less than trivial. A test of Macintosh:Rule 3 (above) calls for the student to understand and use (at least implicitly) the relationships between Macintosh:Rules 1, 2 and 3 in order to arrive at the same conclusion as *trainer* – namely that (given the conditions) the correct product to use is "Adobe Illustrator".

Trainer's expert system roots mean that authoring time can be saved since there is no courseware storyboarding to be done. The sequence of rules relating to a particular topic is important to get right, but this is to do with the integrity and coherence of the knowledge itself rather than with the prescription of a learning sequence.

More than this, though, the possibility of using *trainer*'s logic as an example description language as well as to underpin the inference process affords a further advantage of considerable significance.

With *trainer*, two modes of operation are possible. The use of the logic as an *example description language* allows the system to create a training environment in which the student can browse the subject matter, asking for illustrative examples and submitting to tests of what s/he has seen. The more conventional use of the logic in an *inference process* allows *trainer* to function as an expert system in which the roles are reversed: now the user is effectively "testing" the system and must supply sufficient information (when asked for it) until the system is able to draw a conclusion.

This means that *trainer*, uniquely, can provide a training environment which *survives* training ie it remains useful to trainees after course completion via its ability to function as a consultant.

Student modelling

The student model in *trainer* is a domain-independent overlay model. In other words, a blank copy of the knowledge structure is automatically created and maintained for each student with each rule marked initially as unseen.

Consider Figure 2, which presents a topic map annotated with the number of rules authored for each topic. The student is free to progress through the topic structure, unfolding it in whatever order s/he decides. With reference to the map, a learner can decide to start at any of the Topics A to H. If s/he chooses Topic C, then explanatory materials associated with Topic C:Rule 1 are retrieved. Subsequently, the student can move directly in one of the following directions:

- Topic C:Rule 2
- Topic G:Rule 1

In addition, the learner can indirectly hop to any other topic via a top level menu.

The particular route through the topic map that a student takes represents a learning strategy that may or may not be deliberate. For example, one student may choose a depth-first strategy, starting with Topic A:Rule 1 and then proceeding to Topic B:Rule 1, then Topic C:Rule 1, then Topic G:Rule 1 etc.

Another student using a breadth-first learning style might start with Topic A and work through both rules in that cluster before proceeding to the 3 rules describing Topic B and then to the 2 rules describing Topic C, and so on.

Still another student might move around from one topic to another without accessing all the rules in any one cluster.

Whatever the student's access order, *trainer* notes which rules are actually looked at and marks them as "seen." When the student successfully completes a test, the rule being tested is marked as "understood".

Where a topic is represented by a number of rules, but evidence only exists for a student's understanding of some of the rules, the topic is deemed to be partially understood.

Testing is a learner controlled process once s/he has seen enough of the rule structure to make testing a sensible activity. When a learner elects to take a test, the system consults the student model to identify a rule which has been seen, but not yet tested. Once identified, *trainer* asks a question for which the answer is the result of the execution of that rule. It may be that the rule chosen will first require the execution of other, "lower order" rules on the way to providing the answer. To solve the problem the learner is allowed to:

- ask for information

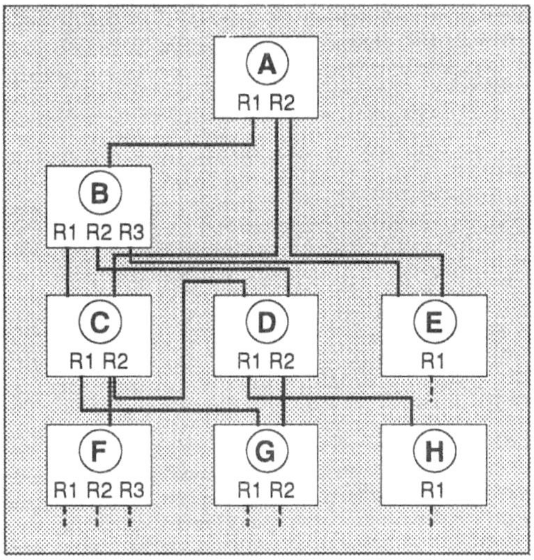

Figure 2. Simple topic map

- provide the final answer
- optionally provide intermediate answers (ie the results of any "lower order" rules)
- receive hints
- re-cap on the given information

The hint mechanism is automatic. By examining the way the learner has so far analysed the problem, and because it has access to an executable rule structure, it is possible for the computer to suggest what information the learner should ask for next, or what answers the learner should be able to provide.

In summary then, *trainer* provides an authoring system for intelligent CBT that virtually eliminates the need for storyboarding, automatically maintains a student model, and is truly generative in its ability to provide examples, tests and hints on the fly.

So how well does it work?

APPLICATIONS

Trainer has been used to build serious intelligent CBT courseware for a variety of domains to the authors' knowledge, including:

- personal income tax legislation for UK accountants (200 rules; 7-14 hours)
- laser printer troubleshooting for technicians (100 rules; 1-2 hours)
- insurance underwriting for private car insurers (250 rules; 9-10 hours)
- physics for civil engineering undergraduates (100 rules; 1-2 hours)

Of these, the underwriting courseware has been formally evaluated by comparison with conventional CBT (N=108) and the taxation courseware has been evaluated via in-depth interviews with selected users (N=2). The results are described below.

KEY AUTHOR ISSUES

Advantages

- Rapid courseware development is possible, with the opportunity to review material from the learner's perspective throughout development.
- Authoring one hour of courseware took only 35 development hours for straightforward, text-based materials to 150 hours for complex materials requiring graphics. Even the higher figure is significantly less than the 200+ hours normally quoted for CBT development.
- It is possible to create rule-based knowledge structures in *trainer* for a wide range of subject matters.

Disadvantages

- There are no high level tools to manage larger rule bases (ie >50 rules) effectively. It is not possible to split and merge parts of a rule base, nor to arrange concepts/topics in author-definable order. Text search and printing facilities are limited.
- It is frequently difficult to explain a rule within the confines of one screen.
- There are no high level tools for creating and manipulating graphics.
- No OR clauses are allowed in the rule syntax, making some rule structures cumbersome to write.

KEY LEARNER ISSUES

Advantages

- There is no difference in teaching effectiveness or learner retention after 6 weeks compared to conventional CBT. (This is an advantage in the context of the relatively small number of development hours needed for *trainer* as compared with conventional CBT).
- Learners found the system particularly valuable as a revision aid — because of its explanation and generative testing capabilities.

Disadvantages

- Learners had difficulty in locating themselves in the knowledge structure. This produced negative attitudes in some learners.

- Certain kinds of generative examples did not work well — often re-iterating what had just been presented.
- Generative testing did not always work well — sometimes the answer was predictable from the question syntax.

EVALUATION SUMMARY

While authoring time for *trainer* courseware is significantly less than for conventional CBT, and the courseware is just as effective. learners do not like the result as much. Dislike seems to have arisen for three principal reasons:

- learners tend to get lost in the knowledge structure
- teaching materials are too integrated with the rule base, leading to tests and examples which are sometimes predictable and unilluminating
- learners need graphics to lighten the learning load — but *trainer* does not easily allow their incorporation and management

In addition, although authors take less time to author materials in *trainer* compared with conventional CBT, the lack of availability of high level tools for accessing, organising and reviewing the developing knowledge structure is a major barrier to the efficient creation and maintenance of rule bases larger than around 50 rules.

ATR trainer, VERSION 2

Reflection on the results of the evaluation studies (as well as much informal feedback) has led to a re-design of *trainer*, culminating in the imminent launch of Version 2.

Version 2 differs from Version 1 in four main areas. These are concerned with:

- Separation of teaching materials from the rule base
- Topic grouping
- More flexible navigation
- Student modelling improvements

Separation of teaching material from the rule base

In Version 1, the teaching material was whatever text and graphics the author had created on one screen to explain the meaning for each rule. This was a good idea in prospect, since it provided the author with standard guidelines, as well as reducing the amount that had to be authored.

In Version 2, the author still creates rules and their meanings. However, these are now used only to provide executable explanations and tests of the teaching materials. The teaching materials themselves can be whatever (and as much as) the author wishes[1]. The overall result is that generative examples and tests produced by *trainer* no longer risk being trivial repetitions of what has just been presented.

To illustrate this, consider the following rule.

relationship **is** wife **if** couple **is** married **and** sex **is** female	**Relationships:Rule 1**

If a learner requests an example of this rule in Version 1, s/he would immediately get the following somewhat repetitive case.

suppose that the couple is married **and** the sex of the person is female **then**, the relationship of that person is wife.	**Example**

The problem is that the rule is being used as an example of itself. However, this same example takes on a much less trivial meaning if it has been chosen automatically from a number of rules about human relationships. In other words, although the rule may be necessary to ensure the coherence of the rule structure, it need not form a formal part of the teaching materials, and may be called on just for its illustrative capacity at an explanatory level.

Topic grouping

In Version 1, more than one rule was often needed to represent what, for the learner, should be a single idea. For example, the following two rules (Loans:Rules 1 and 2) were needed in an early pilot domain to express the single idea that a bank is able to agree to a loan facility if it can do so without exposing itself to financial risk. (Incidentally, note that the second rule has no conditional phrase ie it is a definition.)

proposed-loan **is** acceptable **if** bank-reserves > total amount	**Loans:Rule 1**
total amount **is** (proposed loan + existing loans)	**Loans:Rule 2**

The second rule is needed because "total amount" cannot be defined in the first rule without extending the logic in an undesirable, ad hoc way.

---

[1]The author can choose to be assisted with the structure of the teaching materials by starting with a concept map automatically generated by ATR trainer from the rule base. This concept map can then be edited to produce a topic map containing only those concepts that are deemed teachworthy.

The pedagogical problem with this is that *trainer* (Version 1) treated these two rules as distinct entities that were to be taught separately. In turn this led to a sense of "overteaching" in the learner.

Version 2 allows the author to select rules and concepts that are artifacts of the representational process and which can be taught implicitly at the next higher level ie at the level of a genuine topic. In effect, the rule that computes the value of "total amount" (Loans:Rule 2) is executed when necessary by the system, but never taught as a separate rule from the one that deals with "proposed loan" (Loans:Rule 1).

More flexible navigation

In Version 1, learners were limited in their movements to one topic up or down, and the next or previous rule in the current rule group. This was supported by a navigational aid that located them by reference to the topic above. For example, *"You are currently learning about "object-oriented" which is part of the topic "product"*. This turned out to be inadequate positional information for most learners, who also found it very difficult to return to particular screens they might have visited (possibly hours) before.

In Version 2, it is much easier for learners to re-visit topics either temporarily to reference something, or permanently in order to move on to new material from an old location.

Student modelling improvements

In Version 1, the learner was provided with a constant rough estimate of current topic completion, expressed as a percentage. This was calculated by combining how much of the topic's group of rules had appeared on the screen with how many tests had been completed successfully.

In Version 2, this estimate is calculated in a more sophisticated way. Up to 20% of the completion score is contributed by the extent to which the learner has accessed the teaching materials for that topic; up to a further 30% is contributed by the learner's (modelled) understanding of lower level topics; the remaining 50% depends on the learner successfully completing the tests for that topic.

Additionally, the author has more control in Version 2 of the amount of testing that should be done for each rule, or indeed, whether the rule should be tested at all.

FUTURE DEVELOPMENTS

While Version 2 embodies some important, principled changes, the Kingston ATR design team is already looking at the requirement for *trainer*, Version 3.

There are three main issues driving progress towards a third version:

- A graphical user interface like Windows 3 will mean that navigation can become significantly easier through the use (for example) of multiple scale topic maps. Authors will also be able to take advantage of graphical tools for manipulating knowledge structures as well as creating and integrating (possibly animated) graphic teaching materials.

- The complete separation of the teaching materials from the rule base is also foreseen. This will mean that teaching materials will be organised by a topic map whose structure is independent of the relationships between the rules in the rule base. Authors might suggest a storyboard, but learners will be generally free to unfold and explore the teaching materials in whatever sequence they wish. The rule base will form an executable, explanatory level which can be used as an modelling environment that is mutually accessible by the system (for demonstrating examples to the learner) and by the learner (for demonstrating understanding to the system).

- There is great potential for connecting the student modelling process with the tutorial process in more powerful ways. Here, the likely emphases will be placed on developing and utilising the notion of conceptual closeness to allow the system to suggest new learning objectives to the learner; and the development of *trainer*'s current ability to monitor the learner's "workings out" as a basis for tutorial support. The existing pedagogical strategy of using the student model as a basis for learner choice rather than control of the learner will remain.

SUMMARY

Trainer is an expert system-based intelligent CBT environment that distinguishes itself in a number of important ways — both from conventional CBT and other approaches to constructing intelligent CBT systems. Its central feature is a logic that can be used to underpin both an inference process and example description language. This enables the system to function in both an expert system mode and a tutorial mode respectively. In tutorial mode, the logic is able to generate novel (ie unprogrammed) examples and tests on student demand. Its underlying pedagogical philosophy is strongly learner-centred.

In the light of considerable experience with the system, including a formal evaluation, the system has been re-designed to improve learner and author acceptability. Of principle concern in Version 2 is the separation of the learning topic structure and teaching materials (the teachables) from the rule structure, and improved topic navigation. It is planned to complete this separation in Version 3 where users will also be presented with a graphical interface, allowing much easier navigation and control mechanisms. In addition, it is planned to extend the student model's current capability to monitor student "workings out" to the point where the tutorial process can make direct use of this knowledge.

There is much to be done to refine the shell, improve author effectiveness and increase learner support. However, as a bona fide intelligent CBT system, *trainer* is generative, it uses a student model, it runs on a PC, it is out of the lab and it works.

REFERENCES

Brown, J.S., Burton R.R. & de Kleer, J., 1982, Knowledge Engineering and Pedagogical Techniques in SOPHIE I, II & III, Intelligent Tutoring Systems, Eds D Sleeman & J S Brown, Academic Press: London and Orlando.

Carbonell, J.R., 1970, AI in CAI: An Artificial Intelligence Approach to Computer Assisted Instruction, IEEE Transactions on Man-Machine Systems, 11, 190-202.

Clancey, W.J., 1979, Tutoring Rules for Guiding a Case Method Dialogue, Int. J. of Man Machine Studies, 11, 25-49.

French, P.D., 1987, Intelligent Training, presented at Intelligent Tutoring (Training) Systems '87, Israel. Advanced Training Research, Kingston Business School, Kingston Hill, Kingston, Surrey, UK.

French, P.D., 1990, A Domain-Independent Student Model for an AI-Based Training System, Computers Educ. Vol. 15, No. 1-3, pp 49-61.

Gregory, D., 1990, Affordable Intelligent Tutoring, in: Proceedings of the 7th International Conference on Technology and Education, Brussels 1990, Eds. N. Estes, J. Heene and D. Leclercq, CEP Consultants Ltd., 26-28 Albany St, Edinburgh, Scotland.

Goldstein, I.P., 1979, The Genetic Graph: A Representation for the Evolution of Procedural Knowledge, Int. J. Man Machine Studies, 11, 51-77.

O'Shea, T. 1984, Artificial Intelligence, in: Proceedings of New Technologies in Training Conference, Brintex Ltd.

Sleeman, D.H., 1982, Inferring (Mal) Rules from Pupils' Protocols, in: Proceedings ECAI-82: 1982 European conference on artificial intelligence, Orsay, France).

Wreyford, S., 1990, AI-Based Training for Car Insurance Underwriting, in: Proceedings European Training Technology Event: Training for Competitiveness, Eds F. le Fèvre & B. van Muylwijk, The Hague, 16-18 Oct 1990.

AN AUTHORING ENVIRONMENT FOR TRAINING SIMULATORS

H. Kuiper and A.F. Ambagtsheer

TNO Physics and Electronics Laboratory
PO Box 96864, 2509 JG The Hague
The Netherlands

ABSTRACT

This paper describes an authoring environment to compose lesson material for use with real-time training simulators. A prototype of this environment was realised in the beginning of 1991. The authoring environment is part of a Universal CAI System, that can cover the training process, as for instance tank-crew training or submarine-crew training, for several real-time training simulators. After explaining the context, the different aspects of the authoring environment will be dealt with. These aspects are among other things: breakdown of the course, general requirements, user-interface, modular design, flexibility and prototyping. The authoring environment is checked against the criteria mentioned by Park and Seidel concerning the evaluation of authoring systems. Finally some conclusions will be drawn and some suggestions for further improvement will be given.

INTRODUCTION

Some years ago the TNO Physics and Electronics Laboratory developed a tailor-made Computer Assisted Instruction (CAI) system that was integrated with a training simulator for an anti-aircraft armoured vehicle. Based upon the positive experiences with this system, a project was started to develop a system that had to be applicable to a broad range of real-time training simulators, such as tank-crew trainers, flight-crew cockpit trainers and submarine trainers. The system is called Universal CAI System or UCS (Kuiper, 1990).

Figure 1 shows the context of the UCS in a real-time training simulator environment. The UCS consists of four main components:

- *System Definition*
  This component is used to define the target simulator, which means that all switches, logical states and other variables receive a function name. System definition creates a system reference matrix that is consulted by the other parts of the system.

- *CAI-preparation*
  This is an authoring environment, with which the lessons, subdivided into lesson-modules are created. The lesson-modules contain both scenarios, i.e. the stimuli for the student and the desired behaviour or response of the student.

- *Universal CAI Module (UCM)*
  Being the heart of the system, this module takes care of the real-time training.

Advanced Technologies Applied to Training Design, Edited by
R.J. Seidel and P.R. Chatelier, Plenum Press, New York, 1993

The interaction between UCM and Simulator is a cyclic process. The UCM determines a lesson-module or scenario, with a duration of a few minutes, for the simulator. After the lesson-module has ended, the UCM performs an analysis, i.e. comparing the student reactions with an expert reaction, a judgement and a progress determination, i.e. determining the next scenario based upon the student results.

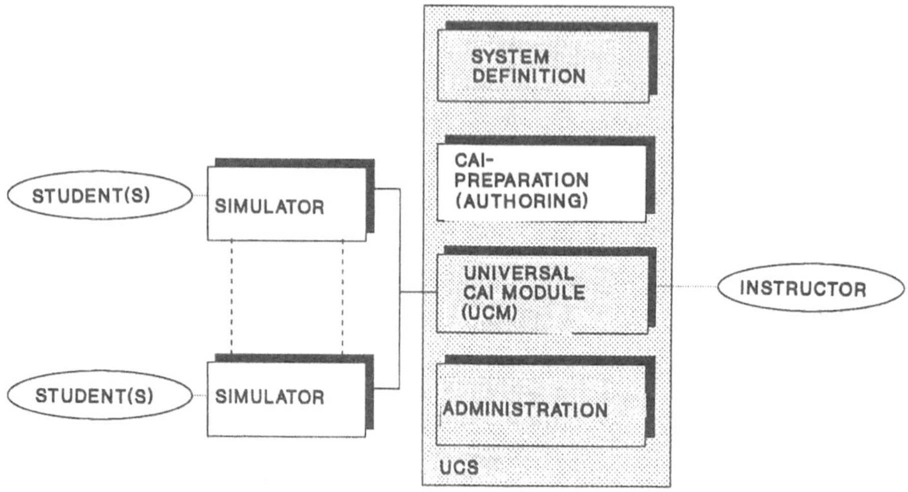

Figure 1 Universal CAI System within a Training Simulator configuration

Administration
Among other things, it keeps record of a database with personal data, results and learning history for every student.

The remainder of this paper will focus on the authoring environment for the preparation of lesson-modules.

THE AUTHORING ENVIRONMENT

Introduction

We will now focus on the basic concepts behind the development of an authoring environment. Starting point is, on the one hand, what we can reach by making a complete breakdown of the course and on the other hand the required lesson material for the simulator.

After determining these basic bricks to be composed with the authoring system, some general system requirements are summed up.

Constituent parts of the course

In order to determine what has to be created by the authoring environment, we first have to make a decomposition of the training program. Generally spoken, after a task and function analysis have been carried out, a number of learning objectives can be defined. The

The course structure diagram is depicted in figure 2.

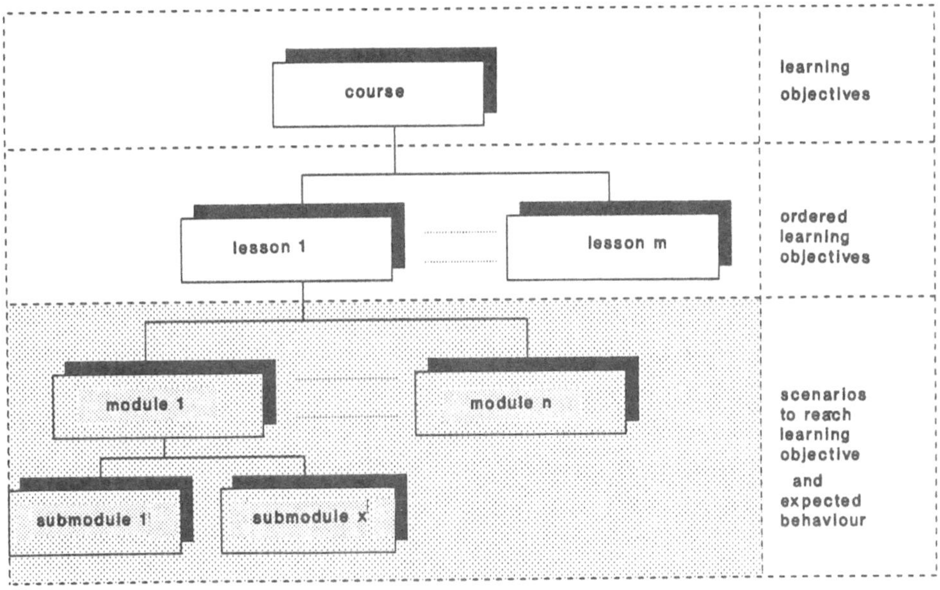

Figure 2 Breakdown of the course elements

total number of learning objectives constitute the specific course. We can divide the course into ordered learning objectives called lessons. A lesson contains one learning objective, for instance, to learn the procedure "target acquisition" for a radar system. We can reach this learning objective by dividing the trajectory into small steps, called lesson-modules. When a student masters a lesson-module he has reached a subgoal leading to the learning objective.

Following the training system model described by Dana (1985), the next step in the training program is the selection of the proper media.

In this case it is a given fact that the training will be performed with a training simulator. The kind of training simulator we have in mind is, as mentioned before, intended for tank-crew training or submarine training. In general, difficult time-critical complex operator actions will have to be learned. Taking the lesson-module as a base-level brick, we will define the contents of this lesson-module.

What the simulator needs is a scenario, i.e. the presentation of stimuli to the student. The Universal CAI Module (UCM), however, needs the expert solution to the scenario, judgement criteria and so on. So here we have the basic needs for the contents of a lesson-module.

Figure 3 is depicting this schematically.

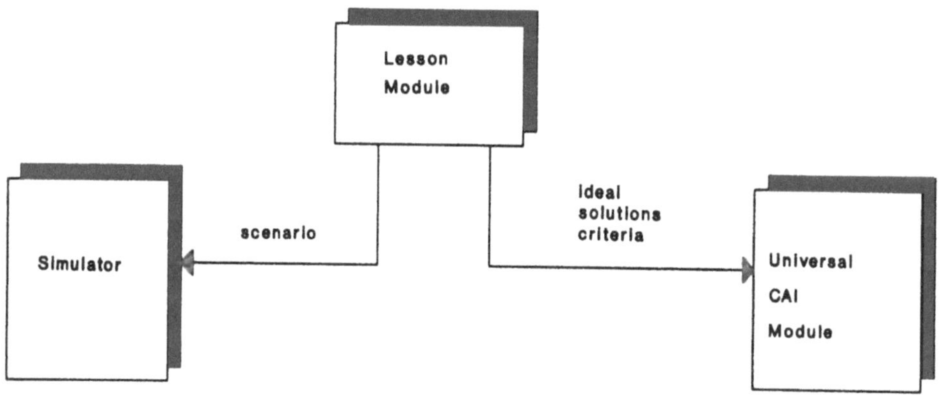

Figure 3 Splitting up a lesson-module

In order to be adaptive toward the student, a network of lesson-modules has to be created in which we anticipate the possible behaviour of the student. The first level is the main-module level. When the student performs well, he gets modules from this level. When the result is doubtful, he may get a repeat-module; in general, a module with the same complexity as the main level but in another form. It should also be possible to give the student instruction on a specific item that caused difficulty during the presentation of the main-module. We call this the corrective level.

So input for the training simulator will be a network of lesson-modules, created with an authoring environment. The network is schematically shown in figure 4.

During the training process, this network will be used to graphically present to the instructor those lesson-modules that have been visited by the student, in other words to show the learning path of the student. This facilitates an easy wayfinding which is important in human computer interaction (Jones, 1989).

Other general requirements for the authoring system

Also, a number of general initial requirements have been imposed on the authoring environment that must create the lesson material:

- *Support*
  The authoring environment has to support the full trajectory of lesson-development. This implies that there has to be some kind of computer assisted support for the author as early as possible.

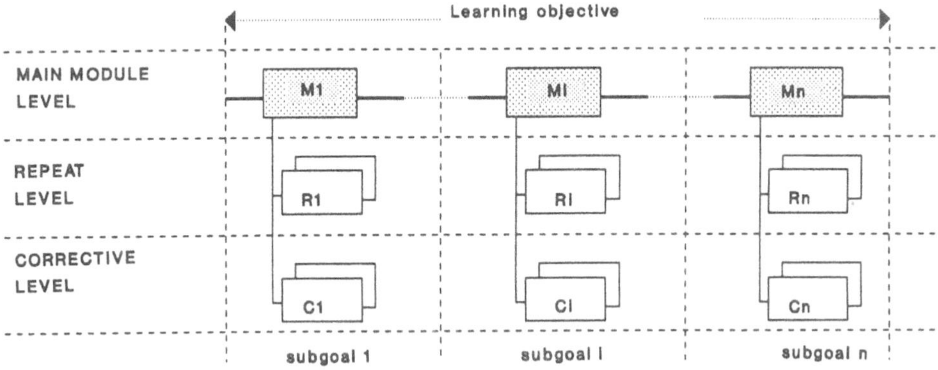

Figure 4 Lesson-module network

- *Universal parts*
  The system has to be designed in such a way, that simulator-system dependent parts are separated from universal parts, i.e. a modular design. This implies that an authoring environment for another training system can be realised easily.

- *Portability*
  The courseware has to be documented in such a way that not only the original author but also other users may get a fast insight into the lessons.

- *Surveyability*
  Courseware has to be presented in a clear and surveyable manner, both on screen as well as on paper. Besides that it has to be possible to make surveys on different levels as far as the structure is concerned using different intersections. Because courseware contains a lot of data, the author must have the facility to apply some structure. For this purpose a library-facility is necessary in which lesson data can be simply stored and retrieved.

- *Verifiability*
  The author should have the possibility to check whether the created lesson can be used by the training system.

- *User-friendliness*
  It must be possible to write courseware without the author having to know any programming language. This means that not only the system has to be simple and

logical from the point of courseware design but also that the author can hark back to a context sensible help-facility. Besides, the user-interface has to be such that both experienced and inexperienced authors can develop courseware.

- *Modifiability*
  Courseware is often composed of modifications of existing courseware. This method of courseware development has to be supported also.

- *Flexibility*
  The author should not be compelled to compose the lesson-modules in a fixed order. Top-down as well as bottom-up lesson-input must both be possible.

Design of a lesson-module

With the help of the authoring system we have to create a network of lesson-modules that can be used by the training system.

This section deals with the lesson-module in more detail.

A lesson-module can roughly be divided into two main parts:

- the scenario, and
- the expected expert behaviour towards the scenario.

Scenario. The scenario describes which perceptional stimuli the student receives at a given moment. For instance, when the student is trained to operate an anti-aircraft tank, the stimuli may be a fault simulation in a radar system, a target appearance in the periscopes or on the radar etcetera. The student has to react to the situation in a proper way. When, for instance, a target appears on the radar screen he has to lock on the target and identify it as being a friend or a foe.

Expected behaviour. In order to be able to judge in how far the student has reacted properly to the scenario stimuli, the expected behaviour, being the behaviour of an expert on the scenario, is also part of a lesson-module.

With a lesson-module we have to anticipate all possible and relevant kinds of errors a student can make when trying to reach a learning objective. First the lesson-module has to contain a unity with which we can record whether the partial-learning objective has been reached (a reference frame). This unity is called main-module. The main-module is always the first module to present to the student.

We have to check whether the student is taking correct action within a given period of time on the scenario stimuli. The key to this check is the placement of timing-windows on the scenario. Figure 5 shows an example of window placement. For example, switches s1 and s2 have to be operated at t1 and t2. The timing window checks whether the student reacted on this precise moment or not. By dividing a timing-window into trajectories we can determine the timing of the student and link a weighting factor to this timing for judgement purposes.

In this way we can place many timing windows in a scenario. To measure, for instance, the students reactions to system-faults, but also to measure periscope operation or radar operation inside the turret of a tank when targets appear.

Other lesson-module contents. Messages for introduction of a module are also part of a lesson-module, as well as feed-back reference messages, judgement criteria etcetera.

PROTOTYPE OF THE AUTHORING ENVIRONMENT

Overview

Before prototyping the authoring environment we may ask ourselves, whether there is no commercially available authoring system that meets the requirements. Most of these systems are developed for interactive CAI with the use of a Personal Computer. There are no systems available that are suited to this kind of training on a simulator integrated with CAI, so a new authoring environment had to be developed.

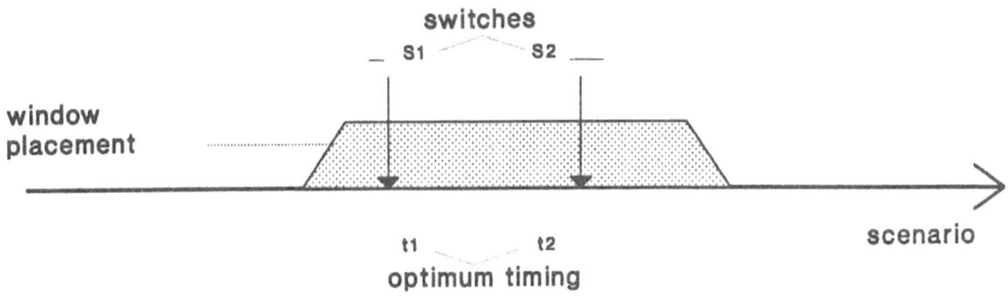

Figure 5 Window placement

A prototype of the authoring environment was built with the standard window system X-View and the programming language C. The prototype uses a WIMP (Window Icons Mouse Pointer) environment to get as much freedom and flexibility as possible for the user.

The authoring environment consists of a number of editors to build the lessons and a number of tools to support the author.

The following editors can be distinguished:

- *Course editor*
  This editor defines the order in which the lesson networks are passed and it enables easy access to the lesson networks.

- *Network editor*
  With this editor it is possible to create a network of lesson-modules containing different instructional levels.

- *Module editor*
  This editor defines what stimuli the student receives at a certain moment and what the ideal reaction to the stimuli should be.

- *Window editor*
  This defines what checks have to be made in the simulator and what score the student receives for his actions.

Some examples

Figures 6 and 7 show some examples of screen-dumps from the authoring environment prototype.

In figure 6 the network editor is depicted. A network of lesson-modules with different instructional levels has been created. The components of the network, the lesson-modules, can be easily manipulated using the mouse.

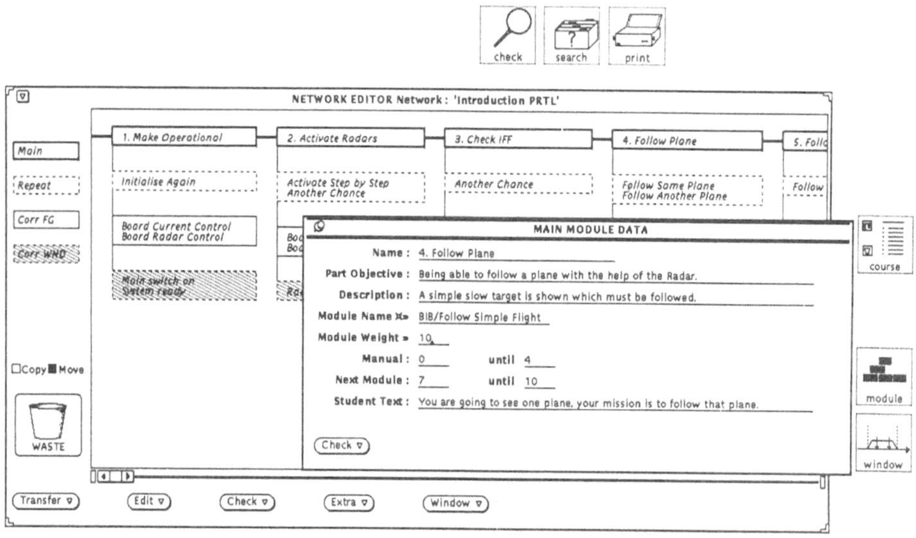

Figure 6 Network editor

By selecting an object from the network (zooming) a template is shown which can be completed with the necessary data. The network can be split up into two parts to see both ends. On the right, outside the window, the other editors are visible as icons. They can be called directly by clicking the proper icon.

Figure 7 shows the lesson-module editor. Inside the window we see an item with an airplane track. When the simulator has a visual system or a radar system, we refer in the scenario to a predefined library of tracks of an airplane or a tank.

The window item refers to the placement of timing-windows in the scenario. The action item refers to the placement of, for instance, simulated system-faults or change of meteorological circumstances (like rain, fog) in the scenario.

Future improvements

Integrating an intelligent help-facility, taking account for both experienced and inexperienced users, will be one of the future enhancements. Also the placement of timing-windows by an expert should be made as much user-friendly as possible, by making it a largely automated process.

EVALUATION OF THE PROTOTYPE

Let us follow some guidelines mentioned by Park and Seidel (1989) for selecting an authoring system to make an evaluation of the features of the prototype. The checklist will

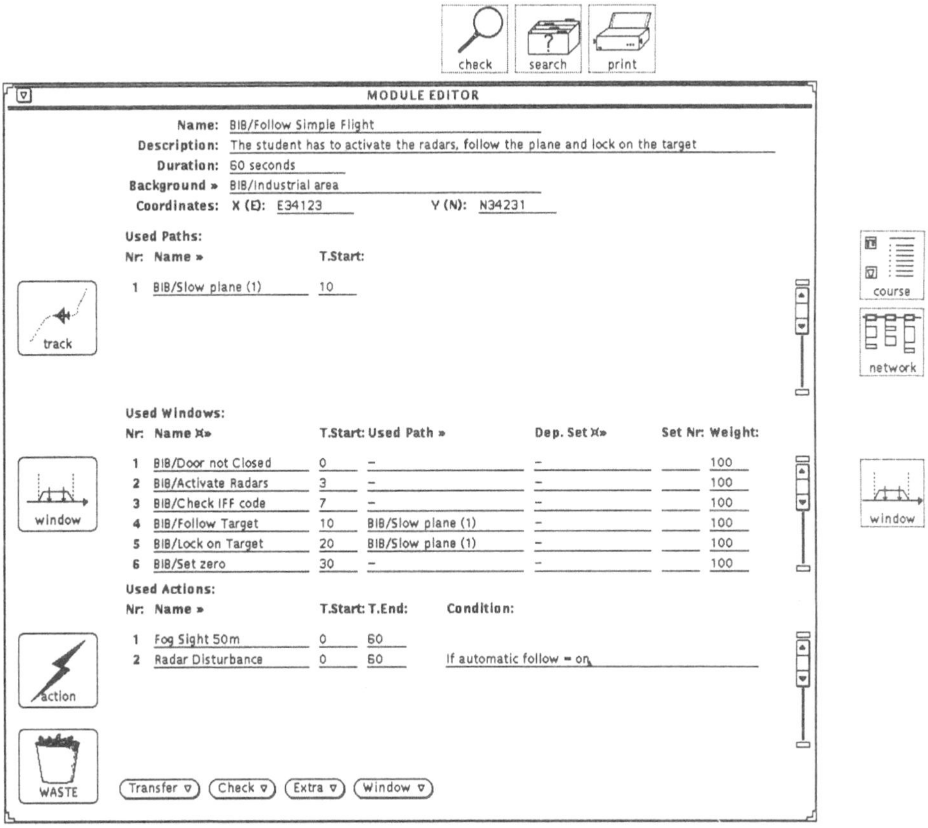

Figure 7 Module editor

be summed up and for each item some comment will be given whether or not the authoring environment supports this feature.

A INSTRUCTIONAL DESIGN AIDS

1. *Aids for laying out the lesson structure*
   In fact, the lesson structure has to be defined on paper. Future developments

include tools for covering the total training trajectory from job specification to evaluation of the training.

2. *Aids for selecting instructional strategies*
This feature has been integrated in a limited way. As stated earlier, at this moment there are three possible levels: main, repeat and corrective level. Future enhancements may include the creation of instructional strategy guidelines that come with the authoring environment. These guidelines will be used by the real-time training component, the UCM, to determine the proper lesson-module for the student, based upon student-history and student-results.

3. *Use of templates for developing components*
Templates are used for the module and the window editor and for headers of some components.

4. *Aids for selecting data collecting options*
All student reaction data are collected during the simulation. With the authoring environment the user can specify what is needed from this data for the specific learning-objective.

5. *Aids for developing presentation displays*
We can briefly comment on this point because we are dealing with a training simulator, which means that on the one hand scenarios have to be developed, and on the other hand feed-back presentation has to be done. For the purpose of feed-back we assume that there is a small video-monitor in the simulator. The first point, the lesson-module editor, is extensively described. The second point has resulted in a text-editor. At this moment there is no graphical editor. A future extension could be the development of such an editor to include graphical feed-back also.

A sound editor is not present in the system. An idea is launched to support the student during the first lessons in a simulator with feed-back in the form of sound, which makes a sound editor necessary.

Feedback can be given in more or less detail, and is selected by the student.

B FLEXIBILITY

1. *Interface with existing software development tools*
There is an interface with all tools provided by the X-View toolkit, like a word processor.

2. *Interface with programming languages*
It is possible to invoke a special program running in the simulator for some complex operations. This program is triggered from within the scenario.

3. *Interface with built-in utilities*
Interface is possible but not present in order to keep the system universally applicable.

4. *Interface with training management systems*
The system can be interfaced with a training management system. An administration system, which can perform this feature, is available.

5. *Interface with hardware peripherals*
The system interfaces with the training system via a network. All hardware

peripherals connected to the network can be used (e.g. printer).

6. *Interoperability among independent hardware systems*
The authoring system is written in a general programming language and uses the X-window system. Both are platform independent and supported by several commercially available workstations.

7. *Online communication and network capability*
The system can run within a network environment like ETHERNET. The authoring environment is built in such a way that different authors can edit (parts of a course) at the same time.

8. *Use of a variety of input devices*
The authoring system provides the user with a WIMP-environment.

C USER FRIENDLINESS

1. *Flexible design and development*
The components of a lesson (lesson-modules) can be manipulated easily, i.e. move, copy, delete. Lesson-module components - windows - can be developed separately, in the order desired by the author. Existing windows can be changed easily. So lesson-module design is very flexible.

2. *Easy development and revision*
Lesson-modules can be easily revised in a user-friendly way by the use of the WIMP-environment. Furthermore, the authoring environment uses a uniform window lay-out and a consistent user-interface.

3. *Built-in templates*
Templates are included for window editing, course and lesson-module headers etcetera.

4. *Easy integration of developed components*
Different components can be developed separately and can easily be integrated just by manipulating the WIMP-facility. Whenever an object has to be filled in, a choice can be made from already defined objects. Once an object is selected, the author can easily zoom in on that object to see its complete definition.

5. *Adaptation to author skill levels*
This feature is not present at this moment. It is part of a study and it may be a future enhancement.

6. *Use of student mode during development*
This feature is under consideration. In the case of a training simulator one may think of running a lesson in the simulator by an expert. The reactions of the expert are collected and timing-windows are placed automatically based upon his reactions. This could save a lot of time because at this moment most of the windows have to be placed manually.

7. *Help features*
Integrating intelligent help features will be part of future development. A tutorial as well as context-sensitive help is part of the study.

8. *User manual*
A user manual will be supplied with the authoring system.

CONCLUSIONS

An authoring environment was prototyped that is universally applicable for creating courseware for several types of real-time training simulators.

The environment is developed in a general purpose programming language and in a standard X-View window system. This enhances its portability to different commercially available workstations.

Largely, it complies with the evaluation criteria set by Park and Seidel. Future enhancements include among other things creating an on-line help system to support different levels of user experience. As stated by Park and Seidel, it is important to cover a large part of the training cycle with authoring tools. So tools should be created to support, for instance, task analysis and learning objective specification. These tools must make a continuous link with the authoring environment.

REFERENCES

Dana, D.K. (1985). The use of computers in training in the British Army in: "Proceedings of the symposium on Computer-based Instruction in military environments", Brussels, 15-17 April 1985, NATO-report DS/A/DR(85)165, pp 99-124

Jones, M.K. (1989). "Human Computer Interaction, a design guide", Educational Technology Publications, Englewood Cliffs.

Kuiper, H. (1990). A Universal CAI System for real-time simulators, in: "Proceedings of MARSIM & ICSM 1990", Tokyo, pp. 293-299.

Park, Ok-choon and Robert J. Seidel (1989). Evaluation Criteria for Selecting a CBI Authoring System, THE Journal, September 1989, pp 61-68

SECTION II
MODELS AND ANALYTICAL METHODS FOR TRAINING

Models and simulations are becoming an extremely important part of our education and training culture. However, the validation of these approaches continues to meet with many questions. This section will address the feasability of using such approaches to determine how to use them as predictors of effective training features as well as their cost effective usage in training devices and simulators. This section has a great deal of cross-feed with Section IV, (Advances in Hardware Technology).

MODEL OF PSYCHOLOGICAL IMPACTS ON MILITARY TRAINING IN SIMULATION

Rolf Otte

Personalstammamt der Bundeswehr
Kölner Strasse 262
5000 Köln 90
Germany

SUMMARY

In the recent years much effort was made to improve the technologies of training devices and training programs. But also attention should be turned to the fact that the human factors have to be adapted by training to optimize behavioral outcomes and performances. Neglecting the specifics of the human interface will cause problems.

A model of psychological influences will be discussed (figure 1 and 2) - after a brief description of the basics in military training and simulation - which can contribute to modulate and improve the performances in military training sequences, especially on simulators.

1.0 INTRODUCTORY REMARKS ON DEPLOYMENT OF SIMULATION IN TRAINING

The growing availability of small and high-performance computer systems for complex algorithmic data processing and system control has added to the improvements of the standards of technical aids in training, especially of standards for simulators.

The simulator as a man-machine system produces a technical and physical synthesis of real systems and a computer-based representation with the help of abstract models. In the field of military training it is particularly important that it should be possible to study man-machine systems through systems analysis and operation research methods with the aim of:

- to derive information for the optimal technical and ergonomic design,
- to achieve time and cost savings in the development and implementation of such systems,
- to be able to vary and check risks and interferences, or also
- to test logistics (include. life-cycle costs) and management,
- to improve the efficient use of such systems.

Hence the importance of the interfaces between the technical-physical systems and those with the task of operating these systems in the context of a real situation or a real scenario, which means that human factors must already have been considered during the phases of the

Advanced Technologies Applied to Training Design, Edited by
R.J. Seidel and P.R. Chatelier, Plenum Press, New York, 1993

system development and have been integrated as an interactive, communicating element. Reviewing the development of recent years, what can be seen is that, on the one hand, technological developments and improvements have proceeded rapidly, whilst on the other hand, it is assumed that the human factors or man's psycho-physical capability can fully utilize these changes without difficulty that only appropriate learning programs and training opportunities need to be applied. Particularly in the fields of military flight training and its continuing training programs where complex and comprehensive synthetic representations of the real world are reflected through simulators, experienced pilots, having been exposed to certain training units and modules in flight simulators, have voiced considerable reservations about the synthetic representation of pilot action essentially because:

- training simulators given their design constraints, and realizations make action sequences only effective in terms of flight training objectives and give only little transfer to pilot actions during flight operations,
- the Trainees (T) at the simulator feel that they are being manipulated,
- conditions in a "real situation" allow for only conditional transfer of acquired or trained behavior,
- synthetic flight situations lack the feeling of flying and awareness of flying,
- the knowledge that "lack of reliability in the operation/handling of the system will not be penalized" is more or less always there,
- in a real system specific personality traits (due care, attention, decisiveness, readiness to take risks, stress resistance, psycho-physical performances under different load conditions) become of paramount importance.

How then can we still justify and explain the immensely high costs for the development of system simulators and the corresponding instructional personnel necessary for the development and use of synthetic simulator representation for military training purposes? What must be the abilities and aptitudes of a human-interactive interface to enable T to apply the acquired knowledge to real-world stressors and achievement situations? Any satisfactory answer to these questions would indeed call for a general discussion of simulator concepts which would go beyond the scope of this paper.

The following will therefore only contain a brief description of psychological impacts on training in simulation (figure 1) which, from the point of view of the psychologist (or human factor engineer), are of interest in the context of man-simulation inter-action, and for which he can contribute methodological and didactical solutions.

This will not be a model for the technical optimization of the training program. If you want that please look at STICHA, SINGER, BLACKSTEN, MUMAW & BUEDE (1987).

2.0 SYSTEM DEVELOPMENT, TASK ANALYSIS AND THE HUMAN - INTERFACE IN SIMULATION SYSTEMS

Internal interfaces in simulation systems (the overall methodological and technical realization, the corresponding software as well as the operator/ supervisor personnel) are determined by three functional areas.

The first function comprises the learning and/or training objectives and the corresponding normative and ideal-type descriptions. These descriptions reflect the requirement analyses, their resultant characteristics and behavior patterns which should be as close as possible to the real system and enable the T after achieving the instruction or training criteria to transfer the acquired knowledge and/or behaviour to the requirements of the real system. This implies an unambiguous definition of content-oriented according to the general theory-of-learning and ergonomic knowledge of psycho-physics and personality as well as according to practical demands; mostly, these must be combined with validation criteria of the greatest possible precision and appropriate feedback.

Model of Psychological Impacts on Training in Simulation

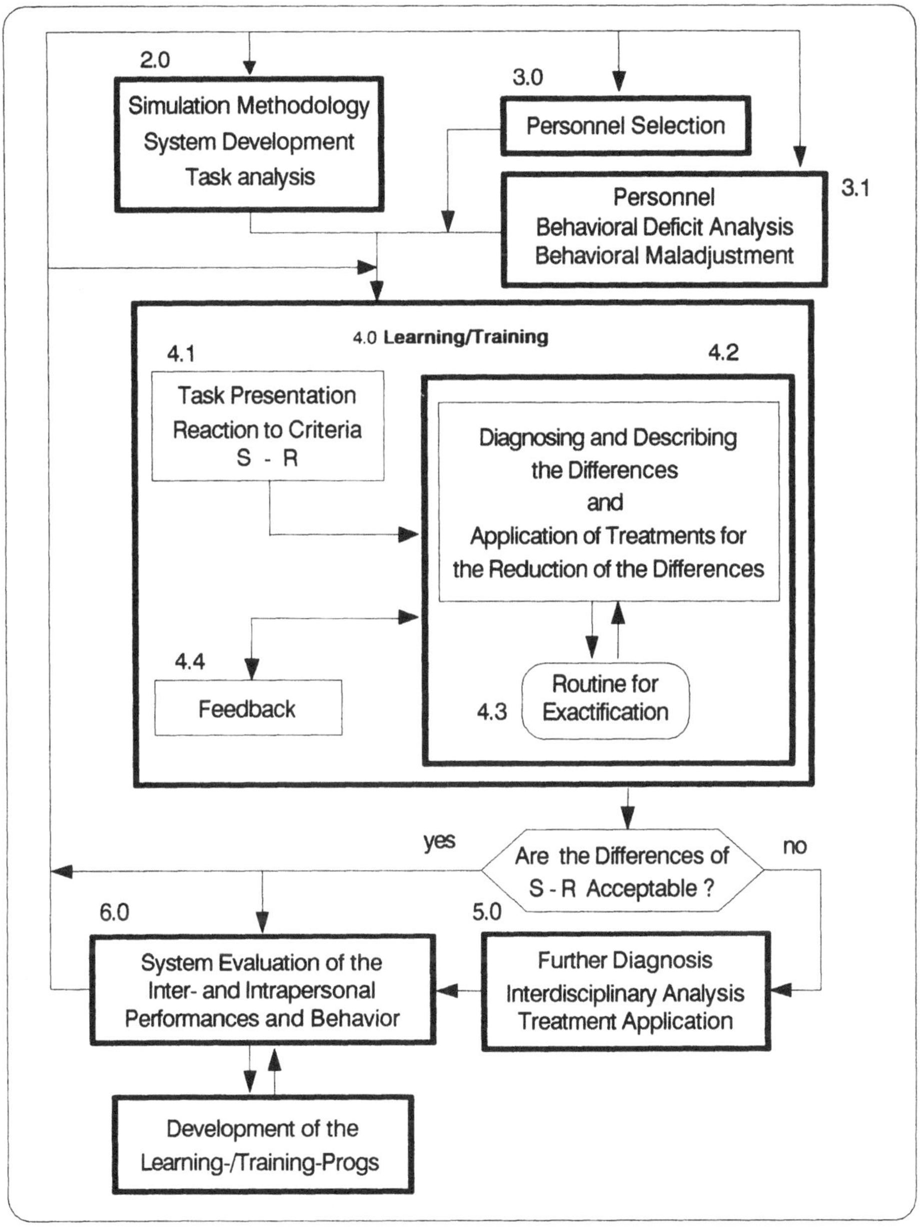

Figure 1. The Model of Psychological Influences and Factors on Training outcomes and Performances

The numbers correspond to the chapters in the text.

The simulation system:

- controls and checks all target and task-relevant information,
- stores data from the interaction process between the simulator and the subject's human factors,
- varies and checks the impacts of conditions which modify the learning/training process,
- charts the corresponding impacts derived from arbitrary interventions of the operating personnel/supervisors.

Miniaturization and the enormous increase in the efficiency of computers as well as automatic control engineering allow for real-time representation, coding and processing of almost all important data. Under these conditions - and from the point of view of the simulation system - the instruction and learning/training process can be optimized and designed to make available the behavior patterns and knowledge to be acquired/trained which are needed to meet the requirements of the real system. This will also augment the probability that whatever has been learned/trained will be transferred and applied to real-world requirements.

The second function comprises the modular structure of such instruction / training sequences which offers the advantage of being able

- to vary the factors impacting on the training process in a systematic and repeatable manner,
- to adjust single sub-components, depending on the requirements,
- to the specific individual background and style of knowledge acquisition and learning.

These possibilities would be excluded in nearly all real systems. There are these variable conditions which permit expansion of the experience background of the T, but also of that of the system designers and/or operating personnel, and improvements with respect to the assessment and elimination of critical system conditions of the real system.

Depending on complexity, function and requirements, we are able to distinguish between system simulators and part-task trainers with the latter used to train certain modes of procedure and defined sequences of actions.

The third function that is relevant to the man-machine interface comprises that of mathematical-logical models with which the simulation system processes, controls and checks any information and with which the T communicates in a synthetic language.

Requirement analyses and the resulting task structures of the simulation system are generally based on cognitive theories or on stimulus-response-control models (ROUSE, 1980; NEWELL & SIMON, 1972).

According to the basic assumptions of these cognitive theories, tasks primarily require interaction between intellectual ability dimensions and decision behavior. This includes complex memory performances, strategies for problem solution, comparative judgments and evaluations and simultaneous sequences of actions.

Tasks derived from stimulus-response-control theories almost exclusively imply psychomotor performances and operative-functional acting. The theoretical assumptions of such models compare the individual to the "optimal control person." Individual deviations and variance are considered as "errors." Checking inputs are compared to defined criteria to permit checking of the quality of inputs and/or responses.

According to the requirements, both types of tasks must reliably and validly represent and cover the most essential elements of knowledge, of ability and achievement dimensions as well as psychomotor skills, and they must, in line with basic assumptions of the simulation model, enable us to predict the expected behavior as accurately as possible. Rational knowledge acquisition and rule-guided knowledge implementation within the task structures are essential conditions for valid predictions. However, the problem of prediction being as accurate as possible is that the negative acceleration of reliability in system simulation incurs exponentially increasing costs and a corresponding technological and software effort as well as additional instruction personnel.

Yet the tasks should not only reflect content aspects of complex action sequences, but, in connection with assessment and evaluation criteria, make it possible to find a mode of evaluating responsive behavior in terms of achievement and to establish the necessary feedback to the T. Synthetic illustrations of symbols and signs (symbols, graphics) will control and check the perceptive, cognitive and psychomotor responses.

Accordingly, the following technological criteria are valid for the assessment of aptitudes and achievement (norms):

- Time Measurements:
  - -- begin and duration of stimulus presentation,
  - -- begin and duration of psychomotor responses,
  - -- begin and duration of possible corrective actions,
- Measurements of accuracy and precision (quality of achievement)
  - -- frequency of occurrence,
  - -- quantity achieved (performance volume),
  - -- consumption of resources,
- Psycho-physiological measurements,
- Assessments of behavior.

Considering the operationalised and hence empirically verifiable basis, these criteria will ensure

- objective and reliable comparisons between individuals,
- statistical analyses and methodical evidence,
- fundamental information about the successful application of instruction/training programmes, as well as,
- valid assessments of the probability of the subject's later success, if panel data and results are taken into account.

These criteria, used as the basis of assessments and performances, are of growing importance as a function of the increasingly more complex use of system simulation and configurations to support and accomplish task entities. From a methodological point of view this seems to be contradictory in some way having "soft data" (assessments of intrapsycho-dynamic processes) as the basis of "hard data" (mechanical hardware and synthetic languages) and vice versa.

Within this complex action context psychological factors such as

- multiple mental load on the cognitive, perceptive and psycho-physical system,
- fear of failure (to commit mistakes) and stress, as well as
- coping with tasks under time pressure

will act as moderators or intervening variables which will have an additional impact that is difficult to check methodically or to explain the absence of the expected behavioral outcomes and performances.

3.0 PERSONNEL SELECTION

Procurement and use of cost-intensive simulators can be justified for many important reasons, one of them being the optimization of training. However, the simulator should not be regarded as a substitute for a real system, but it is only through the integrated and coordinated inter-connection of learning and training at the simulator as well as in the real system that - at least according to the training concept - long-term effective and successful training can be ensured.

The history of the development of simulation systems confirms the fact that efforts have frequently been focused only on the requirements of the system and their technological realization (including the interactive interfaces). At least of equal importance, however, are the requirements to be met by those who will train with these systems.

Within the German Federal Armed Forces the legal regulations concerning professional careers, personnel structure and policy provide the framework for recruitment or military assignment. Psycho-diagnostic and socio-psychological procedures as well as aptitude and achievement tests are used to investigate the individual preconditions of each candidate, to relate insights from several independent assessors and to back up any prediction concerning a future successful career in a given military assignment.

These selection and/or classification procedures (cf. Assessment Center) are designed to ensure that an individual will be assessed as "apt" if the appropriate personal, intellectual and physical conditions have been identified which would satisfy the training course requirements for the military assignment and enable the individual to complete these courses successfully.

In this context - and this applies particularly to beginners - the personal learning and development potential of the individual are taken into account, since specific personality-related structures and achievements can be crystallized, molded, changed (through learning) or kept at a certain level (through training).

The above mentioned criteria for the description of reactive functional behavior are completed by other (as far as possible) operationalised aptitudes which describe the personal, cognitive and physical (health) conditions for a selection or placement decision formalized according to allocation rules. Theoretically, this decision is based on a dynamic aptitude concept which is not only derived from the product of an achievement or an ability, but which takes into account particularly the process-type events of performance behavior. Thus, aptitude is interpreted as the sum total of time-constant and time-variable abilities, performances and personal dimensions which determine both, the process and the product of a observed behavior.

Personal characteristics include dimensions such as

- problem-analysis behavior
- conscientiousness
- judgment
- decision making and execution
- social competence
- group cohesiveness
- learning/performance motivation
- leadership responsibility.
- coping mental and physical workload

Cognitive abilities and functional performances such as psycho-physical stress resistance, concentration, memory are described either by conventional or, if possible, by adaptive test batteries (cf. computer adaptive testing). Although the influence of these personality traits will become only partially relevant for training on system simulators and systems trainers, they are seen as conditions to be met in the sense of personal aptitude being generally applicable to military assignment within the German Air Force (GAF), and hence they will be taken into account accordingly in the assignment to special military training courses.

As in the selection or classification procedures which combine data from several and diverse stations and diagnostic methods by independent assessors into an overall assessment, the methods applied to this condensation of information are of particular, critical importance. Such a decision - especially in case of a decision taken by the assessor board - frequently fails to spell out the extent to which, who, at what time, at what station, applied what abstraction, substitution, comparison or weighting processes which have ultimately had an important influence on the aptitude decision and allocation taken.

If this information fails to be recorded in a manner which permits its content to be appreciated with insight and to be empirically verifiable, the reliability and validity of the results can no longer be examined in the light of learning/training success. There is empirical evidence that the application of sequential decision matrices for forecasting professional success is promising if, during the first separate step,

- "hard" data as objective and quantitative measurements directly generated by the candidate (e.g. data of aptitude or performance tests, psycho-physical measurements such as rectal temperature, heart rate ...) are related by mathematical models and matched (conjoint measurements) in matrices for preliminary placement decisions, and then, in a second step,

- "soft" data (e.g. assessments by independent experts of complex problem-solution, decision behavior, motivation ...) are used to assign ranks in the sense of aptitude categories.

This principle of "soft modelling" achieves a high degree of validity for aptitude prognosis and learning/training success (OTTE, 1985, 1992).

The aim of selection is not to admit (by mistake) any personnel to training who will most probably not be successful. In empirical analyses of training units on a full mission simulator those pilots performed best who - in comparison to a control group - scored above average in the following dimensions:

Prior to training

- high perceptual speed,
- ability of attention sharing,
- responsiveness,
- spatial anticipation and imaging.

The larger group of these pilots demonstrated better training results whenever the pilots showed more marked traits with respect to

- understanding of instructions,
- ability of Gestalt-perception,
- power of decisiveness,
- tolerance to psycho-physical stressors,
- conscientiousness,
- ability to concentrate and to memorize.

A particularly marked high degree of action competence - also under great psycho-physical strain and mental stress - is seen as a good prerequisite for the successful completion of a pilot's training.

In deciding about the selection or the assignment of personnel to go on to cost-intensive training and extension training it is crucial - from an institutional point of view - to avoid, if possible, two types of error:

- to admit, by error, a person to training units on a system simulator who later fails to pass the test in a real situation or lags behind the expected performance level

or

- to reject, by error, an individual who though rejected during a selection procedure, would still be able through training to cope with the real requirements, if he were allowed to do so.

From the point of view of the institution, the error of incorrectly admitting a person who fails to meet the objectives later is the more painful one. This decision does not only burden the selection system, but it also causes costs at the expense of training or of the real system or also causes a delay for an "apt" trainee who could now be trained later and came later to service.

In addition, such erroneous decisions result in delays in the training of "apt" personnel given the limited instruction/training capacities. Findings show that failures of this types are frequently due to causes which could not be diagnosed and/or verified at the time of the selection procedure, so that these erroneous decisions cannot be held a posteriori against the selection system.

However, the other error is more difficult to avoid, because "rejected" persons very seldom get the opportunity to participate - despite their negative results in an selection procedure - in further instruction/ training units.

Empirical evidence demonstrates that especially negative assessments of personal characteristics which are used to justify such rejections, sometimes may no longer be upheld or assessed differently by the same expert when looking at the same data at a later date. Therefore, the ratings of personality traits should not be the only basis of the overall aptitude assessment.

In spite of these not very encouraging results from expert ratings, we should continue to let experts assess in a more global context as a "check" of the objective, quantitative and interpersonally comparable behavior and test data directly generated by the candidate.

Considering the two types of error, analyses of "errors" at all levels are of particular importance. The following considerations assume that development and procedural errors in the design of diagnostic methods have been eliminated as far as possible.

3.1 Behavioral Deficit Analysis and Maladjustment

In section 3.0 I mentioned major variables which are important for assessing "aptitude" and/or for the prediction of any subsequent success with a high degree of probability or lack of aptitude and/or low probability of subsequent success. In case of quantified characteristics, a "not apt" assessment is substantiated on the basis of frequencies and distributions of;

- error scores,
- reduced quantity per time (unit),
- ratio between error and correct responses,
- quality of errors.

To define the quality of responses or behavior we can also use the generally valid description of "human reliability" (cf. SWAIN & GUTTMAN, 1983) according to which individuals (re)act reliably in the sense of the system if they

- (re)act in accordance with the requirements and the system as well as correctly within the expected time span, and
- show no other activities which would interfere with the system.

According to this definition any behavior or reaction could be termed an "error," if somebody

- (re)acts incorrectly,
- fails to (re)act,

- fails to (re)act in the expected or defined manner, or
- though (re)acting correctly, fails to (re)act within the required time interval.

In a systematic way, flawed acting could be classified as follows:

- wrong selection of a particular action from several action alternatives (selection error),
- sequential error,
- time error,
- qualitative error.

An important requirement for an efficient and effective error analysis is not so much the description of error types as the conditions and causes of their origin. Frequently such error analyses demonstrate inadequate personal background, lack of motivation, high risk acceptance, lack of emotional control or insufficient tolerance to stressors.

4.0 LEARNING/TRAINING

Generally, learning can be defined as a change in behavior substantiated by experience. Training, however, is rather the practicing of sequences of actions with the aim of developing or maintaining a specific ability or skill. All abilities and skills require crystallizing, molding and consolidation. Training is closely related to repetitions or exercises. Both concepts are important conditions for the application of knowledge, ability and skills.

The instruction and training units on simulators in the context of military requirements are more aimed at canalizing knowledge and skills and to optimize them in terms of criteria such as accuracy and speed. This applies particularly to complex sequences of actions which are to be repeated especially under different and divergent conditions and which are to be integrated and automated as far as possible within the psychomotor part of behavior.

The development and application of instruction/training programmes are based on the assumption that

- differences exist at the beginning of the learning/ training between the predetermined criteria of a task (TARGET NORM) and the behavior observed (abilities, motivation...) as the ACTUAL NORM,
- this difference between the TARGET/ACTUAL can be reduced through learning/training,
- learning/training processes are necessary conditions for the reduction or maintenance of existing, tolerable TARGET/ACTUAL differences.

Simulators are able to considerably reduce the time required for such training and the costs through suitable configurations in comparison to training in a real system:

- through individually adjusted and isolated repetitions of complex or difficult sequences of actions, and
- through a degree of availability which is practically unlimited because it is relatively independent of any external constraints.

Performance and behaviour parameters can be recorded and analyzed during any selected phase so that the necessary treatments may be applied adaptively.

4.1 Task Presentation and Reactions to Requirements

Depending on type, scope and intention, the different tasks and those with different degrees of complexity according to requirements as well as the sequences of these tasks may be laid down. Every numeric and graphic information is represented in a manner and form which is in line with ergonomic knowledge. Hierarchical and structural presentation of tasks facilitates any subsequent analysis of the flood of data as well as the systematic and integrative assessment of performances. The methodical quality of the recorded data and the corresponding analyses will ultimately determine the present tasks of the system as well as the abilities and performances of the candidate.

4.2 Diagnosing and Describing Differences

The term "diagnosis" within this model designates not only the determination of a certain, observed behavior in terms of an ACTUAL state and the concluding comparison with the fixed TARGET NORM or criteria, but the term also includes the recording and analyzing of all relevant data which have led to the description of the ACTUAL state.

Besides focusing on the mentioned assessment characteristics, the diagnosis of individual behavior comprises at its centre

- systematic information processing,
- processes of problem solution and decision behaviour,
- processes of memory performances

and the determination of their critical values under varied and diverse requirement conditions and relates them to

- time events and intervals,
- quantity and quality,
- frequency of errors and their history.

For most simulation requirements in the military field we find that

- time required for an action and
- accuracy

demonstrate a high degree of positive correlation:

The more time becomes available for a tracking task the more accurately, as a rule, can the task be executed. Linear shortening of response time intervals between successive stimulus configurations (forced pacing) will result in an exponential increase in the rate of false responses. Hence, diagnosis constitutes a comparative statement about functional relationship between ACTUAL states and the corresponding TARGET norms.

TARGET norms are reflected in the technologically feasible and predetermined rating instructions for the tasks to be tackled and provide, in the context of the "man-machine" interface, the basis for the assessment of performance and abilities and for the subsequent classification and selection decisions.

In contrast to these technological TARGET norms, the description of the ACTUAL states as functional norms imply notions such as “motivation,” "abilities" and "performances". These notions cover behavioral aspects which can be compared or evaluated either in isolation or as interacting with defined criteria.

4.3 Routine for Exactification

In the context of instruction/training programmes, the products of instruction/ training processes are reviewed on the basis of the above criteria. The assumption is that there are differences between actual achievements and the predetermined (a priori) criteria which in the light of accuracy requirements must be reduced through special procedures. Based on theory-of-learning principles, this "exactification" routine is to use methods and procedures

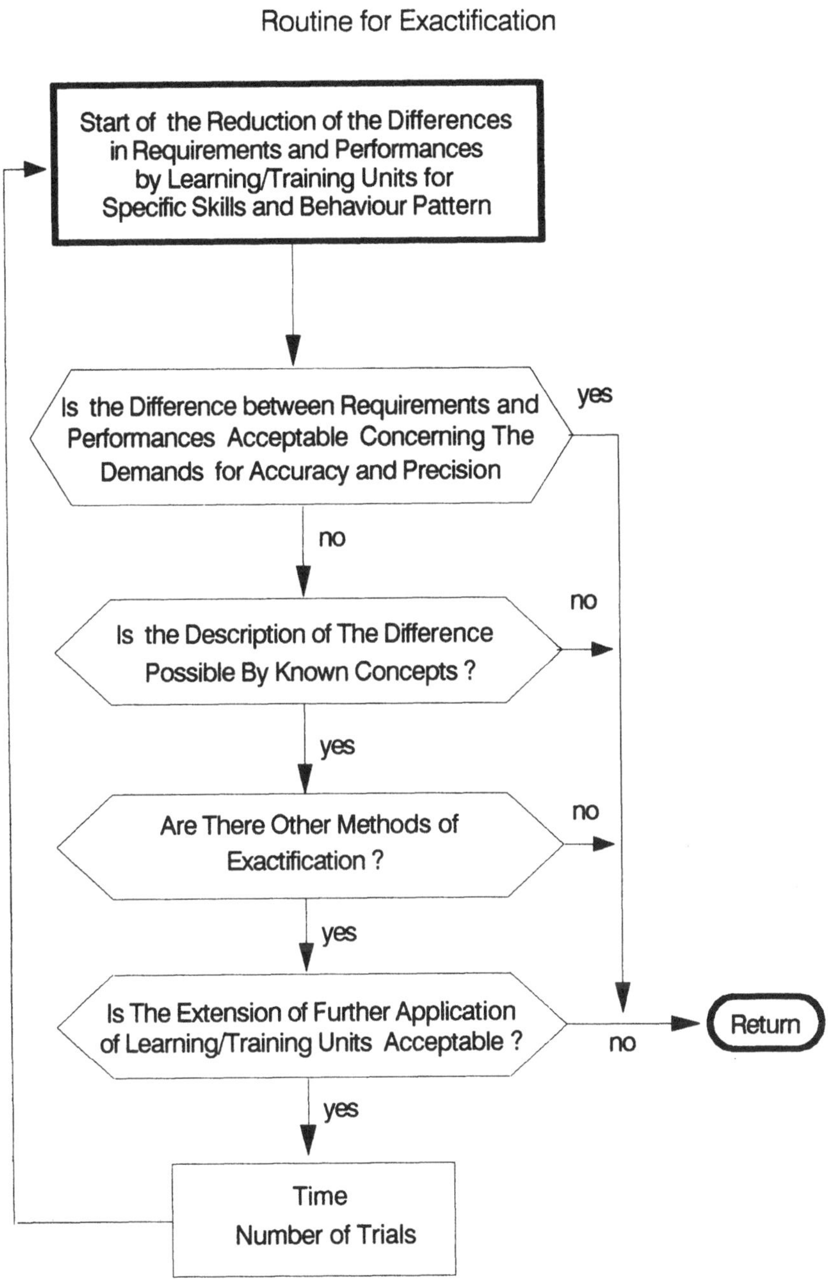

Figure 2. Routine for Exactification of Behaviour Pattern

which permit the isolation of behavioral sequences, and to train accordingly with the help of drills and tasks so that the accuracy requirement with respect to the predetermined criteria and expected behavior can be met.

The term "exactification" describes the iterative process of the reduction of the TARGET/ACTUAL difference. "Accuracy" or "precision" are the guiding principles for the concluding comparison and the corresponding assessment between the predetermined and achievement (as the behavior product). It should be noted that the reduction of the criteria/achievement difference also involves a more precise conceptual description of TARGET/ACTUAL discrepancies. Together with the "exactification" routine we must also make the diagnostic descriptions more precise and improve conceptually the possibly underlying expert system.

4.4 Feedback

According to current knowledge, effective learning depends essentially on type, scope and time of reinforcement. Particularly in learning situations which are implemented on system simulators, (T) is required to understand and analyze complex stimulus configurations and to answer and match them with corresponding behavior modes.

Criteria such as speed and accuracy of the stimulus-response-contiguity are decisively dependent on adequate adaptive feedback hierarchies which must be immediately available to the "errors" or any undesirable behavior. This not only involves reinforcement which follows on the heels of the stimulus-response situation, but the effort is rather aimed at embedding and at understanding the overall relationship between a critical training situation and the desired behaviour, in order to secure long-termed training (learning) effects. Computer-assisted (based) instructions (CAI,ICAI, CBI...) are able to support the training results in every respect.

POLZELLA (1983) discussed different instructions and representations in the case of simulation and proves the effectiveness of knowing about the results in the form of feedback for the purpose of successful training of psychomotor skills. Depending on the complexity of the training (e.g. communication and fire control on platoon level) there is no doubt that a positive social climate and positive social reinforcement by tank crew members or by the instructors will have a great influence on the lasting training success and a positive training transfer. During the training periods direct social feedback by the instructor will confirm the trainee in better following and understanding the instructions or in improving behavioral outcomes and performances, especially when performances of the team (tank crew) were controlled.

Empirical results showed that the team performances of the tank crews of which each member had reached average skill performances, but were highly cooperative in the group, scored significantly higher than the "normal" (randomized) tank crews.

5.0 INTERDISCIPLINARY ANALYSIS AND TREATMENT APPLICATION

If the above mentioned training support measures fail to be successful, e.g. if the differences between ACTUAL states and TARGET criteria are unacceptable, then a comprehensive cause-effect analysis becomes necessary, in order to identify and, if possible, to eliminate any system and/or selection generated errors or restrictions.

Everybody who is involved in the technical realization of complex hardware and software systems is aware of the problem that, though the subsystems are working completely smoothly, these interacting and communicating subsystems as a whole are liable to produce errors which are difficult to identify and which can have a more or less harmful effect on the systems. If technical deficits in hardware-systems may be eliminated by investing more money in better technology or infrastructure, eliminating problems is however more difficult with regard to the man-machine-interface and, particularly, with regard to training.

Lack of training success leads to the presumptuous conclusion that poor motivation or low aptitude levels are the reasons for failures, and these failures could have been avoided by applying suitable placement criteria. This premature conclusion is indeed induced by the fact that the amount of effort to "cure" technical deficiencies is equated with the effort to overcome human inadequacies.

But there are many unforeseen reasons caused by inappropriate methodical and didactical training-treatment-interactions producing unsuccessful training results. Therefore, in complex simulation systems with high demands on physical and psychological stress components an interdisciplinary expert team representing engineering, psychological and - as necessary - medical competence should monitor training phases, analyze deficits of training and implement correction procedures.

6.0 DATA OF A BRIEF TRAINING SEQUENCE ON A PART TASK TRAINER FPS 80

6.1 Description of the device

The part task trainer (FPS 80) is integrated as a part of the training and selection system to identify the suitability of novices for flight training and to provide accurate and objective measurements and assessments of the performances. The system include

- the capability of explaining the required task in a computer based and automatic mode,
- continously monitoring his performances from the beginning of the task presentation to applicant's reaction,
- providing cues and components as required,
- recording and processing data and performance parameters.

The identical test stations include a seat and the standard elements of general aviation aircraft cockpits, such as landing gear lever, flap lever, frequency selection panel, stick, rudder, and throttle and all necessary flight instruments. Three CRT-indicators, part of a fast high resolution color graphic system and ETHERNET-LAN connected, provide the visual system which show a field of view of 36 degree vertically and approx. 120 degrees horizontal field of view. These monitors are synchronized imaging an 3D-visual scenery.

6.2. Tasks

Five different tasks called missions will be available for selection and/or screening purposes. Duration and contents of these missions were selected and designed in cooperation between the staff of the Flight Medical Center, experienced pilots and CAE as the manufacturer of the system. Tasks can be changed or added as required. Generally, the tasks are parted in 3 phases:

<u>Phase 1. Demonstration:</u>

All necessary controls, instruments, and visual displays for the particular mission will be demonstrated and explained.

<u>Phase 2. Practice:</u>

During the performing these measurements will be compared to the ideal flight parameters. Visual feedback will be provided as soon as the trainee leaves predetermined tolerances. Reaching the maximum value will terminate the mission.

<u>Phase 3. Check:</u>

During this phase no feedback will be given to the trainee, but all actual flight parameters will be continously registered, compared to the ideal parameters and stored. If the simulated aircraft will enter an after termination flight instruments and the visual system will return to the initial positions.

The description of the five tasks (missions):

1. Mission FPS-01
   - introduction of visual system
   - take-off and climb,
   - straight and level flight,
   - change of heading, 20, 40, 60 degrees bank,
   - return to base with touch-down, (demonstration only during all phases),

2. Mission FPS-02
   - take-off from base and climb,
   - repetition of FPS-01-manoeuvres,
   - return to base circuit,
   - take-offs, touch-downs with circuit,

3. Mission FPS-03
   - take-off and leaving the circuit,
   - low-level navigation cross-country flight with check, points, identification points such as buildings, roads, and mountains will be displayed,
   - return to base and touch-down,

4. Mission FPS-04
   - take-offs and 90 degree turn to an altitude of 6.000 ft AGL,
   - recovery of unusual attitudes (nose-down, nose-up, with or without banking), requirement: recovery from these unusual attitudes,
   - formation flight,
   - air target chase with firing of gun and/or missiles,
   - touch-down,

5. Mission FPS-05
   - take-off,
   - penetration of a rotating tunnel,(synthetically imaged, geometric figures of differen colours and shapes, the tunnel appears to be infinite the task is to fly the aircraft level with respect to the imaged physical limitations).

6.3 Performance measurements

The following variables are measured and can be used for data analysis:

- pitch, bank, yaw,
- stick deflection: lateral, longitudinal,
- rudder deflection,
- speed, rate of climb, heading,
- deviation from ground track,
- distance to other aircrafts,
- keyboard inputs
- number of hits and misses

Data are gathered, analyzed and stored continously. Further statistical analysis and evaluation are kept variable and can be performed after finishing the missions.

6.4 Results

Figure 3 illustrates the results of a brief training period on the part task trainer of 188 novices with only a few hours of real flight training.

The mass of qualitative and - even more - quantitative data can be reduced by different ways. Here a matrix was chosen which show three areas representing the groups of differen

Tests of cognitive Information processing, instructor assessments

Performance on Simulator	8	7	6	5	4	3	2	1
-49	4	2	2	3			1	
50-54	5	2	1	1				
55-59	7	10	5	1	1		1	1
60-64	3	7	15	8	8		1	
65-69	1	4	1	19	8	2	2	
70-74	1	3	2	8	10	6	3	1
75-79	2		1	4	3	1	2	2
80-100	1			1	5	1	3	2

Trainees of category 3 (less than 40 % of the maximum)
Trainees of category 2 (between 40 and 70 % of the maximum)
Trainees of category 1 (more than 70 % of the maximum)

Figure 3: Flight training on simulator FPS 80

training success. Areas of the matrix can be analyzed to yield certain error types, behavioral maladjustments or incompatibilities of the training design.

This matrix as a heuristic model of "soft modeling" was built by the statistical reduction (factor analysis) of several measurements of the psycho-motoric tracking tasks with simultaneous control of the instruments and the instructor's assessment after a flight training. The axes of the matrix were defined by factor-analyzing these measurements and the corresponding assessments (on the x-axis) and the performances on the simulator (on the y-axis), representing the percentage of the fulfilled requirements. Each of the independent factor-analyses yielded three second-order-factors of which each first factor (second order) recorded at least 60 % of the total variance of the corresponding set of measurements. The categories were built by the method of (LEWIS, 1966).

Important to know is that individuals in one cell of the matrix are parallelized by a huge number of different diagnostic information resources. This gives more information and more opportunities to analyze the training process and the variables involved and to apply additional training in different ways to the individual. Further, these results show - at the first view - the calamity of the general linear model which fails to be an appropriate method for representing all the data and effects. At this moment the process of analyzing the data for further conclusions is on-going.

LITERATURE

LEWIS, A. E. (1966) Biostatistics, New York: Reinhold

NEWELL, A. & SIMON, H.A. (1972) Human Problem Solving Englewood Cliffs, NJ: Prentice Hall

OTTE, R. (1985) Eignungsreihenfolgen für Offizierbewerber (Soft modeling) BMVg-PII4, GMOD, 1985 (unpublished)

OTTE, R. (1992) A Heuristic Model of Integrated Data Reduction Workshop NATO-RSG 15 in Brussel, 1991

POLZELLA, D.J. (1983) Air Crew Training Devices: Utility and Utilization of Advanced Instructional Features AF HRLRep. TR 83-22

ROUSE, W.B. (1980) Design and Evaluation of Computer-based Decision Support Systems In: SALVENDY, G. (ED) Advances in human factors/ergonomics I: Human-computer Interaction N.Y.:Elsevier, pp. 229-247

STICHA, P.J., SINGER, M. J., BLACKSTEN, H. R., MUMAW, R. J., & BUEDE, D. M. (1987) Optimization of Simulation-based Training Systems - Year II Report; Interim Report 87-30, HumRRO, IR-PRD-87-30

SWAIN, A.D. & GUTTMAN, H.E. (1983) Handbook of Human Reliability Analysis Sandia Nat. Lab.,NUREG/CR-1c278, Washington

CHANGING STRATEGIES FOR TRAINING MILITARY UNITS[1]

Franklin L. Moses and Jack I. Laveson

U.S. Army Research Institute for the
Behavioral and Social Sciences
Alexandria, Virginia, USA

ABSTRACT

New technologies and methods are emerging for training military units. The promise of simulation technologies is discussed as an element of a family of effective training methods. Historically, unit training has been provided through a mixture of small-unit drills and periodic field exercises. However, a new variety of threats, an increasing need for rapid deployment, and the availability of more capable equipment make operational and related training requirements more complex. At the same time, there will be fewer resources in the future: the size of the force will be reduced, budgets will decrease, and range and maneuver facilities will fall short of needs. The question addressed in this chapter is how training can meet a changing threat and the need for operational readiness.

INTRODUCTION

How should we progress from current military training methods into the future? To be ready to fight and win, without need for specific training just before deployment, continues to be a goal that requires training at all levels -- individual, leader, and unit. However, resource constraints including equipment, maneuver space, and time are limiting our training options. Our concern is "how we train to fight." While training effectiveness versus costs must be assessed, the critical question is how to do this rationally.

This paper focuses on how to determine the best mix of alternative training methods for large units or groups (i.e., combined arms battalions and brigades, and even joint or unified operations).[2] The emphasis is on simulation and related technologies that are projected to

---

[1] The opinions expressed in this paper are those of the authors and do not represent the official position either of the United States' Department of the Army or the Department of Defense.

[2] Note that "large" units sometimes imply "operational level of war" rather than tactical-level to include logistics. The following definitions are offered:

Combined: More than one branch within a Service as in Mechanized Infantry **<u>and</u>** armor.

be a significant component of future training strategies. The newer technologies need to be integrated with the more traditional training methods such as large scale field training exercises and command post exercises. Tradeoffs among methods should be based on assessment of the effectiveness of each and comparisons among them.

Specialized evaluation facilities for large units are one answer to this problem. The key to these facilities is the use of sophisticated data analysis procedures and display technologies. They can serve as a tool to integrate information obtained from training methods. In addition, information can be played back in various forms to allow evaluation hypotheses to be developed and confirmed. Current facilities that have some display and data analysis capabilities include the Army National Training Center (NTC) at Fort Irwin, the Air Force Operation Red Flag at Nellis AFB, and the Navy Air Combat Maneuvering Range at San Diego, CA.

Although the paper uses the authors' experience with the U.S. Army, the topics have broad implications for large units, whether the forces are single service, joint, or unified. These ideas apply wherever training involves a large area of real estate and large amounts of personnel, equipment, time, and other resources. Four major topics are considered: (a) the need for unit combat training, (b) the integration of simulation into unit training strategies, (c) the assessment of unit training technologies, and (d) the facilities/infrastructure for performing the assessments.

1.0 THE NEED FOR UNIT COMBAT TRAINING

Much has and will continue to be done in the area of individual training, but unit collective training needs emphasis. Consider the changes occurring in the threat (e.g., reduction in Soviet power and growing regional power threats) that apply pressure to modify or redesign unit training programs. Desert Shield highlighted the challenge of training combat units, particularly Army combined arms forces (Brown, 1991). The implication is that both the Active and Reserve components (AC/RC) of the United States' (U.S.) military and its foreign allies could benefit from training and rehearsal at the unit level.

Large unit field training is expensive and complex. It typically requires long-distance travel for personnel and equipment. There are added resources to operate and sustain a force for field training exercises (FTXs) in locales such as the NTC. Damage to the environment and danger to local populations add intangible costs. The need is to increase training effectiveness and efficiency despite more complex operational requirements and declining resources.

Decreasing resources create problems for training. The escalating costs of "drill and practice,"(e.g., flying, driving, steaming, and using "live fire") already have reduced training hours. For example, consider (a) how the cost of live ammunition affects training, and (b) how missiles already have capabilities that exceed most range limits. In the U.S. and NATO nations, the 1990s are likely to produce greater training demands with fewer resources.

Finally, the biggest driver for changing training is that military missions are becoming more complex. This is evident in several ways. The threats are more variable and in locations with no forward deployed forces -- Iraq, Grenada, Panama -- to name only a few. Forces with advanced technology equipment must rapidly deploy in contingency theaters. Furthermore, there is a growing dependence on specialized elements of heavy, light, and

---

Joint: More than one military Service within a country.

Unified: Multiple countries fighting as one force.

special operations forces. The net result is the need to consider how to change training strategies to maintain our military readiness.

2.0 THE INTEGRATION OF SIMULATION INTO UNIT TRAINING STRATEGIES

Evolution of Unit Training

Large unit exercises typically have been infrequent events. Even when resources were available, exercises such as those conducted once-a-year by NATO in Europe required extensive preparation, time and logistics. An alternative to large unit exercises is to use limited training areas with constraints including number of participants, types of service branches, weapon ranges, and so on. Examples of limited U.S. training areas are Twenty-Nine Palms, California (U.S. Marine drills) and Fort Hood, Texas (Army Mechanized). Limited areas provide the opportunity for practice of field maneuvers with some live fire.

During the 1970s, the National Training Center (NTC), Fort Irwin, California introduced computer-supported and instrumented training on a large scale. Real weapon systems equipped with low-power lasers for ammunition provided combat realism. Battle movements and (laser) engagements were recorded for analysis and after action reviews. Weapon simulation helped to make field training more realistic and beneficial by overcoming many constraints of limited training areas.

The Promise of Simulations

The use of more extensive simulations (including models, wargames, and networks) for helping training shows great promise. The challenge is to determine the best way to employ them in a mix of training methods. Simulation, as used in this paper, deals with devices, simulators, and "models" of battlefield activities that include people as an integral part of their operation. Simulations are an analog of the "real thing" and have soldier interactive input.

Simulations can provide the opportunity to improve unit collective performance wherever people need to practice expensive or dangerous activities under realistic conditions. However, its early use for individual training received mixed reviews. For example, the first simulator training of individual aviator skills using the Link trainer ("blue box") was viewed as no substitute for actual flight time even though it was an important training asset. The same is true of more recent simulations of aircraft and vehicle operations that train individuals and small crews, particularly in emergency procedures. Traditional simulation technology has been very expensive, limited to fixed locations, and difficult to reprogram. Between about 1940 and 1980, users viewed simulators as necessary at best and to be avoided when possible.

In the 1980s, at the Defense Research Projects Agency (DARPA), attention was focused on how simulation technology could be improved and utilized for training combat units (for a history, see Alluisi, 1991). Warfare training was the purpose, but similar technology could benefit other unit missions such as fire fighting and disaster mobilization. The essential methods are to use new adaptations of "virtual world" and "networking" technologies.

Both "virtual world" and "networking" are familiar terms, but in combination they may not be. The virtual world uses computerized simulations to provide environments that are representative, but not necessarily duplicative, of any part of the real world. Virtual worlds can include the working environment for commanders, crews, teams, and so on. Networking uses computer communication to link many of the simulations together so they can interact in real-time. Such interconnections among individuals, groups and systems, including actual equipment and simulators, can allow training under battle-like conditions on a common battle area. One example of this is the SIMulator NETworking (SIMNET) concept and demonstration summarized next.

Simulations and the SIMNET Concept

SIMNET has become a premier demonstration of combined arms exercises using simulation and networks. It provides a network of low cost computers, computer graphics, and combat vehicle simulators that demonstrate the concept of tactical training of armor and mechanized infantry units (U.S. Army Armor School, 1989). In addition, an expanded version of SIMNET demonstrated a joint capability by networking helicopters, fixed wing aircraft, and an "off-shore" battleship. Other demonstrations include leader, commander, and staff exercises in command, control, and tactical movement at platoon, company/team, and battalion/task force. Some applications of SIMNET at Division and Corps have been tried. SIMNET is evidence that use of modern computers, graphics and communication technology allows simulation: (a) to be less expensive, (b) to be available for many distributed locations, and (c) to support "free play" among large units on a virtual battlefield.

Of course, improvements can be made to increase realism and utility of the SIMNET concept. Progress and plans already are underway to expand, verify, and validate it in a program called Battlefield Distributed Simulation (BDS). Present efforts include supporting dismounted infantry and obscuration. They also are aimed at improving graphics resolution and detail, accommodating existing aircraft simulators, and so on. An underlying concept is to achieve interoperability among heterogeneous simulations. A "common" language and standards are being developed for interactions (see, for example, Loper, et. al., 1991). A major objective is to combine manned and unmanned simulations to provide a potential for training units within larger entities without having all forces actually networked.

3.0 THE ASSESSMENT OF UNIT TRAINING TECHNOLOGIES: AN APPROACH

Assessing unit training and the success of training technologies always has been difficult. The advent of simulation with its capability to record data faster and in more detail than possible with human observers (e.g., umpires, judges, evaluators) emphasizes the need for better unit evaluation. The only real test is the effect of training on actual battle performance. Yet, as Deitchman (1988, p. 2) put it, "the available knowledge about the effect of training on the outcomes of battles and wars is the stuff of anecdote and history, but quantitative descriptions of the relationships among the multitudinous variables that affect the outcome are usually shrouded in mystery." The solution lies in (a) detailed knowledge of training/mission needs, (b) validated measures of unit performance, and (c) a system for determining training strategies, i.e., the appropriate training methods to use in terms of costs and indicators of performance.

Determining Large Unit Training Needs

Determining training needs starts with a statement of unit mission. From this statement, subordinate units and then individual soldier training objectives are identified. For individual soldiers and small unit performance, this approach has been fine. However, large unit training requires more than just a scaling up of the individual and small unit training process.

Consider how the training process works for smaller units (company and below) where the emphasis is on individual soldier performance. Current training is based on mission along with enemy, terrain, troops, and time available -- METT-T. From these, a mission essential task list (METL) is developed. METL items are compared to training methods (e.g., map exercise, MAPEX; tactical exercise without troops, TEWT; field training exercise, FTX; command post exercise, CPX) to determine which ones to use.

For smaller units, extensive task analysis data are available providing a basis for measurement. Training is viewed by evaluators or controllers who provide unit performance estimates in the form of After Action Reviews (AARs). The AAR addresses what happened during training and what training could have been performed better. Smaller units can be considered as having limited interactions (dealing with only a few other personnel) making analysis straightforward. For large units, extensive task analysis data are not available and interactions are extensive.

Understanding and measurement of large units is not straightforward because it depends on how well all the pieces work together. Interactions have received little attention. When the performances of individuals and small units have been combined to assess larger units, the results have been incomplete because interactions are lost. The problem with large unit analysis is summarized by the following for task forces:

> "Army Task Forces are open ended nonlinear systems composed of subsystems which interact one with the other and with their environment. Task Forces as complex systems are unanalyzable into irreducible parts because the parts as with any open system are constantly being folded into each other by interactions and feedback....The whole system is viewed as an interrelated process system and not necessarily hierarchical" (McFann, Hiller, & McCluskey, 1990, p. 2).[3]

A closer look is needed at a system to get at the appropriate factors and assess them. Clearly, it is training objectives that must drive the selection of training methods. Other factors considered in the selection are a function of mission, time available, and varying levels of training resource availability. An NTC may be necessary for some purposes, or a Command Post Exercise or Tactical Exercise without Troops for others. A simulation network may be the training method of choice for certain needs. The issue is whether the training will satisfy the requirements and whether it is worth its cost.

A System for Determining Training Strategies

To determine a system for training and training assessment requires several Research and Development (R&D) phases shown in Figure 1. The large unit mission is the driver. The phases after that are necessary to assess properly the possible mixes of training methods for large units.

Phase I - Determine Task Clusters. The objective of this phase is to determine how task clusters for unit performance can be developed. Task clusters identify what should be trained, but not how training should be done. A task cluster is defined as a set of tasks or groupings of related tasks encompassing major human functions in combat. Task clusters, thus, are broader than tasks performed by individuals. They address the major operational functions performed by units. Examples include communication, command planning, vehicle maneuver, and weapon operation (Miller & Bonder, 1982).

One example of how to develop task clusters is through task analysis, a logical decomposition process for combat missions:

- A list of generic unit tasks for larger units, such as division, brigade, and battalion, are collected from doctrinal publications and combat models.

---

3 The complexity of the planning process makes it clear why performance measurement is challenging. Simply viewing the intra-echelon information processing requirements of command and control (C^2) functions results in many interrelated tasks and feedback loops (Hicinbotham, et. al., 1990).

- Combat missions such as hasty attack, defend in sector, etc. are decomposed to determine which generic tasks and clusters apply.

- The resulting task clusters are reviewed by commanders and selected staff of large units to confirm that tasks have been identified and properly organized.

LARGE UNIT MISSION (driver)

↓

Phase I
DETERMINE TASK CLUSTERS

↓

Phase II
DEVELOP INDICATORS

↓

Phase III
VALIDATE INDICATORS

↓

Phase IV

MODEL & ASSESS TRAINING STRATEGIES

Figure 1: Research and Development Phases

Generic task clusters (Miller & Bonder, 1982) associated with fire support, for example, are Artillery Fire and Fire Support Allocation.

- Artillery Fire tasks consist of: ordnance selection, target element acquisition and selection (if point fire), decision to fire, delivery pattern, delivery accuracy, target posture, suppressive effects of fire, effect of fire on communications and decision making ability, and time to fire or firing rate.

- Fire Support Allocation tasks consist of: target and mission priorities, preferences for type and amount of support against a given target in a given situation, preferences for types and amounts of fire to be allocated to preplanned missions, criteria for ignoring targets or dropping them from a list of acquired targets, and delays in making and implementing decisions.

Phase II - Develop Indicators. The objective of this phase is to develop performance indicators that can be associated with the task clusters. That provides the basis for comparing training alternatives and, in turn, how to enhance combat performance.

Indicator development requires three sequential steps. The first is an enumeration of possible indicators and their assignment to a categorization such as process measures, outcome measures, and subjective judgment described below. The second is the development of a set of screening criteria considering meaningfulness, relevancy and applicability. The last step is screening the indicators to identify the most promising candidates for matching with task clusters.

The **enumeration step** addresses the collection of indicators. There are no constraints on the indicators (e.g., implementation feasibility, applicability) as the goal is to collect the largest set. Indicators can come from military research, discussions with subject matter experts (SMEs), e.g., commanders, observers, and modelers, and inputs from simulations.

These indicators/measures can be categorized by using:

- PROCESS PERFORMANCE MEASURES - whether or not doctrinal tasks such as defend a sector were performed,
- OUTCOME MEASURES - results of mission performance as measured by objective measures such as enemy and friendly attrition, terrain control and movement, and
- SUBJECTIVE JUDGMENT - interpretation of the goodness or relevance of an action, process, or outcome.

The subjective judgment category is not mutually exclusive from process and outcome. Subjective (expert) judgment is effectively used in the AAR which addresses all battle phases of planning, preparation (e.g., a commander's or observer/ controller's judgment). All three types of performance measures are needed for assessing training effectiveness.

What follows is sample indicators based on mission planning, preparation, and execution. Associated with each indicator is the categorization into process (P), outcome (O), and subjective judgment (S).

Planning
- Iterative Actions Generated (P)
- Likelihood of a Plan's Success (S)

Preparation
- Logistics and Resupply Adequacy (P)
- Responsiveness to a Battle Plan (S)

Execution
- Total Kills (O)
- Loss Exchange Ratio (O)
- Total Ammunition Expenditures (O)
- Fuel Use Over Time (P)
- Battle Position Achieved/Progress Toward an Objective (P)
- Number of Command Communications (P)
- Supporting Fires (P)
- Departure from Phase Line Time (P)
- Line-of-Sight for Supporting Movements (P)
- Volume of Fire/Timing of Fire (P,S)
- Integrated Unit Movements (P,S)

Screening criteria need to be developed. One approach, described by Deitchman (1988), is to assess reasonableness based on the implication for interactions among training and war outcomes, simulations, etc. This and other approaches rely on expert judgment. To capture that judgment, a systematic elicitation approach such as used in the development of a knowledge base is needed.

To be effective, an indicator must meet at least the following criteria: (a) meaningful by being logical and understandable, (b) relevant to the task cluster being assessed, and (c) applicable to a variety of combat operations. It also must be reliable and valid in classical testing terms.

Consider, for example, an indicator such as progress toward an objective in terms of unit movement speed. A sustained (constant) unit movement speed could be considered the mark of a well planned maneuver and be meaningful. However, this assumes suitable terrain (flat versus mountainous) and a strategy requiring sustained movement in order to be relevant. What, if instead, the battle plan calls for a unit to slow down as in a feint. Then, without modification, the constant speed indicator would not meet the criterion for applicable. In other words, an indicator can be considered generalizable if it meets all three criteria.

Screening identifies which indicators selected above can be applied to specific task clusters. SMEs match indicators to task clusters at the attribute level (e.g., ordnance selection for Artillery Fire). The result is a set of potential indicators to be validated.

Phase III - Validate Indicators. The objective of this phase is validation of the indicators. These procedures need to be usable for all types of training. The current focus, however, will be on exercises (e.g., NTC) and simulations (e.g., SIMNET) because they are the most readily available sources of detailed training and outcome data. The data can be analyzed by experts who apply the indicators and then compare their results with (independent) expert evaluations of the battles.

There may only be two levels of performance -- "good" and "bad" or "go/no go" that are discriminated reliably. If additional rating points (e.g., excellent, good, fair, poor) could be associated with the expert judgments, then the indicators could be similarly evaluated. The validation is successful, however, if a list of effective indicators is compiled for measuring task clusters in the planning, preparation, and execution of battles. Indicators that are effective can be assessed against cost data in Phase IV thus resulting in cost-effectiveness values.

Phase IV - Model and Assess Training Strategies. The objective of this phase is to identify which training methods (e.g., simulation, FTX, CPX) can support the task clusters and then which overall strategies are cost-effective. This can be done by developing a

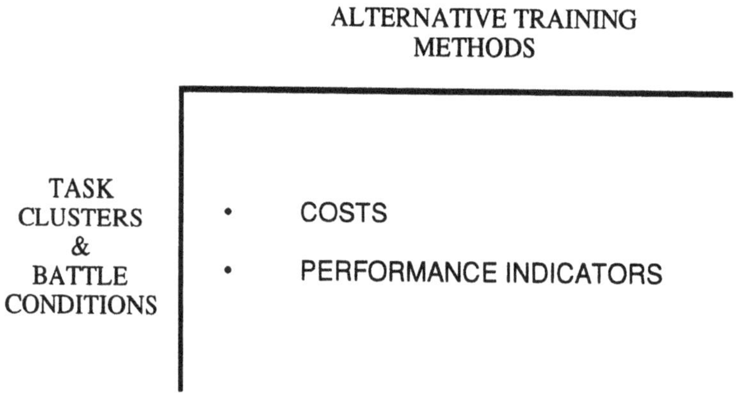

Figure 2: Matrix to Identify Strategies

matrix of task clusters and training methods showing how clusters can be supported (Figure 2). Comparative effectiveness is associated with strategies as a function of costs and performance indicators, where costs include time and other facility/resource factors.

Cost data, while seemingly straightforward to obtain, is expected to be difficult to use. Cost data are not based on task clusters but rather are accounted as total expenses for units and facilities. For example, field exercises include items such as fuel and repairs (e.g., petroleum, oils, and lubricants, and repair parts). There also are cost data such as transportation costs (e.g., moving a unit to the training site and back again) and personnel costs. These costs need to be summarized by task cluster and by training method in order to assess different training strategies.

When task clusters can be linked to multiple training methods, the relative efficiency and effectiveness of different strategies can be determined. In other cases, there may be one and only one method that can support a task cluster with associated costs and training effectiveness. We need to determine what training methods with what benefits at what costs are most suitable.

Ultimately, we would like a methodology that will estimate what the cost effectiveness is, for example, of spending "days" using a simulation facility like SIMNET in addition to using an instrumented facility such as the NTC for field exercises. That methodology may include statistical approaches such as response surface analysis and linear programming to study the combinations of possible strategies. The appropriate methodology remains to be determined.

4.0 FACILITIES/INFRASTRUCTURE FOR PERFORMING ASSESSMENTS

The final question, after having described what training assessment needs to be done, is where and how it should be done. The appropriate data bases and analytic facilities have to have the capability for complex data storage for unit training, its display, processing, and analysis. Historical data on actual combat operations and information on training doctrine should be part of the information. Facility support requires a staff of computer specialists, researchers (such as human performance specialists, modelers, statisticians), and subject matter experts (such as former commanders). While components of such a facility exist, they are just beginning to come together.

Two facilities, one for data from the Combat Training Centers (CTCs) and the other for SIMNET-type data, currently are being implemented. Others should follow. The instrumented data and records from the NTC/CTCs have given rise to the Combat Operations Research Facility (CORF) in Monterey, CA. The CORF is a facility to integrate hardware and software at an analyst's workstation. It will allow access to all data sources in the CTC archive and other relevant databases. Capabilities for the CORF are to include access to and replay of: digitized battlefield operations, AARs, communications tapes, etc.

The newest facility for SIMNET-like data, under installation (September 1991) at the Institutes for Defense Analyses (IDA), Alexandria, VA, can show ongoing battles or complete replays with sound and digitized video. Using a special "stealth" console for the simulation, any perspective is possible -- map overview with friendly and enemy locations, view through a tank's vision blocks, terrain and battle view from different perspectives, etc. There are complete records of events such as movements, engagements, and communications. Experts currently can study and selectively replay videos of battlefield performance. Additional capabilities are expected for evaluating combat data.

These types of facilities show promise to provide state-of-the art ways to study training objectives and battlefield performance. They make it practical to study the combat mission and task requirements in terms of execution, outcome, and effectiveness of performance. In terms of analysis, for example, knowledge acquisition procedures including protocol analysis, structured interviews, and expert system technology could be used to collect data needed for assessing training strategies. Together, advanced data collection, display, and analysis methods will provide personnel with the tools to gain insight and, in turn, to develop more cost-effective training.

REFERENCES

Alluisi, E.A. The Development of Technology for Collective Training: SIMNET, a Case History. Human Factors, 33(3), 1991.

Brown, F. J. A Simulation-Based Intensified Training Readiness Strategy for the Reserve Component (IDA Paper P-2611). Alexandria, VA: Institute for Defense Analyses, December 1991.

Deitchman, S.J. Preliminary Exploration of the Use of a Warfare Simulation Model to Examine the Military Value of Training (IDA Paper P-2094). Alexandria, VA: Institute for Defense Analyses, March 1988.

Hicinbothom, J.H., Zachary, W.W., Knapp, B.G., Zaklad, A.L., Bitter, A.C., & Broz, A.L. Doing Deception: Attacking the Enemy's Decision Processes (Research Report 1550). Alexandria, VA: U.S. Army Research Institute for the Behavioral and Social Sciences, February 1990.

Loper, M.L., Thompson, J.R., Williams, H.L. Simulator Networking: What Can It Offer the Training Community? Military Simulation & Training, April 1991

McFann, H.H., Hiller, J.H., & McCluskey, M.R. Unit performance measurement: Concepts, Techniques, and Examples. Presented at the Military Operations Research Society (MORS) Meeting, Williamsburg, VA, June 1990.

Miller, G.J., & Bonder, S. Human Factors Representation for Combat Models (Technical Report 571). Alexandria, VA: U.S. Army Research Institute for the Behavioral and Social Sciences, July 1982.

U.S. Army Armor School. SIMNET Users' Guide. Ft. Knox, KY: USAARMC, April 1989.

THE OPTIMISATION OF TRAINING SYSTEMS

Alexander von Baeyer

IABG-SLA
Ottobrunn
Germany

INTRODUCTION

This paper has been written as a problem-oriented introduction to an ever new topic, namely the question: What methods are available to design training systems and how good are they?

The approach in this paper is one of methodological analysis. However, the subject cannot be treated accademically alone. The practical use requires a deeper insight into the usability of training design methods under given or new circumstances.

SETTING THE SCENE

The paper is organised in the following way:

- Analysis of the methods and tools to plan, design and conduct cost/effective training

- Analysis of methods to develop cost/effective training technology and simulaton

- Application of these methods and tools to developmental tasks, especially in consideration of the imminent changes in NATO's role as a defence organisation.

Basically, the aim of the paper is to elaborate a larger scheme, into which the training design methods and the cost/effectiveness issue may be inserted. The larger scheme includes Operations Research, which is still a new areea for the training community.

Since steering a military organisation into new directions is always a slow process, changes in the manpower and training domain are even achieved slower. Therefore, means must be chosen, evaluated, used, and optimised, that enable decision makers in the training domain to base their endeavours on knowledge pertaining far beyond their day-to-day-experience. This situation requires OR methods and tools. The application of Operations Research to training is, however, not a standard procedure, although some methods are already in use.

The most obvious application of traditional military OR to training is, of course, the use of war gaming scenarios. However, the topic here is about OR as support for the long-term planing and evaluation of training.

An appraisal of the benefits of OR to training should not take the well known methodology

Advanced Technologies Applied to Training Design, Edited by
R.J. Seidel and P.R. Chatelier, Plenum Press, New York, 1993

and see how it applies to the development and implementation of training systems. Instead, the very notion of "Operations Research in Training" must be newly defined at first. As always, the object of research must govern the methods of research.
My paper therefore begins with a brief clarification of the notion "OR in Training". Then follow analyses of the various applications of a set of methods and tools, which mainly deal with staff development and the planning of qualifications. This is done, because in both areas far reaching decisions must be made, although the relevant data are difficult to obtain and are "soft data" by their very nature.

Operations Research in Training

The notion of OR in training means that staff development and the planning of qualifications must be dealt together and are subjects of a prognostic and analytic approach.

In terms of data quality and research methodology, this means:

- that the majority of data describes human behaviour and is difficult to measure or even to estimate

- that the hard data describing technological and economic facts depend on human data

- that the scenarios depend on so manifold a set of variables, that valid alternatives are again difficult to establish

- that data collection efforts are feasible but constitute organisational problems such as getting the right data into the right hands and pose methodological problems such as controled data collection.

If OR in training is to become a promising endeavour, the weaknesses described must be balanced by a considerable advantage. The advantages of OR to training are:

- the structuring of decision contexts

- the modelling of human resources allocation alternatives

- the cost and effectiveness control.

Therefore, OR's task is to produce:

- a set of models, into which data can be fed and by which the value of data can be analysed

- and those methods, which either help to find the data or produce them.

To sum it up: OR in training is the construction of a blue print that describes the allocation of human resources over a long period of time in terms of data and production rules.

So much for the broader scheme of OR in training. Now back to the narrower topic of this paper: to the optimisation of manpower and training development.

1.0 ANALYTICAL METHODS AND PREDICTION MODELS

In this first part I shall describe approaches used to predict or to simulate the allocation of human ressources. For each approach it will be said:

- why it is neccessary to follow this aproach

- and what is the the state of the art (to my knowledge and in broader terms).

1.1 Manpower and Qualification Requirements, Staff Development

The following considerations about manpower and qualification requirements describe the most eminent problem area in large organisations and tell, how OR can help. Since high tech organisations, like modern armed forces, cannot adopt a "hire and fire"-policy, they must know their manpower and qualification requirements about ten years ahead.

Prognoses and simulations about the right number of people a military organisation must have and train in a far future can be organised along several dimensions according to:

- personnel and/or training areas, e.g. technical personnel, pilots etc

- status groups, e.g. officers, NCOs, enlisted men etc.

- ranks, e.g. captains, sergeants etc.

- occupational specialties, e.g. flightline jet engine specialists TORNADO

- or mixes of occupations and ranks.

The method is the following:

- start with the calculations from a relevant survey of the actual personnel,

- define parameters for

  -- on the job time of soldiers

  -- the age, when soldiers leave the services (and enter the reserve force) according to rank and/or occupational level

  -- re-enlistment probabilities

  -- promotion rates from rank to rank or from one occupational level to the next (e.g. from specialist to first specialist)

  -- the regeneration rate according to the age structure or the occupational structure

- and calculate the dynamic flow of personnel entities from year to year according the just mentioned parameters.

What is achieved with these "prognostic simulations" is information about:

- the active force structure per year and specialty qualifications

- the people to be trained initially per year

- the people to be re-trained from one occupational level to the next

- the force strength of the reserve.

The prognostic simulation of this kind can be completed by calculations of the manpower, technical resources and infrastructure, which are necessary in order to train the yearly number of soldiers required, and by cost projections. This finally results into a total manpower and training simulation.

Industrieanlagen Betriebsgesellschaft (IABG) is implementing these prognostic simulations for various purposes and basically for two levels of prognoses:

- on a gross level using the potential of knowledge based systems and PC's, which we call "Knowledge Based Personnel-Prognosis System" (KBPS)

- and on a detailed level using the full computational power of Micros to simulate the staff development of a whole service (in this case the German Air Force) on MOS resolution(1a).

1.2 New Findings in Learning Theory

Learning theory, though far away from being a systematic and homogeneous corpus of knowledge, is nevertheless an analytic model and serves the purpose to predict the likelihood of any successful and lasting knowledge acquisition. This is why learning theories are treated in this paper.

Basic and applied research on human learning has greatly improved our scientific methods to design training systems in the past. To cite only the most eminent contributions:

- the theory and practice of programmed instruction

- the theory and practice of sensor- and channel-specific presentation (= media didactics)

- the theory of instructional strategies

- and the theory of the aptitude/treatment interaction.

While the foundations of applied learning theory are mostly laid, the actual progress is incremental, but worthwhile to consider. There are currently three interesting areas of learning theory with more or less impact on training design:

- model theories of memory and learning

- practical educational psychology

- and human factors engineering psychology.

1.2.1 Model Theories of Memory and Learning

The first contribution has not yet proven to be practicable, but it shows promise.

The model theories of memory and learning are divided in models of "connections" and models of "dissociations" of memory tasks. Both approaches tend to be empiric and formal at once. Both have not shown their applicability to training design so far, although they were wellcomed with the highest expectations(1). Therefore, the application of the theories of learning and memory must still consider the traditional and well founded neuro-physiological concepts, such as the layer model of memory and the benefits to long term memory of multi-channel presentations.

(1a) See IABG reports B-SZ 1570/07 "Modell der Entwicklung von Personal und Ausbildung" 06. Feb. 1991; and B-SL 3003/01 "Wissensbasiertes Personalprognose System", 26. March 1991.

(1) Douglas L. Hintzman, Human Learning and Memory: Connections and Dissociations. Annual Review of Psychology 41, 1990, p. 109 seq.

1.2.2 Practical Educational Psychology

Findings of recent educationl psychology are more helpful in day-to-day training and should be acknowledged by training designers. Basically, they are the following(2):

- All teaching methods should be adapted to the psycho-social structure of the learner and the group of learners. This means practically that there should be a certain amount of personal relationship between instructor and soldier and that the use of a-personal training methods should be restricted to the necessary training areas.

This requirement is not in contrast to the CBT-philosophy; in the CBT-environment personal relationship to the learner means however someting different than in the conventional classromm instruction (see section 2.1.3).

- The learner should be able to control the sequence of learning tasks himself, whenever the type of instruction and the subject matter allow for such a method. The learner should also receive personal feedback in small steps. If these recommendation can be implemented, they may reduce training time. But on the other hand, they increases the burden of the instructor and military leader.

- The learner should learn how to control and to monitor his learning progress himself. This means that educational skills should not alone be the expertise of instructors, but also - to a certain extent - be to hand of the learners.

- Finally, good instructors and a good social climate make good instruction !

These comon-sense statements are scientifically validated results. Military training institutions have only to comply with them.

1.2.3 Human Factors Engineering Psychology

Human factors engineering psychology is included in this overview, because it constitutes the link from learning psychology to the systems design. The better the systems design is and the more it adapts to the mental structure of the system operators, the better training design can achieve its goals (see also section 2.2.2, which deals with a reason to promote the human factors integration). Human factor engineering psychology has three major domains:

- displaying and interacting with complex visual information,

- mental workload assessment,

- the training of complex skills(3).

In all three domains, there is some applicable progress for the training design task.

<u>Displays:</u>

Since displays of any complex weapon system have to accommodate human performance, the use of displays must correlate with the mental models of the system's functionality. Creating mental models is a central instructional task. Human factors and training design

---

(2) Paul R. Pintrich et al. Instructional Psychology. Annual Review of Psychology 37, 1986, p. 612 seq.

(3) D. Gopher and R. Kimchi Engineering Psychology. Annual Review of Psychology 40, 1989, p. 431 seq. See als Dr. R. Otte's contribution to this workshop, which deals with the problems of man-machine interface in weapon systems design.

match in this important domain. Therefore, the following results are to be understood as prerequisites of successful training.

- The human perception can not process a single dimension with multiple objects at the same time, while it is capable of processing in parallel several dimensions of a single object.

- As well, global features can be processed faster than local features.

- Therefore, integral displays are better than separable displays, "when the data variables are highly correlated and/or when integration of data from a number of sources is required by the operator before a decision can be made".

- However, there are exemptions, such as when the system state is uncertain or when the data are related by correlation or causality. In both cases, separate displays seem to be superior.

Mental Workload:

Mental workload assessment suffers from methodological ambiguities. However, there are results, which can be be used to define the trainability of complex systems. The key distinction is between automatic and controlled modes of mental processing(4).

- As it seems, controlled processes demand high attention, are "slow and effortful", whereas automatic processes don`t demand attention and can proceed in parallel.

- Real-life workload is always a mix of both modes, the automatic and the controlled modes of mental processing.

- Therefore, a necessary task is to determine the optimal proportion of both modes, in order to arrive at a satisfying trainability of complex systems, which in turn leads to the effective qualifications.

Complex Skills

Finally, the training of complex skills has the same problems as coping with information displays. The question is, whether mental models of the complex task can be established for relevant skills. Four tactics are described(5):

- use comon sense and every-day mental models that learners bring into the training situation and base upon them the skill acquisition process

- discover and trace the mental model that develops during skill training and use it to strengthen the new expertise

- introduce good mental models as examples

- take the individual`s current model, discuss it, compare it with other models and place it into the training situation to test it.

1.2.4 Summary

All findings of humam factors engineering psychology, as insular as they are, have one

---

(4)Gopher and Kimchi, loc. cit., p. 441, 444
(5) Loc. cit., p. 446

common trait: This is the quest for task analysis, which is the very basis of every training design. "Systematic research on this issue is sorely wanting"(6).

1.3 Training Requirement Analysis

It goes without saying that analytical methods and predictor models about training requirements and the qualifications needed are necessary tools. Although they should constitute the hard core of any training development philosophy, their systematic foundations are not sufficiently established and researched.

However, methods used in the US "MANPRINT" (Manpower and Personnel Integration), "FOOTPRINT" and "HARDMAN" systems analysis may well be considered as such approaches. These analyses take existing weapon systems, and their logistics, analyze them as for the data required to man a new system or for the similarity of older systems with proposed new developments. They derive information about training requirements for the new system. This historical approach may also yield information to be used in cost/effectiveness analysis.

These methods are mainly data-collection, data-handling and data- processing tools. What they have in common, is that they rely widely on the analogy of data and data structures between old and new structures. What is still needeed, are methods, which produce almost automatically:

- complete lists of tasks of a given weapon system

- related lists of training objectives of a given weapon system and in relation to given occupational specialities

- related lists of proposed delivery systems (courses, media, devices, cost).

IABG has structured the maintenance training of the NATO Helicopter NH 90 along these lines (examples are available on request). As it seems, the maintenance training of the European Fighter Aircraft EFA will be handled in the same manner. And yet, a concise methodology is still lacking.

1.4 Training Development

Training development is another block of methodology, which which should be well established and yet needs a considerable improvement. Training development is the endeavour to procure:

- the right amount of people

- with the qualification needed

- at the right time

- at reasonable cost

- and with the assurance, that people and qualification are sustainable.

How can this be the subject of analytic methods and prediction models?
There are, of course, total concepts of training development for what developmental steps

---

(6) Loc. cit., p. 447

must be done, when and how (most prominent are the Instructional Systems Development/Systems Approach to Training or in short ISD/SAT concepts). Derived from these concepts are a huge variety of "analytic models", which however are rather cookbooks for the training design, especially when they claim to be expert systems and help the subject matter expert (SME) to design training courses without beeing an educational expert (6a).

ISD/SAT rules (and their derivatives) ask for data as basis for decisions. But with the ISD/SAT methodology alone, data are not produced automatically. Therefore, the difficult task remains to be done. ISD/SAT helps to structure the problem, which is only the first step to solve it.

It is, therefore, a most important task to establish a sort of critique of ISD/SAT reasoning by investigating:

- whether the approach of ISD/SAT is a valid approach

- how far the theoretical implications are acceptable and when ISD speculation begins to spoil the developmental advantages of the approach

- what data quality is needed or sufficient.

I would like to create an ISD/SAT family of models. What are then the rules of kinship, and who are the members of this family ?

The rules of kinship are, that:

- technical data are translated into manpower and qualification data

- a future manpower and qualification system is described by production rules, i.e. by "if - then" statements

- the system is highly complex as to the number of interacting system elements and to the amount of interactions per element (nobody has ever counted the complexity of a training system).

The "members" of the family are:

- in the US the procedures AFP 50-58, TRADOC PAM 351-4 (T), ITRO Procedures for ISD, NAVEDTRA 110A(7)

- in the UK the SAT methodology

- in Germany the manuals "Herstellung der Ausbildungsbereitschaft" and the cost/effectiveness analysis procedure "NKU-Am".

Let me turn back to the rules of kinship and enter more deeply into methodological problems. I shall associate with each rule statements about their advantages and the risks:

1.4.1 Data Translation

The data translation is not the problem of an effective data management technique, but lies

---

(6a) See e.g. L. L. Kageff and E. J. Roberts, Synthetic Intelligence for Instructional Design, Proceedings of the International Training Equipment Conference, Maastricht, The Netherlands, 1991, p. 361 seq.
(7) H. B. Sorensen, US Joint Service Systems Approach to Training Design. Proceedings of the International Training Equipment Conference and Exhibition. Birmingham, England April 9 - 11,1990, p. 136 seq.

in the actual translation of a technical-logistical concept into a manpower system. More precisely, the problem lies in the uncertainty of human behaviour. The advantage, however, is that the "two worlds" try to speak to each other by means of an interpreter. The interpreter is always a human being, namely the training analyst. What is needed, would help the training analyst to form a systematic evaluation of standard relationships between technical and logistical components and their impact on manning and training the technical system that could be put into computer manageable formats (this turns out to be the same requirement as addressed in section 1.3).

So far ISD and related procedures have told that and when there should be a translation. Now it must be said, how this tranlation can be achieved and preferably how it can be managed by computers. Automatic translations are still missing.

The risk of this endeavour is hoever, getting lost in the dedale of data. But, this risk is secular to the information society.

1.4.2 Production Rules

The issue of production rules constitutes a greater problem and reaches to the roots of any design system for manpower and qualifications. An appraisal of recent findings in the Cognitive Sciences(8) leads to the following considerations:

The ISD family is an example of using schema models in problem- solving tasks and therefore emphasizes the importance of pattern recognition in thinking. This definition is valid, if experts reason and think in well structured-problem areas. The opposite, however, is the case. ISD-reasoning is mainly done by novices. The problem area is widely unstructred; only precedence cases, e.g., similar weapon systems, can be used as information sources. The ISD-procedures pretend to be strongly deductive and syllogistic; in fact they use strong schemas applied to rather fuzzy problems.

The advantage of ISD as a system of production rules is that the pattern recognition style of problem solving helps to define the data and information needed (see above).

The risk is that the human information processing is done with the wrong problem solving pattern itself. Instead of formalizing the training design more and more, the procedural approach must be revised to handle the complexity. Practical criticism comes to the same result(9).

1.4.3 Complexity

With the complexity issue, the argument is the same as in the preceeding issue. The prevailing strategy of complexity reduction is that of formalisation. This is basically wrong. The complexity is inherent to manpower and training systems. To design them adequately requires a holistic approach, namely:

- a learning strategy of agencies and people
- a long term policy of manning the ISD agencies with long term expert teams

---

(8) Cf. Earl Hunt, Cognitive Science. Annual Review of Psychology 40, 1989, p. 431 seq.

(9) Cf. Avi Kedem et al., New IDF Training Mil Standard for cost Effective Training Choices. Proceedings 10th ITSC, Orlando/FL 1988, p.423 seq.

- and a basic change in the philosophy of military systems development (which no longer is the development of operational systems having a training task, but has to become the development of mannable and trainable weapons).

Since training design in big organisations (like military services) is a task that needs long experience in many interacting areas, it cannot be learnt like a set of applicable rules. Rather, it must be learnt by long practice. It needs competence in holistic thinking (vernetztes Denken); it needs agencies having the capability of synergetics and time to develop special skills. This is the main contribution to complexity reduction, not the creation of military standards.

Therefore, the required change of emphasis, namely from the military engineering to the predominant qualification, renders the development of mannable and trainable military systems less complex right from the beginning.

1.4.4 Summary

To summarise: The methodological analysis tries to do a kind of a "midterm review" of the training design toolkit, actually available. Having done this, firm grounds for further research and in information exchange are ascertained and new areas of collaboration or worksharing are defined. This RSG is an example of collaborarion; an example of worksharing is actually in the process of developing, it is the Research and Technology Project (RTP) 11.1 called "Training Systems Concepts" of the EUCLID programme.

The firm grounds are the existence of rules and channels for the training data collection. However, the real data flow is often far from being satisfactory, especially in the cost/effective domain.

New areas lie in the human information processing abilities as to the collaboration of experts and the optimisation approaches the experts must use.

1.5 Cost/Effectiveness Methodology

The last important methodology I want to treat in this first part of the paper is that of cost/effectiveness analyses. Cost/effectiveness analysis of training is the proof of the practicability of the previously described approaches and many other. My aim now is to elaborate on some practicale aspects of cost/effectiveness analyses: What can be expected from it; what must be done, before these analyses will become worthwile tools; and what is the actual state-of-the-art.

In the second part of the paper I will explain, where certain R and D efforts for training and simulation would gain particularly from cost/effectiveness analysis of training

1.5.1 Expectations and Prerequisites

Cost/effectiveness analyses in training can yield:

- an economic appraisal about alternative training systems to be used, before they are used

- a transparent structure of the alternatives in training as to the decisions about the delivery system and the cost drivers

- a management tool for the design to cost of training systems.

All the expectations must be fulfilled somehow, because designing a training system is looking into the future. Cost/effectiveness analyses in training need two prerequisites to be fulfilled:

- the teamwork of experts in the fields of
  - -- the subject matter in question (technology, operations, logistics),
  - -- staff development,
  - -- education
  - -- and training technology
- a careful training analysis as soon as possible along the lines of the personnel structure policy and/or the weapon systems development cycle.

If the basic prerequisites are met, cost/effectiveness analyses in training may save a lot of precious time and money. They themselves may cost a minor fraction of what they may save.

1.5.2 State of the Art

Asking the cost/effectiveness question in relation to training begins to become a NATO-wide preoccupation. A rather unsystematic start in the mid eighties(10) was followed by a systematic review of available methodologies and cases(11). By the end of 1990, an other workshop was held that monitored the progress achieved so far (RSG 15 Workshop, Huntington) with proceedings that will have been published by mid 1991. The results of this latter workshop may be summarized, as follows:

Data

The need of cost/effectiveness data is generally acknowledged. Producing cost data seems to be feasible according to national cost accounting systems and planning cost overviews. Producing effectiveness data is however not sufficiently institutionalized, exept for the US and eventually for Germany. A promising means of producing effectineness data are instrumented ranges (like the US NTC) or experimental comparisons of simulator data with either real life training or with war gaming simulations.

Whatever data are used, however, a conceptual definition of training effectiveness in the actual case ragarded is of importance, otherwise the raw data of evaluations and simulations cannot be used.

Prognostic Evaluations

The neccessity of prognostic evaluations is acknowledged as a matter of routine in long term weapon system development cycles or in the strategic staff development planning. However, very few case studies were reported, which were really integrated into higher developmental and planning activities. The method of utility analysis (Nutzwertanalyse) is used, when rapid prototyping in software or even hardware are not feasible (see 2.2).

2.0 COST-EFFECTIVE TRAINING TECHNOLOGY

Having specially addressed the methodology of cost effectiveness in training I now turn to the area, where this topic seems to mattter most.

Indeed, it is obvious that training technology and simulation affect the cost and the efficiency of training endeavours. There are some basic considerations, which (if accepted) help to structure the whole issue and prevent fruitless discussions. They concern:

---

(10) See NATO Defence Research Group Panel 7 "Proceedings of the Symposium on the Military Value and the Cost-Effectiveness of Training", Brussels 7th - 9th January 1985, DS/A/DR(85)167
(11) See NATO Defence Research Group Panel 7 "Report of RSG 15 on the Military Value and the Cost-Effectiveness of Training", Brussels 28th December 1989. AC/243(Panel7/RSG.15)D/4.

- a standard electronic medium for the bulk of basic training

- the real costs of siumulation

- the division of labour between the computer and the instructor.

2.1 A Standard Electronic Medium and Training Network

Small instructional technology tends towards a standard electronic medium. It combines the advantages of the so far traditional medium mixes without having the organisational problems of handling a variety of hardware and software.

Obviously, it no longer is asked, what benefits audiovisual and computer based technology have for instructors and students. It is only asked, how to do it and is it cost-effective. Technology, organisation and infrastructure must be analysed, to determine how they can contribute to make training more cost-effective. The role of the instructor needs, however, to be considered seperately.

There are three issues that matter:

- whether the quality of electronic pictures is a limiting factor

- whether the media logistics of a networked and distributed system will be expensive

- if and how the use of the standard medium changes the instructional process and at what costs.

2.1.1 The Limitations of Electronic Pictures

Considering the capabilities of HDTV and video disc storage or even more modern technologies (e.g. the solid state storage, the possibilities of DVI and CD RAM discs), there are practically no areas in the military training, where electronic display technologies would not suffice. High level medical training may be an exeption.

Therefore, the technology of delivering electronic pictures is not a limiting factor in military training and its design.

2.1.2 The Media Logistics

"Media Logistics" deal with producing training software/teachware, bringing it, where instruction takes place, updating it and finally training the instructors or students to use it. An electronic medium should comply with the following logistic support requirements to be cost-effective and useable as a standard device(12):

- It should be used as a stand alone and as a networked device. (This means: All programs and updates must be brought to the student either via disc or any other storage medium or via comunication links (cable, satellite, radio relay)).

---

(12) Cf. Commision of the European Communities: DELTA, Development of European Learning through Technological Advances. Initial Studies, Background Documents. Brussels 1987.See also: R+D in Information and Communication based learning technology. DELTA 90. Brussels July 1990.

- The communication links should be two-way and preferably broad-band links. The backward channel may be a small-band voice channel, if the device is used by students in their units or at home (e.g. for reserve training).

- There must be a central facility that is linked with the student stations and provides documentation and help to the instructors and tutorials to the self-paced learning activities of the students.

- The documentation should be a specially designed didactic information system. This means that all centrally available and distributed programs must be documented to show the learning objectives and tasks they train for: namely the central features in text and in still frame or even in compressed video clips, the previous learning success, training problems and additional information material. The instructor should have the possibility to interact with the documentation center, in order to ask for more or detailed information and to judge the quality of focal pieces of the program.

- A variante of the student device should be equipped with authoring facilities. This should enable the instructor to produce his own video and computer programs, or to alter centrally provided programs (e.g. cut out sequences, mix with other programs, add or alter CBT-functions).

- There should be a systematically planned and evaluated instruction of the instructors. This is an absolute must, or else the value of the training infrastructure described will not be made beneficial. But, there should be a certain amount of student training on using the new medium, too, because this can enhance the self paced learning considerably and is in line with recommendations of the learning psychology (see above).

Can such a distributed training support system, which has to cover a wide range of training applications, be cost-effective ? The answer is "Yes" in two respects and "Not entirely yes" in a third respect:

- The first answer is "Yes", because the electronic student station will become standard technology and will be cheap, if it is introduced NATO-wide. The same is true for the instructor station with authoring facilities. Finally, the network needed to link the students and the central facilities together is a standard military infrastructure. The ISDN capability of a training network, which again is state-of-the-art, would greatly enhance the training effectiveness of the infrastructure. A rough cost overview may show this:

-- Student Station, NATO-wide use = 7.5 K DM, to become cheaper during the next years.

-- Instructor Station, NATO-wide use = 12.5 K DM, and will become cheaper during the next years. Network cost may be sunk cost, because the networks exist already and are also used for other purposes.

-- The cost of software/courseware is high but neutral, because it occurs whatever delivery system is used. The tendency, however, is that the reproduction costs of the standard electronic medium decline, because the programs may be delivered centrally and may be processed further individually by the instructor.

- The second answer is again "Yes", because much of the personnel and infrastructre needed centrally is already available. The personnel, who centrally plan and manage training, actually are engaged in handling the eventual day-to-day chaos. Their endeavours may be directed towards a systematic support system pertaining to most of their normal work. A rough cost overwiev may show this:

-- The cost of centrally available personnel is sunk cost or neutral, because the personnel must be available, whatever delivery system is used.

-- The cost of the instructors is, however, not neutral and tends to be higher, because the instructor`s activities are more demanding (see below) and may result eventually in a higher pay.

- The third answer is "Not entirely yes", because the educational documentation infrastructure does not exist in most countries. Also, extensive training of the instructors and maybe of students is not a common activity in military organisations. A rough cost overview may show this:

-- Instructors should have a two-to-three week introduction into the new medium and a periodic upgrading in training (which may be delivered by the electronic system itself).

-- The basic instructor training is not neutral and may amount to about 7.5 K DM per person.

-- The periodic upgrade in training of instructors will be minute, because this would be part of a centrally performed routine job.

2.1.3 The Use of the Standard Medium in the Instructional Process

The instructional process based on the standard medium with its electronic infrastructure can make training more learner-centered. Therefore, the role of the instructor, who has to serve this infrastructure, will become different from what it is now.

The leader and instructor of a unit will have more responsibility to motivate the students, who will learn in a self-paced mode. Instructors will select, modify and introduce learning materials, and monitor or direct the student's learning behaviour.

The lessons taught by the instructor will become more enriched with audiovisual and computer-based features. Therefore, the instructor will need more time to prepare his lessons. He will certainly need more than the avarage half hour currently devoted to preparing the audiovisual support for daily lessons(13).

Even these skechy remarks point out differences from current training. A networked electonic standard medium renders the training situation more effective for the student and more laborious for the management tasks of the military instructor. Therefore, either the military leadership function will need to be separated from the training management function or the activities related to training management will largely prevail over the military leadership in peace time.

How much the instructor's workload could be reduced by the use of computers, will be the subject of chapter 2.3.

2.1.4 Summary

In sumary, a networked electronic standard medium:

- combines the benefits of a variety of different devices and media,

---

(13) See IABG-Report Rationalisierung in der Bildung und Ausbildung ausserhalb der Bw. B-SL 1412/01, 15.11.1987, especially p.14.

- uses state of the art technology,

- uses available resources that need to be re-allocated to handle the infrastructure, organisation and personnel for such a distributed system,

- requires the role of the instructor to change,

- costs about the same as tradional training, but results in increased training effectiveness.

2.2 The Real Costs of Simulation

The introduction of a distributed system of networked electronic standard media is ultimately not a question of high cost. It does, however, present a problem of organisational change in the military training system. A massive introduction of simulation is going to provoke a deep change in the training philosophy.

To come to the point right away: Simulation will probably be the method most frequently used to do extensive weapon systems training in the future, be it operational training, maintenance training or unit traing above the level of a company. The exeption will be live fire, combat flight missions and field exercises to prepare for war or peace-keeping operations.

Simulation is meant to include embedded training and tactical engagement simulators.

This is, of course, a hypothesis. It is based on the assumptions that weapon systems will become still more sophisticated and complex, and that western societies are less prone to accept military exercises in public.

What does this mean for the cost-effectiveness considerations of training? This question must be answered in two respects: in the short term and in the long term.

2.2.1 Short Term Cost-Effectiveness Perspectives

In a short term perspective weapon systems training and simulator training will still compete with each other. Therefore it will be easy for the simulation to prove its cost-effectiveness. However, investigations of the Comptroller - at least in Germany (14)- have shown that simulators rarely fulfilled the required criteria, for which they have been procured. The savings in amunition or flight hours did not occur after the fielding of simulators. Either the reasons given to procure the simulators were wrong, or the training analysis leading to the request for simulators was incomplete.

Anyway, in a short term perspective and under the prevailing circumstances, wherever the training analysis leads conclusively to the request of simulators, a cost and effectiveness comparison will show:

- that simulation is superior to weapon systems training,

- that mixes of CBT (see 2.1), part-task-training and full-mission simulation provide the best training value for the least possible cost(15),

---

(14) Federal Comptroller Office, report on the use of simuulators in the German Federal Armed Forces

(15) See IABG-Study on Anti Tank Helicopter Training Simulation B-WT 1300/07, dated 13th February 1988; and RSG 15 Report, see note (11) p.132 seq.

- that there are very few exeptions to these rules, only when there are weaknesses in the emulation of training tasks by simulation systems - e.g., combat simulation, truck driving simulation in an complex urban environment(16).

Therefore, the short term perspective does not pose a basic problem to the training and simulation community.

2.2.2 Long Term Cost-Effectiveness Perspectives

Simulation will receive a totally new role in training design and in training itself. This role is twofold, and will be analysed here only as to its impact on the methodologies of training design. One aspect has to do with Simulating Simulators and and the other one with Simulation in the Loop of mission and human factors analysis and trainability assessment (see also 1.2.3 Human Factors Engineering Psychology).

Simulating Simulators:

There is an absolute necessity to conduct cost/effectiveness analyses of training systems, which normally contain simulators or trainers as constituent elements, in order to find out the best training solution. The real task of such analyses is to simulatate the conduct of training done by alternative, but imaginary concepts of simulators or trainers or other media. The usefullness and the LCC of such alternatives are then evaluated with the methodology of cost/utility analysis(17). This method is a cheap "paper and pencil" method with PC support. Its fallacies are the subjectivity of the assesments done by those, who conduct the analyses, and the linear representation of multi-dimensional correlations-of-effectiveness parameters. Its virtues are an illucidation of the complex decision problem and the construction of a transparent structure of the alternatives, e.g. to identify cost-drivers.

This paper-and-pencil method to simulate the conduct of training can greatly be enhanced, when there is the possibility to build rapid simulation prototypes in software or in hardware. The US seems to proceed this way(18).

Simulating simulators is a training design method. This method may be valid, if it fulfils certain conditions of empiric methodology and cost/effectiveness. These are for instance:

- The rapid prototypes must be built according to the real training objectives, or a well described fraction of them. The fraction of training objectives is well described, if it covers a training task to be deliverd by the alternative technology under consideration.

- It only makes sense to produce a simulation rapid prototype, if this endeavour is combined with systematically planned, conducted and evaluated experiments. This makes the simulation rapid prototype subject of a cost/effectiveness analysis strictu sensu (Kosten/Wirksamkeits-Analyse), which is different from the cost/utility analysis, because the effectivenes can be empirically measured(19), and therefore the fallacies of subjectivity and linear reduction due to the cost/utility analysis are overcome.

- Experiments with the rapid prototype must be designed for objective measurement, and the defined learning tasks must be reached by a typical target audience.

---

(16) See IABG-Study on the Cost-Effectiveness of Truck Driving Simulators B-SL 1481/10 (in English); and the summary of the findings in the proceedings of the Huntington Workshop of RSG 15, October 1990.
(17) See note (15).
(18) E.g. simulation for the Advanded Air Defense System.
(19) See note (15).

- The experiments with the prototype should concentrate on presentational cues that provoke a special student behaviour (e.g.thermal image targets and the gunner's identification). Analyses should look at whether the cues are sufficient or not. If they are not sufficient, can the prototype under consideration produce the cues necessary under different circumstances? Or, should another prototype be considered? This is, of course, a lengthy procedure, but it may lead to firm empiric grounds for making decisions.

Simulation in the Loop:

There is another way to use experimental simulation. Simulation in training is now combined with mission and human factors analysis. One can learn from simulation, not only how to simulate in training, but also:

- how much the workload of the operator/maintainer should be,

- and what personnel can do what tasks in what sequence with a minimum of necessary training.

Simulation in the loop of total systems design is a powerful method. This method may be valid, if it fulfils certain conditions of empiric methodology and cost/ effectiveness. These are for example:

- The introduction of a simulation based prototype design phase, in which all experts, including those of human factors, manpower, training and simulation collaborate.

- The existence of a preliminary, though detailed, manpower and training concept and the ability to play freely with all system parameters, especially the human system parameters.

- A plan for a controlled variable system design with particular emphasis on human factors, manpower and training.

Although this new "simulation-aided design" sounds reasonable and feasible under present and forthcoming technologies(20), such a new approach requires changes in the attitude of military organisation towards technological developments and human resources allocation. Therefore: the technology is available; the change of mind must still occur (a mojor step forward may be the DRG Panel 8 RSG 21 on "Human Systems Integration").

Simulation in the loop is cost-effective, because it reduces the risk of developing systems, which may become too complex, and it reduces the workload of the operator, maintainer, and thus the personnel costs. However, simulation in the loop is not for nothing. Paper and pencil simulations may cost one thousandth, soft- and hardware simulations may cost one percent of what they save in terms of LCC.

2.2.3 Summary

What are the real costs of simulation, and what is the benefit, ultimately?

- The cost of simulation is the investment in "brain work" and changes of attitude.

---

(20) For instance the rapid prototyping of cockpit design, done by several companies, e. g. the Canadian company Virtual Prototypes or similar activities of MBB, ESG and others in the field of reconfigurable generic flight simulators.

- The benefits are cheaper and better designed weapon systems with minor investments in manpower for design, training and manning the systems.

2.3 Division of Labour Between the Computer and the Instructor

This section contains a summary of already common opinons about the role of the instructor in a simulator and technology-based training environment. These "rules" are that:

- the instructor is the "animator" of training.

- the computer is the external documentation and storage medium of training data and messages including the delivery of learning cues (e.g. landscape and targets in a gunnery simulator).

2.3.1 Animating the Training

Even with simulation and electronic media or self paced learning, training requires a human interaction with more or less personal involvement of the instructor. The personal involvement, however, is about to change from delivering the training message to arranging for helpful learning conditions, special tutoring and a supportive social climate. This task is more demanding, because it requires psychological and group dynamic abilities. A military leader normally has (or should have) these abilities. What is required in the simulation and computer-based learning environment is, to make full use of these faculties.

Moreover, the lack of reality in simulation-based training must be overcome by an increase of instructional animation, because the presentation of learning cues becomes more abstract than in real life. It is therefore assumed that simulation-based military training creates a relatively artificial environment. (Exceptions to this rule are however the basic infantry and unit training in the field with laser shot devices or the seaman's training onboard with stimulation of the sensors.)

2.3.2 Documenting the Training

There is a good reason, why the division of labour between the computer and the instructor should be re-examined. The computer becomes the incorruptible messenger of the students' real performance. It keeps records of the students's or the group's learning progress. Therefore, the computer can relieve the instructor of all objective analysis of the students' behaviour.

This documentational task is decisive in an artificial environment. Here, much effort should be invested and here is, where the real AI oportunities(21) are in training.

However, there are hidden fallacies, and they pertain to the human reaction to an objective performance measurement: The instructor's new role is then to interpret and evaluate the objective measurements and re-integrate the computer task in the educational interaction.

To put it into one phrase: the computer must not spoil the social climate of the learning environment.

2.3.3 Summary

---

(21) See for example O. Park and R. Seidel, Design of ICAI: AI Techniques and Instructional Pribciples. Proceedings of An International Conference on Military Personnel and Training. Luxemburg, 5 - 7 May 1987.

The division of labour between the instructor and the computer may be summarized in the following overview:

INSTRUCTOR	**COMPUTER**
is responsible for:	
Animating the Learning	Training Messages
Scocial Climate	Objective Measurement
Personal Interaction	Documentation

3.0 TRAINING DESIGN MODELS

Part 3 of the paper creates training design models. They reflect possible changes in the status and role of the armed forces and NATO. They are arranged according to the likelihood of change in the military qualification system and make use of the findings presented in the previous parts of the paper. Since training will remain the most costly endeavour of military organisations in peacetime, its design and implementation will become even more demanding, as money and resources are cut back.
Three design models have been constructed. They are:

- International Conbined Training Efforts

- Streamlining the Delivery System

- Civilian-Military Partnership.

The reason, why this final part is included in this paper is obvious: the three design models will show the practical relevance of the predictor tools and analytical models vis-à-vis real problems to be solved in the forseeable future.

3.1 Design Model A: International Combined Training Efforts

Design model A shows a comparatively easy and proven answer to the manpower and money problems of the future. The model assumes that only the organisation of the delivery system of military training needs to be revised and that the concentration of NATO efforts will be the major goal.

If this is the case, two activities are of focal interest and must be tackled:

How can model A be achieved practically ? It could run, as follows:

Step 1: Re-organisation of the bulk of military training into:

- basic training to be done in training centers

- and continuation training to be conducted in the units;

- both being based on a broad NATO training doctrine and the implementation of widely accepted training systems in commmon areas (which is mostly the case in flight training, already).

The basic training is done in lead nation training centers (like the basic flight training in the US or UK). The continuation training is done nationally. If unit cohesion is not a major training goal, however, lead nation training centers may again be considered for continuation training.

Step 2: Introduction of a harmonized training infrastructure across NATO:

- A standard electronic medium (in hardware and software) will be a minimum prerequisite.

- Training in the lead nation training center must be multilingual and conducted by "native" instructors, when training in English is not feasible.

- The use of cost/benefit analysis for the determination of training facilities that best serve the purpose of multinational training is helpful.

The benefits of this concentrational model would therefore be that it helps to make an optimal use of costly training facilities and environmentally sensitive exercise grounds.

These two benefits should be the measure, used in deciding on lead-nation training centers.

3.2 Design Model B: Streamlining the Delivery System

Design model B shows how to cope with limited human resources, if combined efforts will not suffice.

In this case the focal activity is to re-define major parts of the manpower and training system. Model B could run, as follows:

Step 1 is a very subtle personnel selection system to screen the available manpower and to develop the capabilities of every remaining soldier (using the approach of modern industrial staff development).

Step 2 is a total ISD analysis of the major training areas, to sort out everything that has no military value and is not absolutely necessary to establish combat readiness.

Step 3 is a special analysis of new leadership tasks and instructor workload, because the military leader and instructor will play a more demanding role.

Step 4 is the development of many and highly sophisticated job performance aids, because training time will be reduced.

Design model B assumes that the military personnel and qualification system continues to be an autonomous entity and that it can preserve its identity by adapting to the new constraints. If this is not the case, then design model C may be considered.

3.3 Design Model C: Civilian-Military Partnership

Design model C reacts to a radical change in the role of military forces, as long as defence postures remain at all. One possible answer for training would then be to form a new partnership of civilian and militatry training. If this is the case, it would be worthwile to consider a nation-wide new approach to public-education and military training. This could include at least five steps:

Step 1 is a re-examination of personal attitudes and psychological fittness in a new military environment. This implies the development of a new selection policy, because the boundaries between military and civilian qualifications will become even less strict.

Step 2 is the re-examination of military training in an industrial civilisation including an in depth ISD/SAT approach as to tasks,addressees, remaining military learning objectives and the organisation of training

Step 3 is the in-depth re-organisation of school and unit training as to what training would remain in military hands and what training would be the responsibility of civilian organisations

This leads to Step 4, namely to new definitions of military value of training and effectiveness of training, since combat readiness will no longer be the sole and ultimate measure of training success

Step 5 and the final step is that simulation will be developed to its full capability because there will be very few occasions to exercise purely military activities in a civilian environment. Therefore it will have replace combat exercises, command post exercises, low altitude flights, missile launches, etc.

In this model any analytic method and predictor model of training design would be of great help and would have to prove its value. Military training would have to be reconstructed almost from scratch.

4.0 GENERAL CONCLUSION

In order to summarize, the message of this paper shall be recapitulated: Manpower and qualification development is to become the most urgent task of military organisations now and in the next future. Anayltical methods and tools are available, but need a considerable booster effort in order to help, where they should help. The relevance of the tools and methods can be seen in the following chart.

This paper ends therefore with an overview of the design models and the relevant analysis and prediction tools in matrix form.

DESIGN	Tools				
	MP	LT	TRA	TD	C/E
Design Model A					
1. Organisation of training				X	X
2. Infrastructure					X
Design Model B					
1. Selection of personnel	X				
2. ISD analysis				X	X
3. Leadership	X	X			
4. Job performance aids		X			
Design Model C					
1. Selection of personnel	X				
2. ISD analysis in depth		X	X	X	X
3. Organisation of training				X	X
4. Military value of training			X		
5. Full range simulation		X	X	X	X

Abbreviations:
MP = Manpower and Qualification, see section 1.1
LT = Learning Theories, see section 1.2
TRA = Training Requirement Analysis, see section 1.3
TD = Training Development, see section 1.4
C/E = Cost/Effectiveness of Training, see section 1.5 and part 2

SECTION III
APPLICATIONS OF EMBEDDED TRAINING

The desire for using embedded training as an integral part of military hardware is continuing to grow throughout the alliance. This section will focus on the front end analysis and training value of using embedded training. It is anticipated that discriminations will be possible between where and when to use embedded training for its maximum training impact. Section V (Advances in Cognitive Theory) and Section VI (Advanced Authoring Tools) will have much cross-feed with this section.

EMBEDDED TRAINING

Alain Kada

Army Training Devices Manager, MoD/DGA
Fort d'Issy - 18 Rue du Docteur Zamenhof
92131 Issy Mouline, Cedex, France

ABSTRACT

This document is an overview of Embedded Training applied to the field of Army, Navy and Air Force. After an introduction and a definition of Embedded Training, some benefits and capabilities it can provide are given as well as the aspects to be considered before choosing an Embedded Training approach. Some advice on the circomstances to create and develop Embedded Training are also furnished. Finally, possible Embedded Training roles, forms, modes of operation and types of training are approached.

1.0 INTRODUCTION

The possibilities of applying Simulation to the field of Military Training are more and more attractive. Many different training devices can be proposed for supporting an operational system. But each of them is only a part of the Total Training System that is conceptually diagramed below:

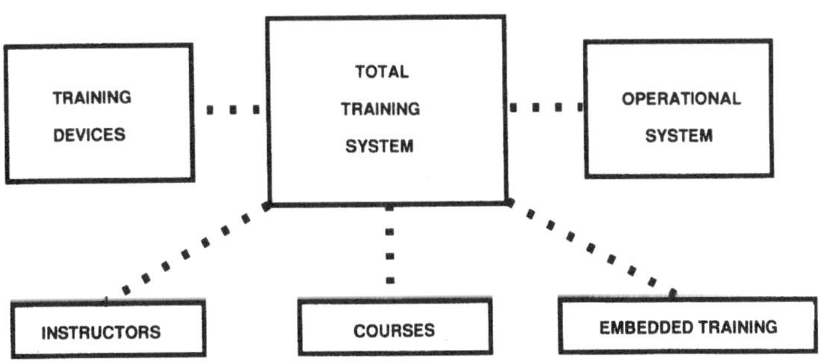

Advanced Technologies Applied to Training Design, Edited by
R.J. Seidel and P.R. Chatelier, Plenum Press, New York, 1993

All the training means can be classed in three groups:

First, those that are completely independent of the operational system,

Second, those that can be considered as strap-on devices, and

Third, Those which are embedded into the operational system.

The “means”, which are completely independent of the operational system, can integrate sub equipments of this system (generally modified) or have only specific simulation components.

For the strap-on devices, they are made with specific simulation components by which simulation can be played on the actual system.

The operational system can be predisposed for training applications, making Embedded Training possible without adding any equipment (or only a few, i.e.:floppy disc).

In this document, only strap-on devices and Embedded Training will be considered.

2.0 DEFINITION

Embedded Training can be defined as training that is provided by capabilities designed to be built into or added into operational systems to enhance and maintain the skill proficiency necessary to operate and maintain that equipment end item.

These capabilities can be provided by features which may be completely embedded within the system configuration by software application or a combination of both software and systems configuration or may be executed by some form of strap-on (e.g.,a video disc player) or plug in equipment (e.g.,a floppy disc) or a combination of embedded and appended components.

The difference between strap-on devices and Embedded Training which may require appended components is explained below.

Strap-on devices are generally not taken into account during the conception/definition phase of an operational system, in opposition to Embedded Training approach requiring appended devices.

Many strap-on devices have been developed for the french Army since 1975 such as:

- DX 143 for MILAN firing post,
- DX 150 for AMX 30 main battle tank,
- DX 133 for VAB-HOT vehicle,
- DX 147 for GAZELLE HOT helicopter.

All these devices have been adapted to existing operational systems. These systems don't "know" when a strap-on device is used. They are unmodified.

That's also the case for a tactical engagement simulator, the "STC-DX 175", which is in service in the french Army since 1985. When the tank is used for force-on-force training, the cable of the operational range finder is disconnected and that's the training device which simulates this equipment. That's done because of laser safety constraints. Nothing else is modified inside the tank. Only some specific training equipments are added. Some examples on these strap-on devices are given in appendix 7.1.

For the future french main battle tank, the LECLERC, the embedded training approach has been selected for force-on-force training only. A training mode will be incorporated into the operational system. However, some strap-on equipments will be necessary such as a transmitter/receiver unit, beacons with retro-reflectors and detectors. But they have been taken into account in the definition phase of this new tank for the integration aspects.

Examples of existing Embedded Training applications are given in appendix 7.2.

3.0 BENEFITS AND CAPABILITIES

PREAMBLE

Embedded Training is both a rather old and a brand new training technology. It is old in the sense that it has been around for some time, first occurring in the Semi Automatic Ground Environment (SAGE) system developed for the US Army in the 1950's, and has been incorporated into some number of systems. It is brand new in two senses:

- More and more "old" systems are now appearing in new versions incorporating advanced technologies (e.g. ,armored vehicles) that can now take advantage of Embedded Training to some limited extent for the first time.

- New high technology systems are now under development that can not only incorporate Embedded Training, but Embedded Training with all the capabilities that a trainer could wish for (e.g., all source analysis system/enemy situation correlation element).

In considering Embedded Training as an option in a system development program, the developers need to begin with an understanding of what is possible and then to proceed to determine what is feasible in that particular end item equipment.

Some capabilities and benefits that Embedded Training can provide are presented below:

- It allows unit accessibility to training;

- It ensures equipment fidelity;

- It provides a training capability that is concurrent with fielding and will support mobilisation.

- Both refresher and sustainment training capability are resident in the unit.

- It relieve the unit training management burden; To the extent that the Embedded Training is designed to provide an assessment capability which is easily accessible by unit training management personnel and does not require on site trainers and evaluators, unit training managers will better know and execute the state of training, and be able to program accordingly, with a minimum investment.

- It has minimal logistics impacts for the end item equipment; Obviously, Embedded Training often imposes some logistics impacts of its own (e.g., wear and tear on generators). To what extent, depends on how it is designed and the conditions under which it is used. It is also possible, however, that Embedded Training may reduce some of the logistics burden imposed by the operational system itself. This is because previous studies have determined that 5% to 65%, with a mode of 13%, of system maintenance actions result from operator and maintainer performance errors. Better unit training through Embedded Training should reduce these error levels,

- It provides training that is standardized across units.

- Training may be more cost effective in meeting essential training requirements that cannot otherwise be provided within cost or safety constraints.

- It provides better job aids (e.g., "help" functions) for the system user. The process for designing tutorials and trainee feedback in support of computer aided instruction and simulation exercises, and the resulting products provide the basis for not only effective Embedded Training but also job performance aids.

However, since Embedded Training requires to use the operational system, the following aspects must also be considered.

- There is a decrease in the operational system potential due to training applications.

- Possibilities offered by Embedded Training may be limited because of the mission operational security.

- The conversion from operations mode to Embedded Training and vice versa must not be time consuming.

- Embedded Training must be more cost-effective than other alternatives, except if it is determined to be the only way to provide needed sustainment training.

For this last point, the life cycle cost of the operational system (including logistics and maintenance) must be estimated early because it is a factor of choice for the Embedded Training approach.

This estimation is difficult to address during system conceptualization. Data to support cost and training effectiveness analysis are typically not available early in the system life cycle, where Embedded Training potential must be addressed; thus, cost tradeoffs are problematical.

Methods for determining cost effectiveness of Embedded Training in relation to other training means have yet to be derived or applied.

4.0 ADVICES

Three admonitions must be held in mind when considering Embedded Training:

• First, Embedded Training should always be viewed as a capability of an operational system within the context of the overall training system supporting the material system. For example, Embedded Training will always exist as a component of the overall training system rather than being the total training system itself (whether used in the unit or the institution).It should be conceived of, designed, and used accordingly.

• Second, consideration of the Embedded Training option must begin early in the system development cycle. In order to realize the most effective accommodation of maximum Embedded Training capabilities within the constraints of the prime system design, one must follow the dictum "earlier is better." The initial question "Does Embedded Training appear to be a viable option for this system?" should first be addressed during the mission area analysis phase of requirement development programs.

• Third, in line with the second admonition, not only must work begin early on Embedded Training, the training developer must be involved from the beginning of and for the duration of the acquisition process.

Since 1987, the US Army policy dictates that an Embedded Training capability will be thoroughly evaluated and considered as the preferred alternative among other approaches to the incorporation of training sub-systems in the development and follow on Product Improvement Programs of all Army material systems. Therefore, the initial definition process for all new and developing systems must include a decision as to whether Embedded Training should be further considered for inclusion in the system capabilities. This policy states that Embedded Training will be included in all new and developing US Army systems unless there are valid and compelling reasons not to do so.

The US Army Research Institute for the Behavioral and Social Sciences and the Project Manager for Training Devices have produced in 1988 a series of ten volumes that includes guidelines and procedures that support the effective consideration, definition, development, and integration of Embedded Training capabilities for existing and developmental systems. A brief description of these documents is given in appendix 7.3.

The major factors that have been identified as impacting on the effectiveness of and the need for Embedded Training in a given system are listed below:

Factor 1: The nature of the tasks and skills demanded by the system concept. What are the requirements for sustainment training?

Factor 2: The feasibility of implementation of Embedded Training

Factor 3: Avoidance of Embedded Training which interferes with operations

Factor 4: The need for training-specific interface hardware requirements

Factor 5: Determine the system availability for training

Factor 6: Minimize the effects of Embedded Training on system reliability, availability, and maintainability

Factor 7: Consider the impacts on system manpower and personnel requirements

Factor 8: cost-effectiveness of Embedded Training (compared with alternative sustainment training capable of achieving the same training goals).

These factors are extracted from the second volume of the documents briefly described in appendix 7.3.

However, training devices may be selected instead of or in addition to Embedded Training due to having greater simulation capabilities or conveniences of a special training environment.

Some advice for creating Embedded Training are given below:

- The features incorporated into the operational system to provide training on the end item equipment should include:
  - -- stimuli necessary to support training,
  - -- performance assessment,
  - -- feedback consistent with improving and reinforcing correct performance, and
  - -- record keeping, to allow management of individual and collective performance trends, improvements and deficiencies requiring additional training;

- Requirements and resources for training must be factored by the systems proponent in concept formulation of the end item/system and pursued throughout the material acquisition process. A training strategy, to include the consideration of Embedded Training, must be included in the initial requirements documents ,

- Any Embedded Training capability, no matter how limited, may be worth the investment when consideration is given to the training requirements and to the available options to satisfy the requirement,

- Know your training requirements and do not yield easily to any reduction in Embedded Training capability. Everyone must yield something in the system development process; but to the extent the training and combat developers, representing the user, know what they want and how to execute it, they are more likely to succeed,

- Consideration of Embedded Training as a major component of the total training system to support a given material requirement must be a continuing, iterative process throughout the early stages of system development. This iterative consideration and decision process may well continue into system fielding and deployment to meet the requirements for decisions posed by frequent need for product improvements and the not-infrequent needs for post-fielding training developments,

- The initial consideration of Embedded Training for inclusion in a given system must be made at the earliest stages of system conceptualization and development. This consideration must take into account at least three aspects of Embedded Training as it relates to what is know about the developing system. These aspects are: the appropriateness of Embedded Training as a component of the training system required to support the defined system, the feasibility and practicability of developing and implementing an Embedded Training component for the system, and the probable cost-effectiveness of an Embedded Training component for this application.

- If an initial decision is made to continue consideration of Embedded Training for the system, the decision process must, as indicated above, be iterated several times throughout the acquisition process. Such iteration is demanded because the developers' information and knowledge about the appropriateness, feasibility or cost-effectiveness of the Embedded Training will necessarily increase as the material requirement proceeds into system concept formulation and subsequent development (e.g., the appropriateness of Embedded Training will become much better clarified when more detailed descriptions of the soldier performance and training requirements posed by the system can be made available for analysis; similarly, the feasibility of incorporating an Embedded Training component into the system will be clarified by more detailed statements of system hardware, software and personnel components). Because of the initial consideration is required so early in the system acquisition, the initial decision will necessarily be based on the usually very limited information dealing with the mission and functional requirements of the system as defined at the mission area analysis stage.

This scanty information base must be related to that which is known or can be determined about how Embedded Training would benefit the training and operational capabilities for this system.

5.0 POSSIBLE EMBEDDED TRAINING ROLES, FORMS, MODES OF OPERATION, TYPES OF TRAINING

Possible Roles

Embedded Training has been traditionally thought of as a training medium for the operational unit environment and as a consideration separate from training devices, the software required to do the operational job, and the design of the soldier machine interface.

The table below is extracted from the second volume of the series which is described briefly in appendix 7.3:

OVERVIEW OF THE INITIAL DECISION PROCESS FOR CONSIDERATION OF EMBEDDED TRAINING IN DEVELOPING OR RETROFITTING ARMY SYSTEMS

PROCESS: Examine known system characteristics in relation to factors affecting probable success of Embedded Training.

QUESTIONS:

1. Are there policy decisions which dictate Embedded Training for knowledge and skill acquisition training for the system?

2. Do the proposed system's critical tasks require frequent sustainment training?

3. Is the development of an Embedded Training component feasible in this system?

4. Is it likely that Embedded Training will be a cost effective training for this system?

DECISIONS: Determine whether to:

1. Continue consideration of Embedded Training into subsequent system definition

or

2. Discontinue consideration of Embedded Training for this system

DECISION CRITERIA:

A. If the answer to question 1 is YES, Embedded Training will be included in the system requirement.

B. If the answer to question 1 is NO, Embedded Training will still be further considered until there is compelling evidence that any one of the other three questions must be answered negatively.

There have been some number of times where that has not been the case. Embedded Training has been used in the institutional environment. The development of software for Embedded Training has, in some cases, contributed to the development of software for both training devices and the operational system itself (e.g., PATRIOT and AWACS). And, especially in the Navy, Embedded Training has been originally developed as a test program for maintenance checks. The US experience with the laboratory Fiber Optic Guided Missile Test Bed (FOG-M), underscored another role that early involvement of Embedded Training considerations can play: that is impact on the design of soldier machine interface.

Alternative soldier machine interface and consequent system design configurations are often feasible; some are more easily and quickly trained. Early involvement of the training developer should be expected to impact on the soldier machine interface design. The fact that development of Embedded Training software can serve a number of roles and provide input to a number of other purposes does not make accounting types very happy. But it is a possible fact of life and makes cost estimation of Embedded Training perse very difficult. The cost of Embedded Training is often not nearly as large as it may first appear.

Forms

Embedded Training is generally defined as "a training subsystem (hardware and software) which is integrated with (but not necessarily fully incorporated within) the overall weapon or tactical system software and equipment configuration. The Embedded Training subsystem provides training and assessment capabilities through the soldier-system interface using software control of courseware and exercises on the operational equipment with auxiliary equipment as necessary."

That which is totally and exclusively a training device and that which is totally and exclusively Embedded Training lie at opposite ends of a continuum. In any event, the choices range from the need for a separate vehicle to the use of removable media (e.g.,floppies or CD ROM discs or video discs) or to totally integrated into the operational software system. The decision as to which way to go is dependent on what the system will easily support operationally and the resultant training capabilities vs. what is needed.

The above Embedded Training definition allows for adjunct or strap-on devices which may not be integral to the tactical system. This potential for application of Embedded Training greatly expands the range and variety of systems to which Embedded Training might be applied. Strap-on devices could conceivably be used with systems which have no (or very minor) computer capabilities internal to the system. It should also be noted that Embedded Training is naturally extendable to retrofitting an embedded capability into existing fielded systems or those late in the development cycle.

Modes of Operations

Embedded Training can be designed to operate in a number of different ways. The following are some of the options identified up to this point:

a. Talking about continua, again, the Embedded Training component can be employed only when the system is rendered combat incapable or, at the other end of the continuum, can be in use while the operational system is in an operational status. The choice depends on both what is design feasible and other considerations like security. Modern electronics and software now permit most systems to be designed such that the soldier can quite immediately transition from the training mode to the operational mode (e.g., a flip of the switch and, at most, a few seconds transition time). Other Embedded Training components have been designed to operate while the system is in a fully operational mode. The later two case have resulted, however, from conscious decisions and subsequent design. There are some cases where the transition time has been unacceptable due to the constraints of older technology and lack of planning.

b. One of the most exciting advents of advanced technology incorporations into systems is the possibility of collective Embedded Training. A first consideration is that demanding "state of readiness" requirements may well be met if it is feasible to allow a subset of the operational asset to employ Embedded Training while other members are in an operationally ready mode. In some cases this will be feasible. Another consideration is the possibility of collective training when other crew members are absent. Again, technology, including artificial intelligence, is increasingly making this a reasonable possibility.

c. Embedded Training does require operation of the end item equipment; the question is how much and to serve what purposes. This issue relates to system hardening, logistics, and unit readiness status concerns. Usage of the end item equipment other that electronic and software usage, does wear out the equipment. But it must be used for training in any event and is in field training exercises, etc. The need is to recognize the requirement for training, to minimize the wear and tear on the equipment to the extent possible, and to harden as necessary. One primary benefit of Embedded Training is its ability to provide training on the equipment when it is in a garrison or other standby modes resulting in much greater training availability and efficiency. This utilizes the electronic and software components, which is RAM beneficial, but also exercises the input control hardware. This hardware may have to be hardened and should be considered and planned for. Another primary benefit of Embedded Training is its ability to provide training in a more active mode (e.g., in flight) that cannot otherwise be provided at any reasonable cost whatsoever (e.g., interactive threat engagement). Here, the choice may be whether or not any reasonable amount of training is to be provided at all.

Types of Training

Embedded Training has been traditionally viewed as a training medium which provides sustainment training. Further, even though it has been billed as embedded "training," in fact, in most instances, it has been embedded practice-period. Given these limited contexts, it has, nevertheless, been viewed by the unit soldiers as being very valuable. Their complaint is mainly that they wish their Embedded Training packages provided more. Again, given technology advances, Embedded Training can in many cases now offer much more. In more and more instances, Embedded Training can now offer:

- True training; that is Embedded Training can be designed to provide feedback, adaptivity, and performance assessment capabilities,

- Current systems technology is now allowing for a combination of computer aided instruction and simulation that will accommodate, not only sustainment training, but also entry level training for military occupational specialty qualified personnel, or acquisition training, and cross training(i.e., it is now possible in some cases to provide tutorials as well as skill development training). However, as always, the requirements for what Embedded Training is to provide must be based on what training is really needed, not just on what technology can now provide.

Another aspect must be considered: the contrast between operator views during garrison or peacetime operation of the system and operator views in battlefield operation of the system is frequently major. In peacetime, or in garrison practice, some systems provide a display which is very limited in variety and may be considerably different in terms of number, density, or nature of stimulus objects (targets) from that which would be seen in a wartime situation. If, as is frequently the case, battlefield stimuli do differ sharply from stimuli provided in normal peacetime or garrison use of the system, there is an implicit need for presentation of these battlefield stimuli to provide sustainment training of operators and commanders. Such a system cannot provide the operators with realistic practice or training against "realistic" events (targets) without some degree of simulation capability which will provide a basis for meaningful sustainment training.

To the extent that electronic networks are designed into systems and/or artificial intelligence can create surrogate crew member and combined arms components, Embedded Training at levels above the individual level is possible.

6.0 REFERENCES

Implementing embedded training: volume 1 of 10: overview. Butler, PA: D.L. Finley (Army Research Institute), D.S. Peckham (Project Manager for Training Devices), H.C. Strasel (Hi-Tech Systems, Inc.), APRIL 1988.

Implementing embedded training: volume 2 of 10: embedded training as a system alternative. Butler, PA: H.C. Strasel & F.N. Dyer (Hi-Tech Systems, Inc.), J.T. Roth (Applied Science Associates, Inc.), I.N. Alderman & D.L. Finley (U.S. Army Research Institute), AUGUST 1988.

7.0 APPENDIX

7.1. EXAMPLES OF STRAP-ON DEVICES

- DX 143 FOR MILAN FIRING POST

- DX 150 FOR AMX 30 MAIN BATTLE TANK

- DX 133 FOR VAB-HOT VEHICLE
- DX 147 FOR GAZELLE HOT HELICOPTER
- STC-DX175 TACTICAL ENGAGEMENT SIMULATOR

7.2 EXAMPLES OF EMBEDDED TRAINING APPLICATIONS

- BQT DEVELOPED FOR THE ROYAL NETHERLANDS NAVY
- T-BQT DEVELOPED FOR THE ROYAL NETHERLANDS NAVY
- OPPAS DEVELOPED FOR THE ROYAL NETHERLANDS NAVY
- FOTOTAS DEVELOPED FOR TORPEDO TEST AND EVALUATION
- TARSIM DEVELOPED TO SIMULATE RADAR TARGETS

7.3 BRIEF DESCRIPTION OF 10 VOLUMES ON "IMPLEMENTING EMBEDDED TRAINING" PRODUCED BY THE US ARI AND PM TRADE

TITLE	DESCRIPTION
1. OVERVIEW	overview of embedded training, the entire product set, and how and when to use what
2. EMBEDDED TRAINING AS A SYSTEM ALTERNATIVE	guidelines for making a decision about whether or not to further continue consideration of embedded training
3. ROLE OF EMBEDDED TRAINING IN THE TRAINING SYSTEM CONCEPT	guidelines for making an early estimation of training system requirements and determining the potential allocation to embedded training
4. IDENTIFYING THE EMBEDDED TRAINING REQUIREMENTS	procedures for iteratively determining the embedded training requirements as information becomes progressively available.and detailed
5. DESIGNING THE EMBEDDED TRAINING	procedures for iteratively developing the instructional, software, and hardware design of the embedded training as the system design envolves
6. INTEGRATING EMBEDDED TRAINING WITH THE SYSTEM	guidelines for integrating embedded training into the design of the system itself including both system software and hardware design
7. EMBEDDED TRAINING TEST AND EVALUATION	guidance for preparation of test issues for the test and evaluation management plan and conduct of in-plant tests and technical test/user test
8. INCORPORATING EMBEDDED TRAINING INTO ARMY UNIT TRAINING	guidance for developing the user documentation provided to unit personnel using embedded training

9. LOGISTICS IMPLICATIONS	guidance to assist the definition of the support requirements imposed on the system by inclusion of embedded training
10. INTEGRATING EMBEDDED TRAINING INTO ACQUISITION DOCUMENTATION	guidelines on which of the above volumes to use in preparation of specific acquisition documents. Detailed guidance on how to prepare procurement documentation with generic models of an Statement Of Work and embedded training specification

U.S. TECHNOLOGICAL INITIATIVES FOR COURSEWARE PORTABILITY

J. D. Fletcher[1]

Institute for Defense Analyses
Alexandria, Virginia USA

INTRODUCTION

Suppose for the sake of illustration, automobile ownership was complicated by requiring every one of the 40-50 makes of automobiles commonly found on our streets to use a unique type of gasoline, so that a Buick would run poorly, if at all, on fuel intended for a Chevrolet, a Dodge would run poorly on fuel intended for a Lincoln, and so on. What sort of impact would this state of affairs have?

Certainly, the technical and logistical demands on car buyers and owners would increase. They would have to learn in considerable detail what fuel their cars needed, where appropriate and reliable sources of fuel were located, which brands of fuel for other automobiles they could use in emergencies, and which they should never use. One can imagine user groups growing up around particular automobiles and sponsoring labyrinthine discussions of which fuels could be used after making what technical adjustments in which engines of their automobiles.

The burdens on gasoline suppliers would also increase because of complications in arranging fuel deliveries, maintaining adequate supplies, and serving customers who, in turn, would require both training and ongoing technical support to keep their automobiles running. Properly trained gas station attendants would be harder to find and more expensive to hire. Legal issues would certainly arise over which fuels dealers had a legal right to sell to whom under what sorts of licensing agreements. These complications would be solved at considerable cost, which would be passed on to the already beleaguered car owners.

Technical innovation in the development of automobile technology would be hobbled. Developers and suppliers would be discouraged from making technical innovations because research and development investments would be returned only by the owners of one make of automobile. Moreover, many new cost-effective features in both automobiles and their fuels would prove to be incompatible with existing fuel requirements. Policies to ensure compatibility could be established, but these would require increased coordination among manufacturers, who, in turn, would need increased control over the market, again at the expense of consumers, to implement their policies.

---

[1] The views expressed in this paper are those of the author and do not necessarily represent those of the Institute for Defense Analyses nor the U.S. Department of Defense. Both the content and expression of these views were substantially improved by comments provided by Scott Lewis, Philip Dodds, and Paul Chatelier to all of whom the author is grateful.

Advanced Technologies Applied to Training Design, Edited by
R.J. Seidel and P.R. Chatelier, Plenum Press, New York, 1993

In general, the impact on the market for automobiles and fuel would be disastrous. Automobile ownership would be far more expensive, cumbersome, and technically demanding for everyone involved, and the growth in automobile technology would occur at a rate slow enough to discourage its most enthusiastic supporters.

The analogy to be drawn is obvious. Software is the fuel of our computing technology. Portability, which in this discussion means the ability to operate the same software across many different hardware platforms, (1) reduces the costs and increases the usability of both computers and their software, (2) strengthens the market for both, and (3) helps institutionalize their use in many diverse applications.

Without portability, different platforms will require different versions of the same application software, and in instructional applications they will require different versions of the same interactive courseware. Without portability for the interactive courseware we use in military training and education, routine and widespread use of this promising technology will be seriously handicapped. As suggested by the analogy with automobile fuel, these handicaps include excessive cost, difficulty of use, technological inertia, and unnecessary complexity in courseware design, development, delivery, and implementation. They will be removed if our interactive courseware is designed to be portable.

Interactive courseware can be portable if developers use standard practices to create it. This is to say that portability requires the establishment and the use of standards, and that the topic of portability is also a topic of standards -- military standards, government information processing standards, national standards, and international standards. Implementation of these standards represents a significant opportunity and requirement for cooperation at all levels. This cooperation will improve the quality and reduce the costs of the education and training we provide at all levels of instruction. This paper discusses what we mean by courseware portability, what we have done about it thus far, current and foreseeable benefits, and some recommendations.

PORTABILITY

This discussion concerns the portability of interactive courseware (ICW). Interactive courseware is a term that is increasingly used in the United States as a catch-all for such applications as computer-assisted instruction, computer-managed instruction, interactive video instruction, and tutorial simulation. Any of these can be presented by such hardware as a computer alone, a computer with a videotape or videodisc player, a computer stimulating the actual equipment used on the job, or a computer using compact disc technologies such as Compact Disc Interactive (CDI) and Digital Video Interactive (DVI). As defined by the Department of Defense (DoD), interactive courseware is computer-controlled courseware that responds to individual student input in determining the pace, sequence, and content of instructional presentations. Courseware by itself refers to all training materials, including the curriculum database and all disks, tapes, books, charts, and computer programs, necessary to deliver an interactive courseware program.

The key distinction between an interactive courseware program and all others is the provision of interactions that tailor the instruction to the needs of individual students. The desirability of individualizing instruction has been noted throughout the history of instruction -- perhaps as early as the 4th Century BC (Corno and Snow, 1986). By tailoring instruction to individual needs, each student receives the level of detail, pace, difficulty, and remediation and the sequence of topics and interactions needed to learn the material efficiently within the limits imposed by time and access to instructional resources. Interactive courseware programs can provide individualized instruction within our current group-oriented institutions. This advantage, combined with the fact that interactive courseware programs tend to be developed as stand-alone, autonomous modules, make them promising candidates for smooth transport of training across many different application areas, instructional settings, and hardware platforms.

Portability means different things to different people. Transportability, transferability, convertibility, and related terms are used to discuss the same concept. A restricted, but straight-forward definition was presented by Dahlstrand in 1984, who defined portability as "the ability to move an application from one computer to another unchanged and get the same results" (p. 17). The DoD definition is not much different. DoD Instruction 1322.20 defines portability as "The capability to run courseware and associated application programs without modification on a delivery system other than the one for which they were originally designed" (p. 3-1). Interoperability is sometimes used in place of portability, but it generally means portability without re-compilation. Portable applications may or may not have to be re-compiled to run on different systems, interoperable applications do not have to be re-compiled.

THE REQUIREMENT FOR PORTABILITY

Over the next several years, the U.S. military Services will invest many millions of dollars to develop interactive courseware. To support this investment, they will acquire a variety of computers, operating systems, peripherals such as videodisc and compact disc units, and all the interface hardware and software needed to communicate within and across these systems. The courseware will be used in all military instructional settings. It will be used by the Reserves and National Guard, in formal courses in military schools, in active duty operational units in garrison and at operational sites, in combined arms and joint staff training, and in educational settings like the Service War Colleges, the Service Academies, and civilian institutions providing Reserve Officer training.

About a dozen different system configurations, at least four different operating systems, and over a dozen authoring systems are commonly used for developing and delivering instruction in the U.S. military Services. Courseware is being produced independently for each of these systems, and in many cases development is proceeding as rapidly as possible to establish a well-anchored position before system standards are introduced. Most of this courseware would have to be "re-purposed," or reprogrammed, to operate on any system platform other than the one for which it was originally developed. Portability will encourage this development to continue while increasing opportunities to share its productions.

Courseware portability will increase the sharing of ICW within each military Service by providing ICW programs that can be used with little, if any, reprogramming across a full range of within-Service instructional settings including initial skill training, advanced skill training, garrison training, job-site training, officer education, and Reserve Component training.

Courseware portability will also increase the sharing of ICW across the different military Services. All the Services teach basic courses such as electricity, electronics, hydraulics, and wheeled vehicle repair. It is not unreasonable to suggest that materials for these courses, or some portion of them, be usable by all the Services. This sharing is neither an administrative nor political impossibility. Joint training is now offered in 62 inter-Service courses or skill areas (Military Manpower Training Report, 1990), and the need for it will increase as training budgets decrease.

Additionally there are actions being taken to encourage, if not mandate, the DoD to share its instructional materials with organizations and users outside the DoD. The Training Technology Transfer Act enacted by Congress in 1988 is intended to encourage transfer of instructional software from government agencies to non-government applications that support the education, training, and retraining of industrial workers, especially workers in small business concerns. As the major developer and purchaser of instructional courseware in the federal government, the DoD is under increasing pressure to provide for the transfer of its instructional materials to non-DoD users. Fletcher, Bosco, Wienclaw, Ashcraft, and Boycan (1991) have outlined issues and processes involved in such a transfer. Portability is prominent among these issues. It will provide a technical foundation for the DoD to meet the requirements of the Training Technology Transfer Act with minimum impact on the resources needed to perform its operational missions. Other federal agencies that adopt the DoD portability procedures should benefit similarly.

Finally, transfer of DoD materials to military and non-military markets is becoming internationally desirable. The establishment of DoD portability procedures that cross international boundaries will be a significant step in this direction. The possibility of opening international markets to courseware developers should motivate greater investment and development in ICW products and technology, improving their quality, lowering their costs, and thereby benefiting both military and non-military users.

THE OPPORTUNITY FOR PORTABILITY

Portability is technically achievable. It is more of an administrative and political issue than a technical issue. As suggested above, portability requires the establishment of policy supported by standards. However, a major issue to be resolved in establishing portability policy is deciding what should be standardized, which is both a technical and an administrative issue. Four possibilities are standards based on (1) hardware, (2) operating systems, (3) authoring software, or (4) the interfaces between these components. Standards based on hardware would specify that all interactive courseware use the same generic hardware specified by its operating and performance characteristics. Standards based on operating systems would require all interactive courseware to use the same operating system, again with specified operating and performance characteristics. Standards based on authoring software would specify that all interactive courseware use the same authoring software. Standards based on a standard interface would establish common techniques for application programs, such as courseware programs and their authoring software, to communicate with and control hardware peripherals.

Despite their conceptual simplicity, there are at least four problems with the first three of these approaches, which would standardize hardware, operating systems, or authoring software. First, all three depend on a common elements specification. Unique features that do not occur in all instances of hardware, operating systems, or authoring software are not likely to be covered by the standard. Unique capabilities in any one of these three areas may languish for many years before being included in the standard, which, in the interim, may be routinely bypassed by exceptions created for users who need access to these capabilities.

Second, changes in the standard must provide upward compatibility for hardware, operating systems, and authoring software already in place. If standards specify hardware, operating systems, or authoring software, then revisions for upward compatibility can accumulate in both quantity and complexity until the standards that incorporate them become unworkable.

Third, the one-size-fits-all philosophy does not work in practice. Different interactive courseware may require different hardware, operating system support, and/or authoring software for good reasons involving both costs and performance. Requiring a single hardware configuration, operating system, or suite of authoring software that is intended to satisfy all users inevitably costs more and satisfies no one.

Fourth, a balance must be struck in standards between simplicity and flexibility. Standards based on hardware, operating systems, or authoring software do not allow flexible inclusion of new features without substantially increasing the complexity of the standard itself. A standard with a better chosen level of specificity, will push the trade-off between simplicity and flexibility much further back than is possible with a standard based on hardware, software, or authoring software.

The DoD is pursuing an approach that is based on standardizing communications between application programs and the devices they must control. This approach meets the criteria suggested by the above considerations. It can cover unique elements, it allows upward compatibility and migration to new systems without requiring substantial modifications in the standard or its structure, it allows developers considerable freedom to exercise their creativity in choosing and developing cost-effective combinations of hardware, software, and operating systems, and, among all the alternatives considered, it strikes the best balance between flexibility and simplicity. Because the interface specified by this standard is generic and intended to apply across many applications and many different physical devices, it is called a virtual device interface (VDI).

THE OBJECTIVES OF PORTABILITY

The current state of affairs and the objectives of the DoD portability effort are illustrated in the progression shown in Figures 1a-c. Figure 1a illustrates the situation without portability. In this situation, application software must be uniquely developed for different operating systems and hardware platforms. The interface between the authoring software and the operating system is unique and usually proprietary for each combination of authoring software, operating system, and hardware platform.

The initial goal for DoD portability is illustrated by Figure 1b. It may be described as system level portability. Its goal is for application software to run on different operating systems and hardware platforms with no modifications to the application. This goal is rarely reached to the point of requiring no modifications, but a proper scheme of portability minimizes these modifications and facilitates their implementation. The goal has largely been reached by the effort described here through specification of the virtual device interface.

A second, more ambitious objective for portability is illustrated by Figure 1c. This objective may be called device level, or plug-and-play portability. Its goal is not only for application software to run on different operating systems and hardware platforms with no modifications to the application (system level portability), but also to permit free substitution of hardware devices, perhaps supplied by different manufacturers, in hardware platforms with no modifications required in either the application or the operating system software. This goal has been achieved elsewhere, for instance by the high fidelity audio community. Records, tapes, and compact discs can be played without modification across a large variety of hardware devices supplied by an equally large variety of hardware manufacturers. Device level portability has been more difficult to achieve in the computer world, but it too, like systems level portability, can be accomplished through the specification of an interface, in this case a device handling interface.

It would have been possible to produce a device level specification directly without a systems level specification as an intermediate step. There were two reasons for the intermediate strategy chosen. First, the DoD needed to produce a specification quickly, and a system level specification is quicker and less risky to produce than a device level specification. Second, a system level specification is more likely to be widely and easily accepted by the interactive courseware community than a device level specification. In this sense, the strategy was successful. The specification was produced quickly, and it has met with wide and often enthusiastic acceptance. The risk in the strategy is that the effort may be too successful and fail to progress beyond a systems level specification. Notably, a plug-and-play technical working group has been formed and is scheduled to produce a device level specification in the next 18 months.

Figure 1a. Standard architecture without portability.

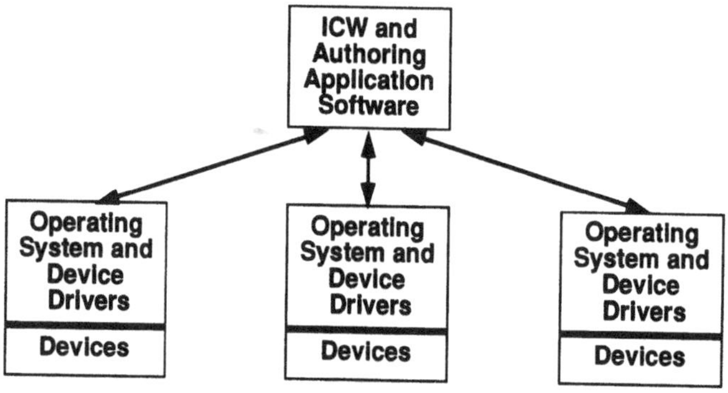

Figure 1b. Standard architecture with system-level portability.

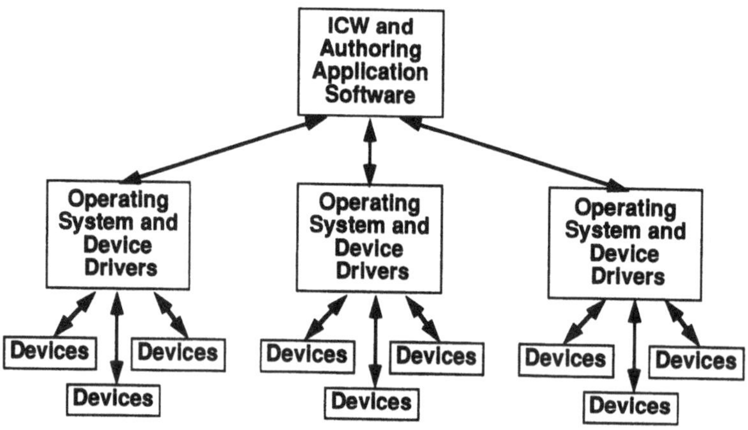

Figure 1c. Standard architecture with device-level portability.

The standard that was produced grew out of close cooperation within the DoD between its Training Policy and Audiovisual Policy Offices. Its development also depended on cooperation with an industrial organization, the Interactive Video Industry Association (IVIA) which is now called the Interactive Multimedia Association (IMA). These organizations adopted a specific strategy and course of action to establish a systems level virtual device interface standard for the portability of interactive courseware. As Lewis (1991) discusses in more detail, this strategy produced an initial standard that could be quickly developed, easily implemented, and widely accepted. The initial standard was developed by a working committee in the IVIA in cooperation with DoD. Version 1.1 of the standard was published by Dodds, Lewis, McFarling, Mistrot, Snowman, and Spiegelberg in 1990 and was incorporated in Appendix D of MIL-STD-1379D, "Military Training Programs," dated 5 December 1990. Conformance with this standard is required by DoD Instruction 1322.20, "Development and Management of Interactive Courseware (ICW) for Military Training," dated 14 March 1991.

The focus of efforts to accomplish portability has changed in the last few years. Earlier discussions by commentators such as Brown (1977) and Dahlstrand (1984) focused on achieving portability through the use of standard higher order computer languages. Dahlstrand emphasized that portable programs should be in source form and that data files should be in text form (stored as formatted characters). More recently the focus has broadened to include both higher and lower levels of concern. Collis and De Diana (1990) present a seven-level model of portability factors in the life cycle of interactive courseware. The model includes higher level technical, educational, social/cultural, and organizational factors as well as lower level factors such as algorithms and data. The virtual device interface approach pursued by the DoD does not address the syntax or form of the higher order languages that might be used for authoring interactive courseware and it does not address the syntax or form of the (presumably) lower order language in which the operating system and its device drivers were written. Instead it aims somewhere in between these two levels by specifying how they should communicate about the specific functions that the lower level provides and the higher level accesses.

Basically this approach concentrates on what we want done rather than how the hardware does it; its orientation is functional rather than procedural. In practice, most interactive courseware programs and authoring software address operating system subprograms, generally called 'drivers,' that provide the desired functionalities and access to hardware devices such as visual displays, videodisc players, monitors, and cursor controllers. If communication between these drivers and the code generated by authoring software can be standardized, we will establish a significant degree of portability.

To see in more detail how this approach works, we need to start with a shared view of the applications and systems software used to support interactive courseware. Figure 2 shows interactive courseware (ICW), authoring software, an operating system, and device drivers as successive layers of a system of software, which is a common way to describe such systems. Three notions are assumed in such a description. First, each layer may only address the layers immediately above and below it. Second, such communication occurs in a clear, well-understood fashion; the interface between any two layers is standardized. Third, software developers are free to pursue whatever approaches they wish within their layer of concern -- only the interfaces between layers are constrained by standards.

In this view, everything above the Operating System could be described as an application, or as application software. Figure 2 shows a typical ICW application. It contains:

• An ICW data base, which consists of courseware data such as student and course parameters, questions or other items, text, graphics, audio, video stills, and video segments.

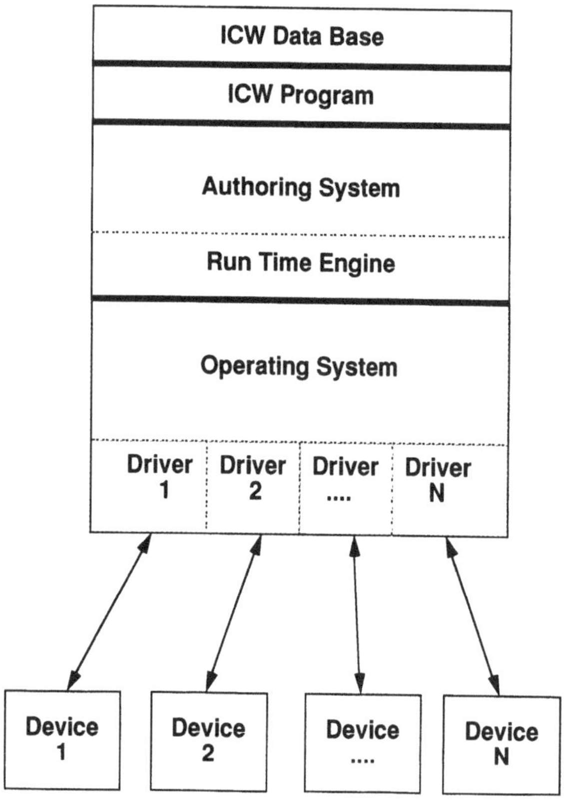

Figure 2. Standard architecture for interactive courseware.

• An ICW management program, which consists of the courseware logic -- the software controlling the selection, sequence, style, and content of courseware data presented to the student.

• Authoring software, which is used to translate the ICW program and data into a form that can be scheduled and controlled by the operating system and be executed by the hardware. The run time engine is that portion of the authoring software that is needed for execution, but not development, of the ICW program.

As Figure 2 suggests, communications between the authoring software and the operating system are rarely seen or accessed directly by courseware developers. The authoring software generates communications in the appropriate form and these are transmitted across its interface with the operating system when it translates the courseware into statements that are scheduled and executed by the operating system, device drivers, and the hardware at run time. Readers used to MS-DOS operating systems should be advised that the operating system in this case is being treated in a generic manner and that it includes other system services such as the Basic Input/Output System (BIOS) which may reside on chips -- in Read-Only Memory -- as firmware.

The virtual device interface (VDI) approach standardizes communications between the application software and the operating system with its associated hardware. It is shown in Figure 3, which is the same as Figure 2 except that it interposes a new, small layer of software, called the virtual device interface, between the application software and the operating system. This VDI layer is unique to each operating system and hardware configuration, but it is the same for all authoring software for a given class of platforms. It allows any ICW application using the authoring software to run on any operating system and hardware platform combination that includes the VDI layer, without reprogramming the application. It allows developers flexibility to do whatever ingenuity and imagination permits, and it allows the ICW application to operate wherever the VDI is provided. The costs for this combination of flexibility and simplicity are the development costs to standardize the calls made by the authoring software run time engine to the VDI, the development costs to write the VDI for the operating system with its hardware configuration, and the run-time costs to service to service the VDI code at run-time. In experience thus far involving 17 ICW courses, we have found the development costs to be 1-4 person weeks, even for poorly designed authoring software (Dodds, personal communication, 19 July 1991). The run time costs have been negligible and cannot be detected by users.

The lack of a standard, device level interface for the operating system device drivers is what requires the VDI to be unique for each class of hardware platforms, such as an Intel-based MS-DOS system accessing a specific set of peripheral devices. Figure 4 illustrates an architecture that provides device level, plug-and-play portability. It incorporates another standard interface, a device handler interface (DHI) to be interposed between the operating system and its device drivers. In this case the VDIs would be nearly, if not exactly the same within a more general class of ICW platforms, such as an Intel-based MS-DOS system accessing any set of peripheral devices. The additional costs for this combination of simplicity and flexibility are the development costs to write the DHI, to modify the operating system to include it, and for peripheral device (videodisc players, light pens, keyboards, monitors, etc.) manufacturers to prepare device drivers for the operating system that can communicate successfully with the DHI from below. The run-time cost is just that required to service the DHI code, but as with the VDI overhead, this cost should be negligible.

Work by the DoD on device level, plug-and-play portability illustrated in Figures 1c and 4 is underway but not yet completed. The systems level portability illustrated in Figures 1b and 3 has been specified and incorporated in a military standard, MIL-STD-1379D, Appendix D, and referenced in a DoD Instruction, DoDI 1322.20.

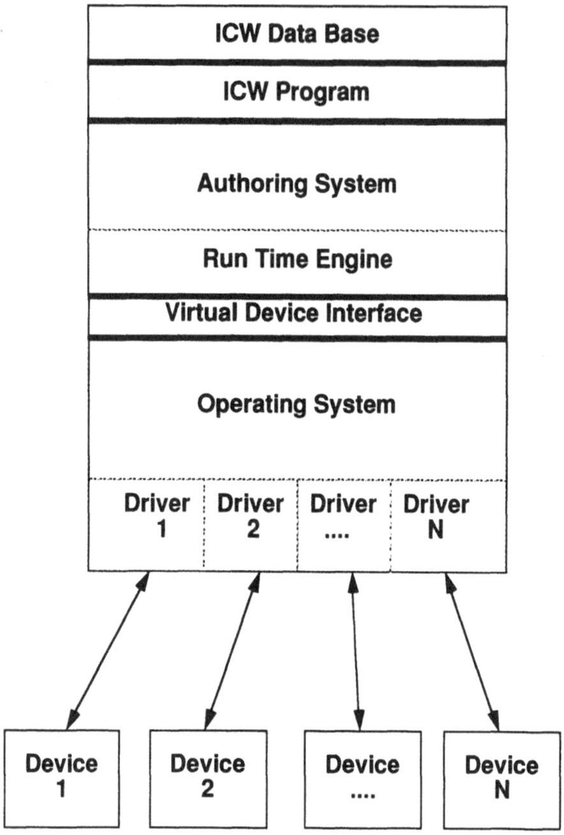

Figure 3. Architecture for system-level portability.

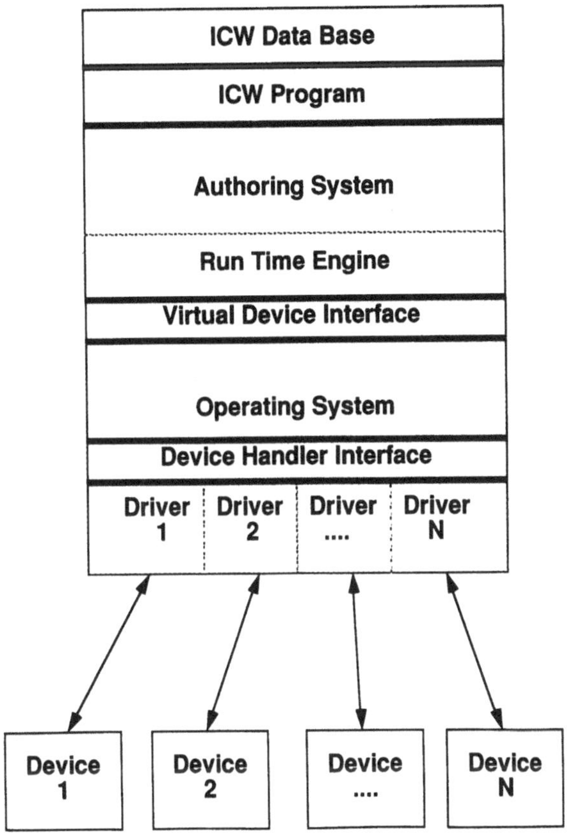

Figure 4. Architecture for device-level, plug-and-play portability.

In brief, the VDI approach is intended to provide the portability needed to satisfy management criteria of cost, effectiveness, and control and developers' needs for maximum flexibility in satisfying the requirements of specific instructional applications. It provides a standard way for ICW applications (courseware and authoring software) to interface with the operating system routines that control hardware. Communication from above (from the applications level to the operating system level) and from below (from the operating systems level back to the applications level) is standardized. Outside of these requirements, all other components of the system can go about their business as usual, unaffected by the standard.

Use of this approach means that most ICW programs already in place do not have to be changed to conform with a virtual device interface standard -- code generation for the authoring software may be the only component of the interactive courseware system that needs to be modified. If the authoring software compiles free standing code, the ICW programs may have to be recompiled to conform with the standard, but this should not be a major undertaking, and DoDI 1322.20 establishes procedures to minimize its costs by requiring that the software needed for compilation be stored and kept available. From a hardware standpoint, no changes in functionality are needed for conformance with the standard.

THE CURRENT STANDARD

The current standard defines an interface between application software, such as interactive courseware and authoring software, and operating system software, which controls access to the resources of the hardware system on which the application software operates. It defines the commands issued by the application software and their expected execution by the hardware, primarily the system peripheral devices. However, the standard is not itself software, it is a specification for software.

There are two forms of the interface definition. One form uses characters, an ASCII interface; the other uses bits, a binary interface. The ASCII interface uses character strings and file I/O with an MS-DOS installable device driver. Of the two approaches, it is less efficient and requires the procedure parameters used to control devices to be translated to ASCII rather than allow them to be used directly as literal values. However, the ASCII interface is easier to read and understand, it is more English-like, and it can be used by any language that can access an MS-DOS file -- effectively any language available on MS-DOS.

The binary interface uses tokens, i.e., numbers, to provide an efficient way to invoke hardware functionalities in MS-DOS using a software interrupt. It is not supported by every language, and it requires more sophistication on the part of its users. However, it is efficient, requires less memory than does the ASCII interface, and eliminates one layer (ASCII to binary) of parameter translation. The two forms differ in user requirements and system performance but are equivalent in purpose and basic functionality.

Seven design criteria were established for the standard that was developed. It was to:

(1) Be compatible with most programming languages.
(2) Function as consistently as possible both for single operating systems and across different operating systems.
(3) Be upwardly compatible to new hardware and new system capabilities so that the standard can be upgraded as required by technological developments without affecting existing applications.
(4) Allow both simple, easy-to-use functions for doing simple tasks and sophisticated functions to support the most demanding multimedia applications.
(5) Provide commands that do not depend on the capabilities of any one operating system.
(6) Not require a specific system software implementation or structure.

(7) Be as inexpensive as possible to implement in terms of memory and performance.

The current standard is generally successful in observing these criteria. It assumes the following as the minimal platform on which it is to be implemented:

- Intel 80X86 architecture.
- MS-DOS version 2.0, or higher, or a functionally equivalent operating system with IBM PC/AT compatible ROM BIOS.
- IBM PC, IBM AT, Microchannel, or Enhanced Industry-standard Architecture (EISA) bus.
- Graphics and video overlay using CGA, EGA, or VGA graphics or two separate graphics and video monitors.
- One or more videodisc players or functionally equivalent video sources and one or more XY-input devices (touch screen, mouse, light pen, bit pad, etc.).

This platform meets the requirements of the standard. On a more practical level, a platform that would support ICW programs in a manner that satisfies nearly all performance requirements would be as above but would include such additions and enhancements as the following:

- At least 640KB of random access memory, a hard disk, and at least one 5.25-inch 1.2MB or 3-inch 1.44MB floppy drive.
- MS-DOS version 3.3, or higher, or a functionally equivalent operating system.
- I/O ports as required by the videodisc player and XY-input device, and at least one IBM AT-compatible parallel port.
- Fully VGA compatible graphics with CGA/EGA emulation in overlay modes and with NTSC or PAL overlay capability.
- AT keyboard, at least one button on the mouse, and Laservision compatibility in the videodisc player along with a capability to play either constant angular velocity or constant linear velocity videodiscs.

In short, what is needed now to make existing interactive courseware portable in accord with this system is an MS-DOS computer, CGA, EGA, or VGA graphics, conforming system software, and conforming run-time courseware. The requirement for conforming run-time courseware means that authoring software must be modified to issue the standard's interface commands during code generation. The content of existing interactive courseware should not require any modification.

One way to determine what functionalities are addressed by any version of the standard is to look at its "service groups." Currently there are four service groups: (1) general system commands; (2) visual management commands; (3) videodisc commands; and (4) XY-input commands. Although the specific commands in each of these groups could be used directly by an ICW program, that is not the intent of the standard. The intent of the standard is that these commands should be issued by conforming authoring software either as ASCII strings or as binary tokens. Authoring software developers are free to present these commands either verbatim or in higher order forms to their users.

The specific commands in the standard are organized and listed under their service groups and briefly described in the following:

(1) General system service group. This group is the only one of the four that must be included in all conforming implementations. It concerns general VDI issues and manages information about VDI software, its configuration, and its state. The commands currently in this group are:

syINIT	Initializes VDI management and the sy service group.

sySTOP — Releases all resources used by the interface and makes them available for other uses.

syGETSTATE — Identifies the service groups for which the system was configured, the version of the standard it supports, and other information on the VDI software release.

syCHECKERROR — Returns the number of the last error, if any, and the command that caused it. This command is particularly needed to detect errors that occur later rather than in immediate response to application commands.

syQUEUE — Stores commands in an internal queue for later execution. It is especially useful for collecting commands that have critical timing requirements and must be executed in close sequence. The queue can be turned on and off, cleared, and executed.

syERRORMSG — Returns an upper-case ASCII text error message that describes the error that occurred. (This is an extended, optional command)

The command, syERRORMSG, is the only extended command in the current standard. Others may be added in the future. An extended command is implicitly a candidate for the standard, but not yet included. An application can test for its implementation by issuing the command and checking for the error message. Even if the command is not implemented, a conforming application should be able to process the command to the extent that it issues the correct error message if the command is not implemented. The application program must then provide its own work-around scheme to deal with the absence of the command implementation.

(2) Visual management service group. These commands concern management of the display screen. They control the graphics display, video display, visual signals, video modes, and graphics modes. They distinguish between logical and physical colors and assume some number of logical colors taken from a palette of physical colors. For instance, using a VGA controller a system can display 256 colors simultaneously, and visual management commands therefore assume 256 logical colors for VGA mode. The commands currently in the visual management group are:

vmINIT — Initializes visual management and sets visual management parameters to known values.

vmGETSTATE — Gets information about current state of visual management including current settings of parameters and available resources such as the number of logical colors and number of video sources.

vmSETGRAPHICS — Sets the graphics mode and the position of graphics relative to video displays. This command also controls VGA emulation of CGA and EGA modes.

vmSETPALETTE — Sets the amounts of red, green, and blue components in a specified logical color.

vmGETPALETTE — Returns the amounts of red, green, and blue components in a specified logical color.

vmSETVIDEO — Chooses the video mode (NTSC, PAL, or neither) and the video input source if more than one is available.

vmSETTRANS — Sets logical colors to transparent or opaque and turns physical transparency on and off. The command also enables temporary override of specified transparent colors.

vmFADE — Sets fade and dissolve levels for computer generated (overlay) graphics, video displays, and/or the relative level of one to the other. Specified changes to specified levels of intensity occur over a specified time. Conforming visual management software can be developed for video hardware without fade circuitry as long as an on/off capability is supported by the hardware.

(3) Videodisc commands. These commands initialize, control, and query the status of videodisc players connected to the system. Current technology uses two types of videodiscs -- constant angular velocity and constant linear velocity -- which differ in the way information is stored on the videodisc and in the functionalities they provide to the user. Constant angular velocity technology stores the same number of frames on every track of the videodisc. It provides more functions (e.g., still frame, scan and frame search, frame-by-frame step forward and backward) and less storage than constant linear velocity technology. The videodisc commands handle both technologies, and treat constant linear videodisc functions as a subset of constant angular velocity functions. The commands currently in this group are:

vdINIT — Initializes videodisc players and sets parameters for their management by the system. This command must be issued for each videodisc player used by the application.

vdGETSTATE — Returns information about the status of a specified videodisc player. This command can be used for purposes such as determining if the player door is open or closed, the current frame number, the state of background play or scan, if the remote control unit is on or off, the player speed, and if the player is parked or spinning.

vdPLAY — Executes videodisc play sequences. The sequences may include any combination of starting frames, ending frames, chapters, directions, and speeds.

vdSCAN — Starts the videodisc player playing either forward or backward from its present position at its maximum speed until it is interrupted by another command or the player reaches an edge of the videodisc. This command is more likely to be used during development than during routine operation of an application.

vdSEARCH — Causes the videodisc player to turn video off, immediately scan to the specified frame or chapter number, and freeze.

vdSTEP — Causes the videodisc player to move forward or backward one frame without blanking the screen.

vdSTILL — Causes the videodisc player to immediately stop at the current frame. With constant angular velocity videodiscs, video remains visible, with constant linear velocity videodiscs the player automatically blanks video.

vdSET — Sets various videodisc player values such as the default logical player number, state of the audio and video channels, index and chapter number displays, and disc spin/park status.

vdPASSTHRU — Communicates directly with a videodisc player to access player features that are not supported by other commands in this service group. This command should not be used in portable applications. It is provided as a convenience to developers who want to use the standard command set for portable applications and who do not want to switch to a different command set for access to non-portable player features.

(4) XY-input device service group. These commands concern XY-input devices (mouse, touchscreen, light pen, bit pad, etc.). They define a uniform way to obtain information from these devices and define coordinate spaces. Many of these devices operate in stream mode, which is to say that they make position and selection information available on a continuous basis. The standard treats all XY-input devices as stream-mode devices with continuously available data. Multiple physical devices may be mapped to a single logical device. The commands currently in this group are:

xyINIT — Initializes XY-input devices and sets parameters for their management by the system. This command must be issued for each XY-input device used by the application.

xyGETSTATE — Returns information about XY-input devices such as the scaling of the coordinate space, what XY-input devices are available, whether their cursors are visible, and the number of buttons they have.

xySET — Scales the XY coordinate space, sets the default input device, turns the cursor on and off, and sets the current XY coordinates

xyGETINPUT — Gets X-position, Y-position, and information on button presses. The standard supports multiple (1-32) button devices, but applications should assume single-button devices for maximum portability. The standard reports only if a button has been pressed; it does not handle touchdown, liftoff, or intensity.

FURTHER DEVELOPMENT

Further development will concern at least six areas. First, the standard will grow to encompass more varieties of interactive courseware. The current standard focuses on interactive videodisc courseware. This focus was chosen because a considerable body of courseware is now being developed for presentation by interactive videodisc systems, and a significant portion of DoD interactive courseware could be covered by the initial standard. Also, many functions of interactive courseware are incorporated in interactive videodisc courseware, so an appreciable portion of the functions that eventually must be covered is already present in the initial standard.

Second, the initial standard will become a device level, "plug-and-play," standard. The progression from non-portability to systems level portability to plug-and-play portability is shown in Figure 1a-c. The current standard is systems level, as illustrated in Figure 1b. It assumes some specific functionalities, and it provides full platform independence across a variety of delivery systems commonly used in the DoD and elsewhere. However, if a single device in an interactive videodisc system is replaced with another, the current standard cannot guarantee that conforming ICW will continue to operate on the newly configured system. Plug-and-play will allow such replacement, and it will increase the ability of peripheral device suppliers to compete freely in the market.

Third, new technological opportunities will be addressed by the standard. These include digital audio, audio management, digital video, graphical user interfaces, and CD-ROM.

These technologies will be addressed in new service groups that are likely to be added in the next 1-3 years. Other new opportunities will continually arise, and provisions have been made by the IMA and the DoD for modifying the standard to address them.

Fourth, the treatment of graphics in the standard will progress beyond its current hardware/firmware level. The current standard cites EGA/CGA/VGA graphics, which is a temporary measure subject to all the problems that accompany standards based on rapidly evolving hardware. Much of the technical work needed to accomplish this expansion of the standard is being completed outside the IMA and DoD portability activities. Many portability needs may be satisfied by the incorporation in the standard of virtual scalability which may be accomplished soon and appears to be technically straight-forward.

Fifth, the current standard will be expanded to cover a variety of operating systems. The current standard focuses on interactive courseware written for MS-DOS. Again, this is reasonable and timely since most DoD interactive courseware operates on MS-DOS. However, the final version of the standard will incorporate POSIX, which may become the government's standard operating system, the many versions of MS-DOS, other PC-based operating systems now available to DoD users, and operating systems based on open system architectures (such POSIX) that are now emerging. This expansion and establishment of the necessary migration path have been discussed by Schneeman (1991).

Sixth, the standard may, in the long run, progress from platform independence to authoring software independence. The current standard provides platform independence which means that courseware can be operated on a variety of delivery systems -- it does not provide courseware that can be modified across a variety of systems, which will be possible given authoring software independence. If a courseware developer wants to adopt a graphics display from one courseware package and modify it for another, the standard now in place cannot guarantee that this can be done, although its platform independence provisions will facilitate this process by substantially increasing the amount of material available for modification. However, authoring software independence is still needed and desired by courseware developers, and it should be addressed as the standard evolves.

PORTABILITY POLICY

Two key formal actions have been taken to establish the portability standard in the United States Department of Defense. First, the standard was incorporated in a formally established military standard. Military standards are created to establish technical requirements for processes, procedures, practices, and methods. MIL-STD-1379D, which includes the ICW portability standard, is the principal DoD standard for training. It establishes: (1) procedures to follow when developing training programs, including guidelines for writing contracts and delivery orders for training; (2) requirements for using DoD Computer-aided Acquisition and Logistic Support (CALS) for training documents; and (3) requirements for using the VDI approach described here.

Military standards are intended as tools for military acquisition. Their existence does not imply any requirement that they be used. DoD directives and instructions are expressions of DoD policy and required procedures. The main goal of DoD Instruction 1322.20 is the cost-effective use of interactive courseware for military training, and it applies to all interactive courseware developed by or for the DoD.

The instruction sets five policies for the development and management of DoD interactive courseware programs and materials. They are the following:

(1) ICW programs are to be designed to promote portability, following the standard DoD programming protocols developed under the DoD portability initiative and other technical requirements prescribed in MIL-STD-1379D.

(2) Payment of royalties, recurring license, or run-time fees, use taxes, or similar additional payments for ICW and associated materials developed for or by the DoD are to be eliminated.

(3) The Defense Instructional Technology Information System (DITIS) is established to provide an inventory and maintain a catalog of DoD ICW programs for use by all DoD components.

(4) Reproduction master materials must be archived for the life cycle of each ICW program.

(5) DoD Components must ensure the availability of all materials necessary to modify ICW programs throughout their life cycles.

In cooperation with the DoD, the National Institute for Standards and Technology (NIST) has adopted the initial standard as the foundation for a Request for Architecture issued to solicit suggestions and recommendations from industry for further development. The product of this DoD/NIST effort will be offered for consideration and adoption as a federal, national, and international standard.

Opportunities for international cooperation in the NATO communities are obvious and needed. Technical review of the standard by ICW developers in NATO countries will determine its feasibility for international adoption and, if it does prove feasible, provide a foundation for its acceptance either specifically or in principle. Such a review will also begin discussions among the military organizations of the participating countries on the technical and administrative means to provide ICW portability and increase the use of this promising new technology for military training and education.

BENEFITS OF PORTABILITY

The main benefit of the DoD portability initiative is to substantially increase the availability and use of interactive courseware. Evidence on the effectiveness of this instructional approach has been accumulating for over 30 years. This evidence has been summarized in reviews starting with Vinsonhaler and Bass, who in 1972 reported a median student increase in achievement of about 40% for interactive, computer-based approaches compared with more conventional approaches, continuing through the studies of Orlansky and String (1979), who found an overall time savings of 30% in the use of these approaches in military training, and most recently found in the meta-analyses of Kulik and his associates. These meta-analyses combine the results of many evaluation studies in a quantitative manner and express their overall findings in standard deviation units. C-L. Kulik and Kulik (1986) found an average increase in student achievement of 0.26 for computer-based instruction used in higher education (this result is roughly equivalent to increasing the achievement of 50th percentile students to that of 60th percentile students). C-L. Kulik, Kulik, and Shwalb (1986) found an average increase in student achievement of 0.42 standard deviations for computer-based used in adult education, which is roughly equivalent to an increase in student achievement from the 50th to the 66th percentile.

In a meta-analysis that was directly relevant to the interactive videodisc approaches addressed by the portability initiatives discussed here, Fletcher (1990) found an overall increase in achievement of 0.50 standard deviations (roughly an increase from 50th percentile to 69th percentile achievement) for students in military training, industrial training, and higher education associated with the use of interactive videodisc instruction. Fletcher also found that across the 13 cost ratios reported in the studies covered by his review, the average ratio of interactive videodisc instruction costs over the costs for more conventional instruction was 0.36, indicating that strong cost-effectiveness arguments for interactive videodisc instruction may exist even though direct experimental examination of this possibility is needed before it can be viewed as conclusive.

It is also interesting to note how much interactive courseware may be available for transfer, provided that it can be made portable. Although the Defense Instructional Technology Information System (DITIS) has only recently been established by DoD Instruction 1322.20, it contains a limited, preliminary listing of interactive courseware programs that are now used by one or another of the Services in the DoD. There are 4644 interactive courseware programs used by DoD already in DITIS. About 70 percent of these programs are implemented on the PLATO computer-based instructional systems. The remaining 30 percent are implemented on various desktop computer-based instructional system. Although this listing is preliminary, the subject matters covered by these interactive courseware programs appear to be representative.

These preliminary DITIS records were examined for applicability to widespread DoD and non-DoD uses. Specifically omitted from the analysis were programs that dealt only with Service-specific equipment or procedures, course management tools without accompanying training packages, and non-training computer programs and application software. Following this initial screening, two criteria for applicability were used to screen the remaining interactive courseware programs: (1) existence of similar jobs in other Services and in the civilian sector; and (2) existence of a vocational area in other Services and in the civilian sector that requires the skills that are taught. Using these criteria, 2718, or about 59 percent of the interactive courseware programs were found to be applicable. This proportion may be higher than more comprehensive analyses will find, but the absolute number of ICW programs that would benefit from DoD portability efforts is still likely to be substantial.

These programs were then divided into two groups: (1) generic programs that deal with basic literacy and general education; and (2) more specific programs that deal with job-related knowledge and skills.

Several lower order categories were found in the interactive courseware programs that deal with basic literacy and general education. In general, 835 programs were found that deal with various topics in general education, including all basic verbal and mathematical literacy skills, subject area-specific preparation for advanced education, learning skills (including learning strategies and reference skills), and personal development skills necessary to function as an adult in modern society (e.g., maintaining a checking account). In addition, 197 programs were found that dealt with foreign language training. These programs cover 22 foreign languages.

As far as programs related to job- or vocation-related skills are concerned, the DITIS records fell into three major categories: Technical Training, Managerial Education, and Professional Skills Update and Enhancement. A total of 1883 interactive courseware programs fell within these categories.

The category of Technical Training includes 1060 programs covering instruction in the knowledge, skills, and abilities necessary to perform non-professional jobs (i.e., jobs that do not require training in the liberal arts or sciences or advanced study in a specialized field). The category of Managerial Education includes 270 programs covering instruction in the skills necessary to direct and administer groups of people and/or processes within an organization. The category of Professional Skills Update and Enhancement includes 553 programs covering instruction to maintain, update, and improve job-specific knowledge, skills, and abilities related to specialized fields requiring advanced education and training.

This cursory and preliminary survey of the DoD inventory suggests that there are ICW programs that are candidates for transfer within the Services, across the Services, from DoD to non-DoD organizations, and perhaps across internationally allied military Services. More information is needed before comprehensive analyses of current DoD interactive courseware programs and the cost savings of portability can be completed. Precise data will emerge as the DITIS data base is filled and as the ICW community gains more experience with the portability standards now in place. However, if the costs of re-programming even a few ICW programs can be avoided, the costs of developing and implementing courseware portability will be recovered.

The issue is not just one of costs. Promotion of ICW and wider realization of its benefits across the DoD are likely to be a significant contribution of the portability standard. Portability should provide more efficient use of the resources that we allocate to military education and training and increase the quality and quantity of the human performance that is an essential and inseparable component of all DoD activities.

Overall, portability will provide for buyers and users of ICW:

- More ICW available off the shelf.
- Increased competition among providers and competition that keys more directly on the costs and instructional effectiveness of the courseware produced instead of its underlying computational requirements.
- Reduced investments of money, manpower, time, and facilities in courseware acquisition. More courseware will be available and it will be more widely useable.
- Less duplicate funding of course development, since less re-programming will be needed.
- Increased interchangeability, reliability, and maintainability of courseware since its development and production will be more widely standardized.
- A well-defined evolutionary path into an open systems environment.
- Greater preservation of the producers' investments in courseware development over time.
- More flexible accommodation of future technical improvements.
- Improved operational readiness of the Services due to more efficient and effective training and education.

Portability will provide for developers and suppliers of ICW:

- A greatly increased marketplace and installed base.
- Access to previously closed markets.
- Reduced development costs.
- A clear path for the evolution of architecture enhancements.
- Reduced stocking and distribution costs.
- Overall increases in the adoption of interactive courseware.

And in general, portability will:

- Lower the per-unit costs of ICW.
- Lower instructional systems procurement costs.
- Increase the use of advanced instructional technologies.
- Increase training efficiency.

SOME ISSUES

Lecarme, Pellissier-Gart, and Gart have stated that "To suppose that a program can carry on an interactive dialog with a terminal is often the surest way to make it unportable" (1989, p. 16), and the defining core of ICW is its interaction with students. In attempting to ensure the portability of ICW, we have not chosen a simple task. Substantive issues remain to be settled. Among these are the following:

- Continued support. We have discussed establishing a portability standard in the same way we might discuss building a bridge, as if once in place and given some maintenance it would continue to perform its function indefinitely. That is not the case. A standard must not only be sustained but continually improved, reshaped, and reconsidered from newer perspectives. The portability standard is a process as well as a specification and the process must be maintained. Development work may to a significant degree be carried out by the IMA, and NIST efforts may produce the necessary committees and coordination, but the process will continue to need advocacy and support from the DoD. Resources and responsibilities for this sustained effort are now only loosely established, and the systematic analyses of costs and effectiveness needed to justify advocacy for portability are far from complete.

- Promulgation. The standard, which began largely under DoD urging and support has grown well beyond DoD. The DoD is still a major developer and user of ICW, but it is only one participant in what has become a large market. The IMA portability effort is now supported by nearly all the major suppliers of interactive multimedia hardware, software, and courseware. The problem is not widespread acceptance, but the opposite. Because the new participants have little vested interest in the current version of the standard, which is incorporated in MIL-STD-1379D, new versions of the standard may leap so far ahead that the current version, and the ICW programs that conformed to it, will be left without a practicable migration path to follow.

- Certification. The portability standard will have little effect if developers and users who wish to conform to the specifications and provisions of the standard have no means for determining if they or their suppliers and contractors have successfully done so. Certification is needed, which in turn means that a resourced facility with clearly defined responsibilities for certification must be established and maintained. This facility does not yet exist, the scope and nature of its responsibilities has not been determined (for instance, do we need certification of hardware, operating systems, authoring software, courseware, or some combination of all of these), and the technical procedures needed for certification have yet to be devised.

- ICW hand-off. A key objective for the effort to establish courseware portability was to increase the sharing of instructional materials within the military Services, across the military Services, and between military and non-military organizations both inside and outside the government. Helping to ensure the technical feasibility of instructional materials sharing is an important component, but technical feasibility will be of little use if the materials themselves cannot be obtained. An organization is needed to establish and maintain an inventory of portable ICW programs. The materials will have to be physically stored with all necessary environmental safeguards. They must be maintained in good working order, and updated and modified promptly by the organization responsible for marketing them. Software, hardware, and an in-house technical capability must be available in order to provide this quality assurance. Users will also require technical assistance in installing the materials for their own applications, and this assistance will have to be provided via telephone hot lines, technical bulletins, user group seminars, and the like. An organization to perform these functions could be established and would probably become self-supporting, but the work to do so needs to be undertaken.

CONCLUSIONS.

Interactive courseware portability is primarily an issue of software. At the systems level discussed here, portability is mostly concerned with code generation and the run time engine of authoring software. It is an issue that individuals concerned with instructional technology must address, but it is not an issue of instructional content or instructional design. It is an instructional production issue, that is best associated with the production and delivery component of a systems approach to instruction.

Provisions for ICW portability will have a substantial impact, both direct and indirect on the costs to develop and use ICW in military and non-military applications. By widening the market for ICW and significantly increasing its potential life-cycle return on investment to developers and users, it should both decrease the per unit costs of ICW and promote its widespread use. If early indications of ICW cost-effectiveness turn out to reflect genuine and substantial efficiencies in instruction, ICW portability will increase the quality and quantity of human performance available to our military Services and thereby contribute substantially to their readiness and effectiveness.

Portability does not come free. There are development costs to be borne by suppliers of authoring software systems and, once device level portability is established, suppliers of ICW peripheral devices. There are also run-time processing and memory costs of ICW platforms that support the virtual device and device handling interfaces required, respectively, by system level and device level portability as described in this paper. However, these costs are likely to be minor in proportion to the development and run-time costs that are already invested to support ICW and trivial compared to the cost avoidances that will result once conformance with the portability standards is established.

The costs to maintain and further develop the ICW portability standards described here are a more serious matter than the development and run-time costs that suppliers will have to bear. The early DoD initiative to establish ICW portability has been turned over to the Interactive Multimedia Association and its participating members. They are likely to support further development of the standard for the near future. If the National Institute for Standards and Technology (NIST) is successful in fusing the IMA and DoD products into a Federal Information Processing standard and later into a standard supported by the American National Standards Institute and the International Standards Organization, its long term survival will be assured. This process has grown beyond direct DoD influence and support. There is little to do but encourage those aspects of the standards process that directly support DoD training and audiovisual policies and that ensure graceful migration of DoD products to newer versions of the standard.

There are opportunities for NATO cooperation in the establishment of ICW portability for military training and education. Review of the work completed thus far to determine the feasibility and desirability of developing a NATO-wide standard for portability would be appropriate and helpful at this time. The most effective approach for undertaking these reviews may be to attempt implementations of the standard in existing and new ICW programs. This hands-on approach with systematic records kept of the time, costs, and critical issues required for these implementations would advance the current state of the art. A central purpose of this paper is to encourage relevant organizations in participating NATO countries to undertake these reviews.

REFERENCES.

Brown, P.J. (1977) Software Portability. Cambridge, United Kingdom: Cambridge University Press.

Collis, B.A., & De Diana, I. (1990) The impact of different portability factors during the life cycle of an educational software adaptation project. Journal of Research on Computing in Education, 23, 306-317.

Corno, L., & Snow, R.E. (1986). Adapting teaching to individual differences among learners. In M.C. Wittrock (Ed.), Handbook of Research on Teaching (Third Edition). New York, NY: Macmillan Publishing.

Dahlstrand, I. (1984) Software Portability and Standards. Chichester, United Kingdom: Ellis Horwood Ltd.

Dodds, P., Lewis, S., McFarling, D., Mistrot, H., Snowman, G., and Spiegelberg, J. (1990) Recommended Practices for Multimedia Portability, Release R 1.1. Washington, DC: The Interactive Multimedia Association.

Fletcher, J.D. (1990). Effectiveness and Cost of Interactive Videodisc Instruction in Defense Training and Education (IDA Paper P-2372). Alexandria, VA: Institute for Defense Analyses.

Fletcher, J.D., Bosco, J., Wienclaw, R., Ashcraft, J.K., and Boycan, G. (1991) Assessment of Department of Defense Job Skill Enhancement Programs (IDA Document D-939). Alexandria, VA: Institute for Defense Analyses.

Kulik, C-L.C. and Kulik, J.A. (1986) Effectiveness of Computer-Based Education in Colleges. AEDS Journal, 19, 81-108.

Kulik, C-L.C., Kulik, J.A., and Shwalb, B. (1986) Effectiveness of computer-based instruction in adult education. Journal of Educational Computing Research, 2, 235-252.

Lecarme, O., Pellissier-Gart, M., and Gart, M. (1989) Software Portability with Microcomputer Issues. New York, NY: McGraw-Hill.

Lewis, D.S. (1991) Introducing the IMA recommended practices for multimedia portability. In P.V.W. Dodds (Ed.), IMA Compatibility Project Proceedings, Washington, DC: The Interactive Multimedia Association.

Orlansky, J. and String, J. (1977). Cost effectiveness of computer-based instruction in military training (IDA Paper P-1375). Arlington, VA: Institute for Defense Analyses.

Schneeman, R.D. (1991) Interactive video software portability: Migrating a MS-DOS application to POSIX. In Proceedings of the 29th Annual Technical Symposium, Advanced Information Interfaces: Making Data Accessible. Gaithersburg, MD: National Institute of Standards and Technology and Washington DC Chapter of the Association for Computing Machinery.

U.S. Department of Defense (1990). Military Manpower Training Report: FY 1991. Washington, DC: U.S. Government Printing Office.

U.S. Department of Defense (1990). Military Standard 1379D: Military Training Programs. Washington, DC: U.S. Government Printing Office.

U.S. Department of Defense (1991). DoD Instruction 1322.20: Development and Management of Interactive Courseware (ICW) for Military Training. Washington, DC: U.S. Government Printing Office.

Vinsonhaler, J.F., & Bass, R.K. (1972). A summary of ten major studies on CAI drill and practice. Educational Technology, 12, 29-32.

DEVELOPMENT AND EVALUATION OF INTELLIGENT TRAINING SYSTEMS FOR AIR TRAFFIC CONTROL

G.C. Morrisroe and S.L. Dickens

Logica Cambridge Limited
Betjeman House, 104 Hills Road
Cambridge, UK

INTRODUCTION

Intelligent training systems (ITS) concepts aspire to provide advanced features in the computer-based learning environment. Aspects of intelligence can provide systems with adaptability and flexibility in the structure and presentation of instructional material. Modelling of the user's knowledge and experience allows a greater degree of individualisation of training. Representation of teaching strategies allows the system to tailor its presentation to the individual. The dialogue structures which can be maintained by the system allow users much more control over the direction of the dialogue.

With such potential and so many advantages, it would be expected that ITS would be exploited in many application areas. However, the goals of ITS may take rather longer to be realised than expected. In this paper we present a series of projects which chart the progress to usable and useful ITS systems, and we comment on the issues of importance and the essential elements to ensure good intelligent training systems.

BACKGROUND

Logica's first ITS project was concerned with the evaluation of Expert Systems techniques for use in training. Funded by Royal Signals & Radar Establishment (RSRE), it was awarded in 1983. It lasted two years during which time a number of evaluation prototypes were produced.

It was clear that knowledge-based systems (KBS) with their explicit representation of knowledge and models of expertise could be of significant value for instructional

Advanced Technologies Applied to Training Design, Edited by
R.J. Seidel and P.R. Chatelier, Plenum Press, New York, 1993

purposes. However, as a result of this study we concluded that the use of expert systems and knowledge bases in themselves do not provide sufficient facilities or functionality for training. Instructional systems require additional flexibility and expressive power not commonly found in most KBS implementations. The requirements for advanced instructional systems could be stated broadly as having:

- some knowledge and expertise in the application area
- some understanding of the type and level of the individual student
- methods to manipulate the conditions of learning

Based on this approach, the project produced a prototype architecture with the now usual components of dialogue manager, student model and domain knowledge in declarative and procedural forms. This formed the conceptual basis for our subsequent work in ITS.

The subject knowledge contains an explicit representation of the facts, relations and rules in the application domain. Broadly, this knowledge base covers the areas of declarative knowledge and procedural skill.

The student model is an individual fine-grained representation of the student. This model is built up over the course of interactions with the system and contains information on the level and skill of the student.

The difference between the model of competence represented in the knowledge base and the student model defines the training requirement. This requirement is interpreted by the teaching strategy component to determine the content and structure of the next instructional event or sequence. The interactions in which the student engages are used to update the student model .

In 1986, we began a programme of work in air traffic control (ATC) training systems for the UK Civil Aviation Authority (CAA). The CAA believed that in the medium term KBS technologies showed promise in ATC training applications.

This paper is concerned with the development of a series of prototype ITS for ATC. Each of these training systems embodied ITS concepts, but addressed a different type of training need, from intelligent courseware to simulation training. The evaluation of these projects offers a perspective from which we discuss the essential elements involved in ITS development.

AERODROME CONTROL SIMULATOR

The overall aim of this project known as CAI was to demonstrate the use of KBS technology in the context of ATC. This aim was refined into an objective to develop a generic knowledge-based instructional framework which would address the transition between theory/knowledge and practice/skill in ATC training.

The area of ATC selected for the training system was aerodrome control. Aerodrome Control is the first practical (simulation) exercise encountered by students at the College of ATC (CATC); the practical sessions follow on from a course of lectures on theory. Students try to implement what they have learnt in an existing rudimentary aerodrome simulator under the guidance of an instructor.

Training Requirements

The training need on which CAI concentrates is the strategic management of traffic and information passing aspects of Aerodrome Control. Students must apply their knowledge of the wide-ranging theory on aerodrome layout, rules and limitations by which the traffic is constrained. The complete task of Aerodrome Control also incorporates the skills of Radio Telephony (RT) and flight strip management which are not addressed in the CAI system.

Application Area

Aerodrome Control is a complex environment in which a range of traffic interact. There are six basic traffic types each with their own ATC requirements:

- Mobiles (eg fire engines)
- VFR Outbound (eg light aircraft)
- VFR Inbound
- IFR Inbound (eg commercial aircraft)
- IFR Outbound to the FIR
- IFR Outbound to Zone

Each traffic type behaves differently and has its own procedures. In addition, aerodrome control contains both airborne and ground based behaviour. The airborne and ground constraints are quite different; taxying aircraft can stop and wait; airborne aircraft can only be given directions using predefined routes.

System Overview

The CAI interface comprises four main areas:

- message formulation, input and reception
- vehicle movement display
- electronic strip board
- commentary area

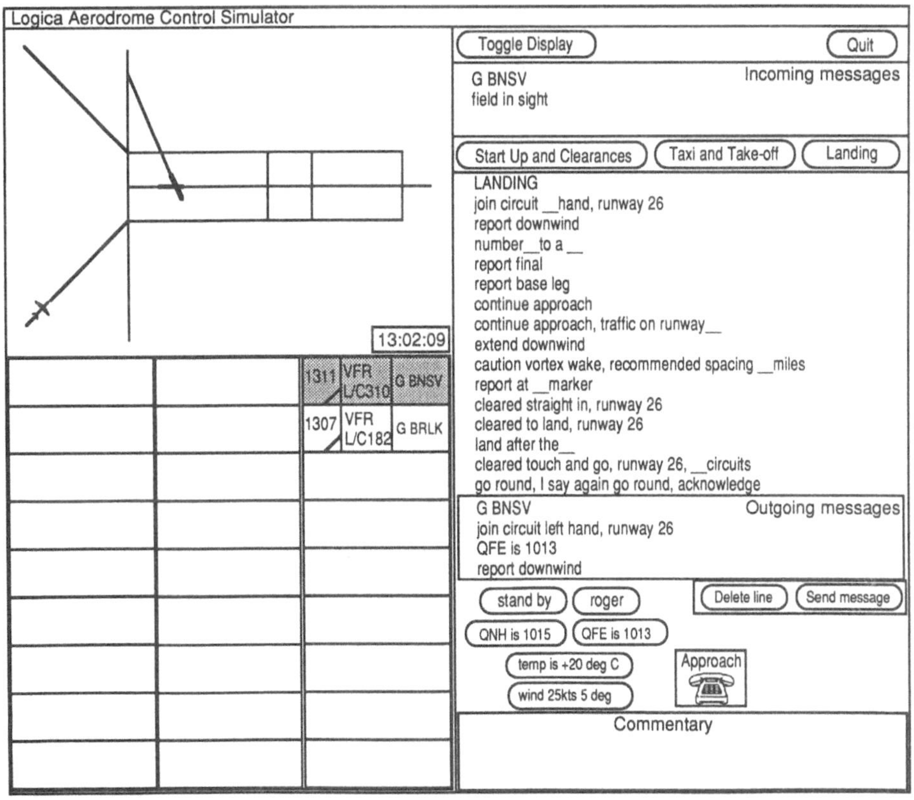

Figure 1 Shows the aerodrome control simulator screen in zoomed-out view

The message formulation and input area provides the repertoire of commands and instructions which can be sent to the various ‘actors’, ie vehicles involved in the scenario. The student formulates messages by nominating the target actor by pointing at the appropriate flight strip and selecting one or more messages from one of three lists depending on whether the message is concerned with startup and clearances, taxying and take off, or landing. Some messages are templates which require specific values to be provided depending on the current situation. For example, an aircraft can be told to 'taxi to holding point X via the Y taxiway'. The student must select the appropriate values for the holding point and taxiway, eg 'taxi to holding point 26 via the southern taxiway'. The message is sent and the vehicle replies with an acknowledgement. Actors send text messages to the student at appropriate times, and these are displayed in the incoming messages area.

Active traffic in the aerodrome traffic zone and the aerodrome itself are represented by two views: the zoomed-in view shows the runways, taxiways and airport aprons and buildings, the zoomed-out view (shown in figure 1) shows active traffic and routes (circuits and standard VFR and IFR approaches and departures) within a ten mile radius of the aerodrome.

The knowledge-based tutor generates the scenarios. At first it introduces the simplest type of vehicle, ie a mobile, and then increases the numbers and the types of vehicles as the student progresses. This allows the student to become familiar with each particular procedure. The introduction of more than one vehicle or vehicle type brings into play the aerodrome constraints. The scenario generator introduces higher levels of traffic and greater variation of aircraft types. The scenarios are not engineered 'on the fly' to emphasise a particular training point; poor decision making by the student will inevitably result in increased task complexity.

If an instruction sent to a vehicle is wrong the requested action is not carried out, and the student is given immediate feedback in a commentary area. All such errors are recorded in the student model and used in remediation at the end of the scenario. In remediation the problem is identified and a statement of the correct action or sequence of actions is given. The scenario generator takes account of the current state of the student model when deciding on the traffic complexity for the next session.

The demonstrator does not include the full repertoire of behaviour and messages possible, but it does allow students without the prerequisite skills in Aerodrome Control to begin to formulate messages to control the movement of traffic in and around the aerodrome. The system allows students to recognise and formulate messages without necessarily being able to recall the exact terminology of RT. The message strings themselves encourage good practice by affixing additional messages to the main phrase, eg ' cross runway 26, *report vacated*'.

Knowledge-Based Tutor

The knowledge-based tutor (KBT) performs three main functions:

- generates scenarios based on student model
- simulates behaviour of actors
- monitors student behaviour

Scenario Generation

The rôle of the scenario generator is to provide the information required to set up and run a scenario. This involves generating a traffic complexity specification (TCS).

A TCS defines the total number of vehicles in the scenario and the type of each vehicle. In addition, the specification associates call-signs, aircraft category, approach and departure directions, landing and take-off runways, and start and finish aerodrome locations with each vehicle as appropriate.

The scenario generator produces the specification on the basis of a syllabus, teaching strategy, and student model. The student modelling component is discussed further in the section on student monitoring below. The syllabus and teaching strategy are described here.

The syllabus is defined as a list of 'activities' ordered according to their level of difficulty or complexity, starting with least complex and finishing with the most complex.

Activities are defined as combinations of one or more vehicle types. Vehicle types are ordered according to their perceived level of difficulty, as are combinations of vehicle types. For example, mobile traffic is defined as the least complex activity, VFR inbound the next least complex, mobiles and VFR inbound together the next least complex, and so on. The most complex activity is therefore defined as the combination of every type of vehicle. The syllabus defined in the system is in no way definitive, and can be changed with minimal effort.

For each activity defined in the syllabus, there is an associated 'activation', which represents the student's current level of expertise for that activity. Activity activation is derived from procedure (or vehicle type) activation. Procedure activation is in turn derived from concept activation, likewise the student's current level of expertise for each concept. Procedure and concept activation are discussed in the section on student monitoring.

The teaching strategy employed by the system dictates which activity is selected for the scenario. The activation for each activity in the syllabus is determined until one is found that falls below a predefined threshold. This becomes the chosen activity. The teaching strategy incorporates a 'progressive parts' training philosophy; individual 'units' in the domain are combined at increasing levels of complexity until the full domain is presented, ie the last activity in the syllabus.

Having determined the type(s) of vehicle to be simulated, the number of vehicles for each vehicle type needs to be established. This is achieved by mapping the activation for each relevant procedure onto the appropriate 'procedure graph'. A procedure graph associates numbers of vehicles with activation thresholds. The numbers of vehicle for each procedure in the chosen activity gradually increase until a peak is reached, when they begin to decrease. The 'shape' of the graph can be defined arbitrarily. In the current system, the shape of the graph is linked closely to the activity threshold described in the previous paragraph. The activity threshold represents the peak in each procedure graph. Before the activity threshold is reached each procedure is taught on its own, ie without any other vehicle types. Once the threshold is reached procedures are taught in combination with other procedures. Thus, it is sensible to increase the number of vehicles as the threshold is approached, and gradually decrease them as other vehicle types are introduced.

The final stage in specifying traffic complexity is to assign appropriate attributes to each vehicle to be simulated. Attributes are assigned by randomly selecting a 'value' from a predefined set of possible values. For example, the call-sign for a mobile is randomly selected from a list of possible mobile call-signs. Selecting attributes such as start and finish locations, approach and leaving directions, and aircraft category is performed in a similar fashion. It should be noted that the number of predefined call-signs effectively defines the maximum number of vehicles that can be simulated.

Simulating the behaviour of actors in the scenario is a prime function of the KBT. Actors are defined as pilots and drivers of simulated vehicles and Approach Control. Simulating the behaviour of actors involves generating appropriate messages from them to the student (ie the Aerodrome Controller) and generating appropriate routes for vehicles on and around the aerodrome.

Vehicle Representation

Each vehicle in the scenario is represented explicitly within the KBT. Each vehicle has the following attributes: call-sign, vehicle type, vehicle category, route, expectations. The call-sign is that assigned by the scenario generator. Vehicle type is one of the possible vehicle types defined earlier. Vehicle category is only relevant to aircraft and is one of: light, small, medium, and heavy. Mobiles have the dummy vehicle category 'mobile'. A vehicle's route has the following characteristics: type, current route, current position, and target location. A route can have one of three types: normal, take off, and default. A ground vehicle follows a normal route, an aircraft cleared for take off follows a take off route, and an aircraft on inbound approach or in the circuit follows a default route. Route types are used when applying safety rules. Approach Control is represented by a special 'vehicle'.

When a message for a particular actor is received, the current state of the vehicle representation is examined and appropriate actions are taken. This process is described in the following section on events and expectations.

Events and Expectations

The concept of events and expectations is fundamental to the function of actor simulation.

Each simulated actor has a set of expectations about which messages and instructions it should receive next. Events are the messages/instructions that are actually sent and received. The expectations of an actor change according to the events that occur; existing expectations can become obsolete and new expectations can be added. The behaviour of actors is determined by the appropriateness or otherwise of events with respect to expectations.

When the KBT receives an event the expectations for the relevant vehicle are examined to establish whether any are relevant to the event. If there are no relevant expectations or the event has unacceptable consequences, an error has occurred and appropriate action is taken. The mechanisms for establishing event acceptability and error diagnosis are described in the section on student monitoring.

For each relevant (active) expectation the following process takes place. The active expectation is removed from the set of existing expectations. Any new expectations which become expected following the activation of the relevant expectation are added to the set of existing expectations for the vehicle. Any existing expectations which become 'obsolete', ie those that are no longer expected, are deleted from the set of expectations. In addition, any 'internal' existing expectations immediately become events, ie they result in some side-effect. Internal events are defined as those emanating from actors and result in text messages being sent to the student.

A vehicle can be removed from the scenario when it has no more expectations. The scenario itself comes to an end when all vehicles have an empty set of expectations.

Actor Message Generation

Events representing messages from actors to the student need to be converted into appropriate text. Each message type has an associated 'template' text string into which appropriate parameters can be inserted. Some message types have no parameters, eg reporting position and closing down. For those with parameters, the event indicates the value of each parameter.

Student Monitoring

The student monitoring function comprises event acceptability, error diagnosis, and student modelling.

Event Acceptability

The acceptability of an event is determined by a two stage process: acceptability with respect to the specified vehicle, and acceptability with respect to a subset of aerodrome safety rules.

The acceptability of the event with respect to the specified vehicle is determined by applying the relevant expectation mechanism to the event. If this stage succeeds, ie the event is acceptable with respect to the vehicle, the event is checked for acceptability with respect to safety. Otherwise the error diagnosis mechanism is invoked.

A subset of aerodrome safety rules is implemented, namely:

- clearances of vehicles to ground locations on the aerodrome
- separation on take off
- separation in circuits

Determining the acceptability of an event with respect to safety involves checking the state of the aerodrome after the event against the above sets of rules. If any conflicts are created, the event is unacceptable and error diagnosis is performed.

Error Diagnosis

The function of the error diagnosis mechanism is to determine the reason for the unacceptability of an event. The types of error diagnosed are as follows:

- irrelevant events
- inappropriate events
- dependent events
- illegal vehicle clearance
- illegal aircraft separation

Irrelevant events are those addressed to a vehicle of a type to which the event does not belong, eg instructing a mobile to take off. Inappropriate events are relevant to the particular vehicle, but occur at the wrong time, eg instructing an aircraft to contact Approach Control before it has left the apron. Events containing illegal parameters are also classified as inappropriate, eg instructing an aircraft departing to the north to take a left turnout (ie head south). These types of error demonstrate the coarseness of the error diagnosis component.

Dependent events are those that should not happen before one or more prerequisite events have occurred. For example, before being cleared for take off an outbound aircraft should have been given clearances and QNH. If a dependent event occurs too early, an error is generated. An error is generated if a route resulting from an event, eg taxi instructions, contains locations that are either occupied or cleared for use by another vehicle. This applies to ground vehicles. For airborne aircraft a separation violation error will be generated if an illegal clearance is given. Minimum separation distances are defined for each aircraft category. Failure to adhere to these rules will cause an error. These rules are only applied when an aircraft reaches the downwind and final legs in the circuit. A second separation violation error occurs when an aircraft cleared for take off is in conflict with an aircraft having recently taken off.

The student is informed of errors in two ways. During the scenario, the student is given some commentary about the error. Commentary consists of one or two lines of concise text, informing the student of the error in terms of the event and the error type, as defined above. At the end of the scenario, the student is given more lengthy error information in the form of remediation. Remediation is presented at a fairly high level. If irrelevant and/or inappropriate events occur during a particular procedure, remediation is given on the procedure in terms of the events that happen and their order of occurrence. Remediation is given on dependent events in terms of the events that should precede the dependent event. All clearance and separation errors are given the same remediation. The circumstances under which a clearance or separation error occurs are explained.

Student Model

The student model is a dynamic (ie constantly updated) representation of the perceived level of knowledge of the student. Each student known to the system has a separate student model, which is saved at the end of a session and reloaded when the student next uses the system.

Knowledge in the student model is represented in the form of concepts. Each message the student can send, effectively each event the student can cause to happen, is a single concept.

The level of knowledge for each concept is represented by a 'score'; the higher the score, the greater the level of knowledge. Each time a concept is used correctly its score is increased, and when used incorrectly it is decreased.

The scoring procedure is the only analysis carried out on the student model during a scenario. A more sophisticated analysis is performed during the scenario generation phase.

The student modelling component provides the scenario generator with concept and procedure activations in order that the latter can derive activity activations.

Concept activation is derived by applying a weighting function to the score for the concept. The weighting mechanism enables concept scores to be normalised with respect to their relative importance in aerodrome control. For example, one might judge that taxying is more important than telling an outbound aircraft to contact Approach. A greater weight is thus applied to the score for taxying than to that for contacting Approach, in order to reflect this distinction in the activations for these concepts. Procedure activation is derived by obtaining the product of the activations of all concepts associated with the procedure.

Evaluation

The CAI system was evaluated by two teams: a technical team of KBS experts and training team of aerodrome control instructors. The technical team considered the system was a useful application of KBS techniques to training, although they had their reservations concerning the genericism of the system. The instructors were impressed by the coverage and representation of the application area. They also liked the challenging scenarios of traffic that the system could provide. But they had serious reservations about whether the system was usable as a part-task training system for aerodrome control. Also they felt uneasy with the method of message formulation and passing, and thought that radio telephony should be integrated within the system. Finally, they had reservations over the simple representation of aircraft and mobile in the zoomed in and out views.

We learnt that it is not enough to produce a good technical demonstrator system. Suspending consideration of the training requirements whilst concentrating on technical aspects runs a high risk of the system not being taken further.

MULTI-MEDIA TRAINER FOR ATC THEORY

In this second project, called AVIATION1, the aim was to deliver a PC-based prototype ITS which demonstrated a credible approach to ATC training together with basic authoring tools which would provide the capability for users to implement course material, and a specification for advanced tools which would replace the basic tools at a later stage.

The system is interactive with the student steering the content and rate of delivery of the teaching material. The student is able to access all the teaching material in the system in a 'free browsing' mode, or can be led through the material in a tutorial mode.

The system supports a mixed initiative dialogue allowing the student to decide what is to be taught next, or to let the system suggest what is the most appropriate subject to be taught. These suggestions are based on the student model, ensuring that students are only taught subjects for which they have the necessary prerequisite knowledge. Whilst presenting particular teaching items, the ITS also presents a list of other related pieces of information from the pool of teaching material to which the student has free access.

As well as providing a course delivery system, the AVIATION1 project investigated the issue of authoring tools, which allow instructors to generate their own teaching material. Although significant research and development has been dedicated to the delivery of ITS courses, little work has been invested in the authoring of such courses. The AVIATION1 project provided rudimentary authoring tools, using the experience of developing and using them to gain an insight into what facilities more comprehensive authoring tools should provide.

Training System Requirements

The training requirement derived from the need for more air traffic controllers in the UK. The CAA is recruiting and training air traffic control cadets in increasing numbers. Whereas in the past, a typical ATC cadet had either direct experience or an active interest in aviation. Now, cadets often have less appreciation of general aviation. The training course structure leaves little room for teaching these general aviation subjects.

The AVIATION1 project examined ways in which this extra teaching could be performed alongside formal classroom instruction such that each individual student's training requirements were satisfied. The AVIATION1 training system was designed and implemented as a personal tutoring workstation, with the ability for the students to use the system as often and for as long as required.

Conventional computer based training approaches had been considered by ATC training staff but rejected on the basis of their lack of suitability. We were, in effect, adding intelligent aspects and flexibility to a CBT approach.

System Description

The AVIATION1 system was designed to provide structured training on the basis of a pre-specified collection of subjects and for that training to be tailored to the needs of the individual student. There was a requirement for the student to be able to initiate dialogues in terms of the subject being taught and the order in which subject material was presented. In addition, a 'reference library' of facts and figures was required for the student to investigate when necessary.

The presentation of training material needed to be as wide as possible to make the tutoring more engaging. The AVIATION1 system provides facilities for displaying text, static and moving graphics, and video stills and sequences.

Instructors needed the ability to monitor students' progress through the course in terms of the subjects and lessons that the students had covered, objectives met, individual pieces of training material seen, and tests completed.

We now describe the data structures, components, and behaviour of the system designed to meet these requirements.

Data Structures

The data structures contain the course material that has been entered by the instructor. In addition, there is a knowledge and teaching 'language' which is used to describe what should be presented. For each student a record of his/her progress with respect to the current course is maintained.

Syllabus

The subject domain is divided into a number of topics which are arranged in a hierarchical form in a syllabus. An example of a syllabus is as follows. The main subject, theory of flight, has three sub-subjects, aircraft parts, control surfaces, and movement about axes. The aircraft parts subject itself has two sub-subjects, fuselage and wings, and the control surfaces subject also has two sub-subjects, elevator and aileron.

Modules

Attached to each subject is a list of modules. Modules represent lessons that are presented to the student. A module has one or more teaching objectives which define what the module is intended to teach. A module also has a list of prerequisite objectives that must be satisfied before the module is presented. It can also have a set of excluded objectives. This allows the instructor to prevent certain objectives being taught, even if the system considers them appropriate. A module can have a specified preferred teaching strategy, which will be used to teach the module.

If there is no specified teaching strategy the course teacher uses a default strategy. There is also a set of teaching points from which the teaching material to present in the lesson is chosen.

Teaching Points

A teaching point is an individual piece of teaching material. It can be any of text, line or moving graphics, or video still or sequence. A teaching point can be either a piece of exposition or a test. A teaching point is labelled by one or more teaching primitives which describe the content of the teaching point. For example, if the teaching point contains a piece of text describing the fuselage in some detail, the relevant teaching primitive would specify that the teaching point is a 'narration' on the concept 'fuselage' to the level 'detail'. Teaching points are collected together and presented in scenes; a lesson comprises a number of scenes. A scene comprises a small number of teaching points.

Concept Net

The final part of the knowledge base is the concept net. The concept net is a representation of the ATC domain in terms of individual concepts and relations that hold between concepts. This representation is devoid of any instructional interpretation; it is intended as a factual model of part of the ATC domain. For example, the concepts jet aircraft and turbo-prop aircraft are related to the aircraft concept by the 'type of' relation type, and the wings and fuselage concepts are related to aircraft by the 'part of' relation type.

Student Model

For each student using the course teacher there is a record of his/her interaction with the course teacher which is maintained from session to session. This student model records what modules have been covered, objectives met, teaching points seen, and tests completed.

System Components

The course delivery system is made up of teaching strategist, dialogue manager and presentation manager components.

Teaching Strategist

The teaching strategist is responsible for deciding what subject to teach next. This is achieved by comparing the syllabus with the model for the current student to establish which subjects and modules within subjects have not been covered. Once a module has been chosen, the lesson is generated. The teaching strategist generates an outline lesson by applying a set of rules to the objectives specified in the chosen module and produces a much larger set of lesson objectives implied by the teaching of the main module objectives. The rules expand objectives into sub-objectives using the concept net as appropriate. For example, if the main objective is to discriminate between types of flap, some of the sub-objectives would be to describe each type of flap in turn. Types of flap are obtained from the concept net by establishing which concepts are related to the 'flap' concept by the 'type of' relation.

Dialogue Manager

The dialogue manager takes the lesson the course teacher generated by the teaching strategist and generates a scene for each objective in the lesson. A scene comprises one or more teaching points. Each objective is converted into one or more teaching primitives by means of a set of rules. This set of rules maps the teaching semantics of the objective description onto the content semantics of the primitive description. For example, the 'discriminate' objective is mapped onto the 'difference' and 'similarity' primitives. Appropriate teaching points are then obtained by searching for pieces of material labelled by the set of generated primitives.

Presentation Manager

The presentation manager is given the scene and presents each teaching point in turn. Once the scene is complete, the presentation manager passes back the scene 'results' to the dialogue manager. The results specify which teaching points have been seen and which tests completed. The dialogue manager can then establish whether the requirements for the current scene have been met. If they have the next scene is generated, otherwise a new scene is generated for those aspects of the scene that have not been satisfied, possibly using new material. A lesson is complete when there are no more objectives to satisfy for the current module.

System Behaviour

The AVIATION1 course delivery system comprises three main facilities:

- a tutoring component
- a browsing component
- a student support facility

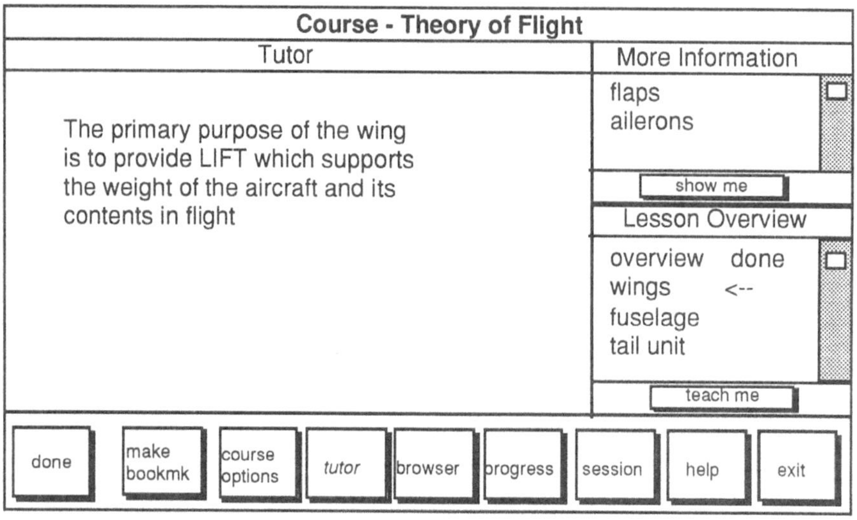

Figure 2 Shows the 'Tutor' Screen

The tutoring component is that part of the system that takes the student through the syllabus in a controlled and consistent manner, while allowing the student some freedom to determine the course of the dialogue. Using the tutoring component the student is able to receive tailored instruction on some aspect of the ATC domain. Students can select lessons to be presented or ask the system to choose a lesson to present. The actual content of the lesson is determined by the teaching strategist / dialogue manager on the basis of the current state of the student model, as described in the previous section.

The browsing component gives the student access to reference material relating to the teaching material presented in the tutoring component. The student is given unlimited access to all information held in the library of information and no particular route through the information is enforced. This facility is intended for use as an adjunct to the tutoring component for students to find out more detailed related information or look up various facts such as aircraft performance data that are not be remembered easily.

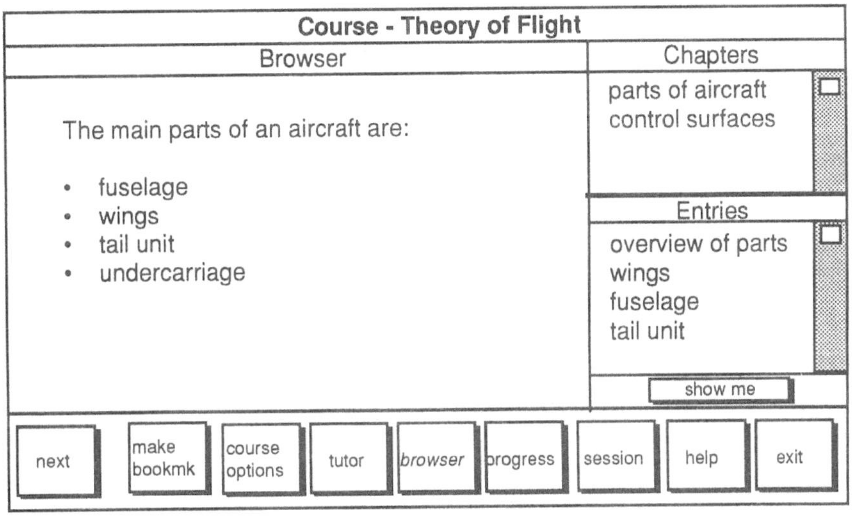

Figure 3 Shows the 'Browser' Screen

The student support facility provides the student with facilities for gaining fast access to certain pieces of teaching material (bookmarks), obtaining a report on his/her overall progress through the course (progress report), and a report on what the student has covered in the current session (session history).

A 'bookmark' facility enables the student to retrieve teaching material in a rapid manner. Each individual piece of material can be marked by the student, and these bookmarks are retained from session to session. The student can examine his/her own bookmarks, select one for display, and the appropriate piece of material is presented.

While the student uses the course teacher, his/her model is constantly updated. This model includes information about what lessons the student has covered, which objectives have been met, which pieces of material have been seen, and which tests have been completed. The model is used by the tutoring component to decide which subject to teach and which material to use. At any time during the session, the student is able to examine the model, which is presented in terms of a progress report.

The student is also able to obtain a report on what he/she has done during the current session. This report is obtained from that part of the student model that references the current session.

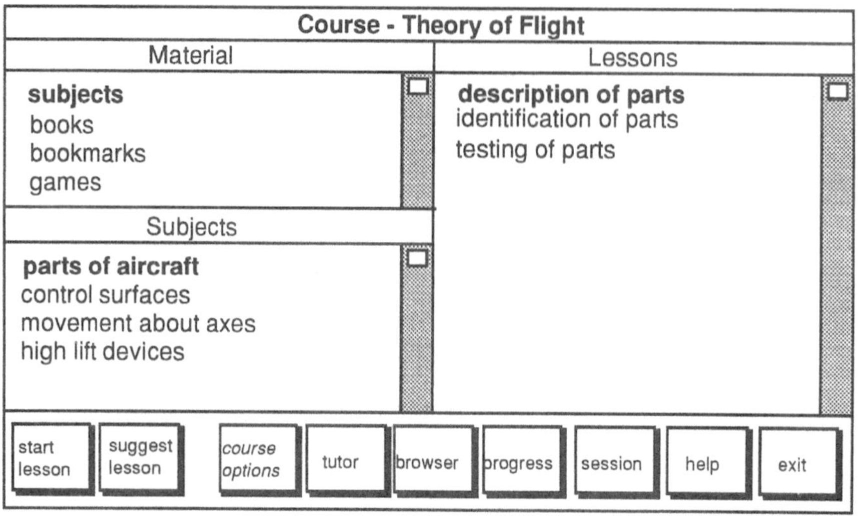

Figure 4 Shows the 'Course options' Screen

Instructional control is provided by the 'suggest' button on the course options screen, where the course teacher decides which lesson it would like to teach, the 'done' button on the tutor screen, where it decides what piece of teaching material to present next, and the decision to give a student access to the games facility.

Student-initiated control is provided in various ways. The student is able to:

- decide which part of the training system to use, ie subjects, books, bookmarks or games ('Material' window on the course options screen)
- choose the subject and lesson to be taught ('Subjects' and 'Lessons' windows on the course options screen)
- choose the order in which material within lessons is presented ('Lesson Overview' window on the tutor screen)
- ask for additional information on the current topic ('More Information' window on the tutor screen)
- select a desired chapter and/or entry ('Chapters' and 'Entries' windows on the browser screen)
- create bookmarks on selected pieces of training material which can be accessed easily on demand ('make bookmk' button on the tutor and browser screens)

Feedback is provided by the training system in the form of student progress reports, session histories and responses to students' test answers.

Evaluation

AVIATION1 was envisaged as a system that trainees would use outside the normal coursework and their performance would not contribute directly to any formal assessment. In addition, the system would provide an extra teaching resource available to instructors, complementing the various teaching strategies already in use. The system should also provide the facility to integrate the lessons learnt in discrete areas of instruction into meaningful real world ATC scenarios.

Trials were carried out to examine how trainees used the system and evaluate how the system matched the goals of the development project.

During evaluation the ATC cadets liked the system very much and found it quite easy to use, but when pressed further they said that the subject area was not formally examined and therefore was of low interest. Their ideal way of using the system would be short bursts of training on the small scale models. Further discussion revealed a relatively stable list of subjects which might be usefully incorporated, namely, altimetry, aspects of meteorology and aviation law.

Following discussions between users and designers, it became clear that the rôle for the system to deliver complete structured courses was not established. What emerged was the need for a system which taught aspects of ATC that were not easily explained on the blackboard or in books, but that were required for practical exercises on simulators, to assist in the vital transition from theory to skill in the control of aircraft.

The ground rules for such system models were defined as:

- an area of known difficulty in ATC training
- should take no more than 30 minutes to master

The new requirements identified became the foundations for the second phase of the project known as AVIATION2.

MODELS-BASED TRAINER

The training delivered by AVIATION2 cuts across different curriculum areas. It brings together various aspects of the theory in a series of interactive models, to provide trainees with a deeper understanding through self-exploration, of the topics they have been taught, to assist them in developing a better picture of the overall subject of ATC.

This project took these new and well defined requirements and defined the areas to be tackled, namely altimetry, meteorology (DALR/SALR), take-off run requirements and flight rules availability. We were now allocated the training specialists in these areas to define the models which were to be included. Alongside the main training system, tools were developed.

The AVIATION2 system consists of two main parts, the delivery system and the authoring system which enables instructors to describe the teaching material in the system. No programming knowledge is required since AVIATION2 converts the material into the code required for the delivery part.

Small-Scale Simulation in AVIATION2

AVIATION2 focuses on small-scale modelling of a specific ATC scenario. Each simulation contains a model of some aspect of ATC. The model contains variables whose values combine to produce a consequent effect in the scenario. Each time a change is made to a variable the consequent effect is shown. Underlying the model is a set of rules or formulæ which determines the consequent effect given the current values of the variables. It is possible to set a variable to a constant value and then subsequently release it to the trainee for manipulation when certain preconditions have been met. This enables the model to be presented in a progressive parts manner where each successive stage in the model introduces more complexity until the whole model is manipulable.

These models can be used to promote understanding of theoretical issues presented in the classroom by allowing trainees to construct their own hypothetical scenarios and observe the consequences. In particular, trainees can be encouraged to look and see, discriminate, and generalise. These models can also be used for testing understanding.

Example Model

In this section an example model is presented. The basic mechanics of the model are described along with the supporting functionality that the AVIATION2 system provides.

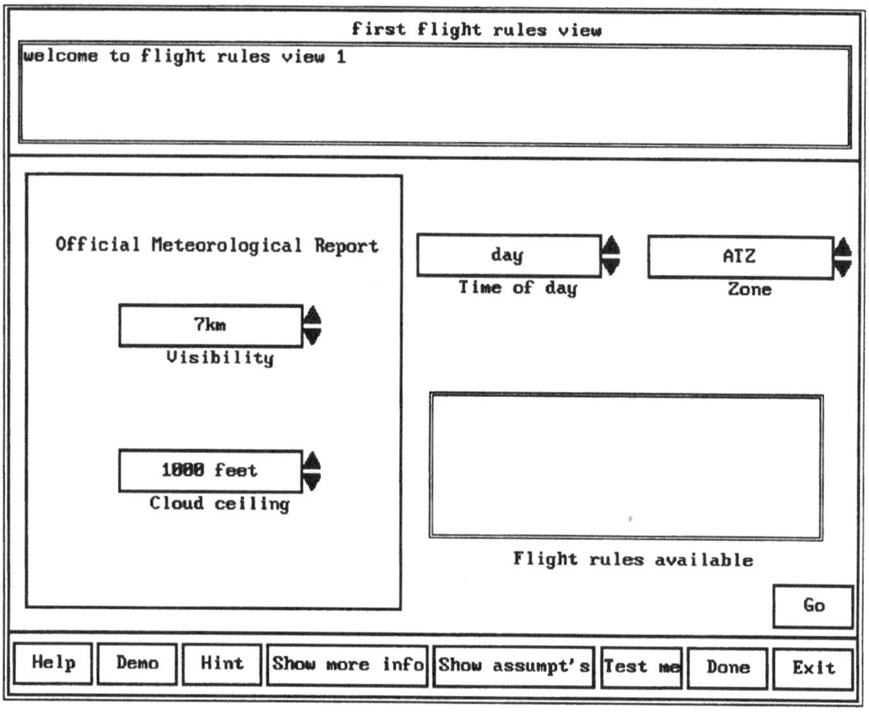

Figure 5 Shows the 'Flight Rules Availability' Model

The example model covers the availability of the three types of flight rules under which aircraft can operate: Instrument Flight Rules (IFR), Visual Flight Rules (VFR) and Special Visual Flight Rules (SVFR). There are four factors that govern which flight rules are available: time of day, cloud ceiling, flight visibility and the type of airspace, ie zone type, in which the aircraft is operating.

Each variable that the trainee can manipulate is represented in the model by an arrow button. The trainee can change the value of any of these variables by selecting the up or down arrow on the appropriate variable. Each time a value is changed the new set of available flight rules in displayed in the box provided. This set is generated by the model's underlying set of rules.

There are other ways in which variables can be represented in a model, ie dials, sliders and graphs. Dials and sliders are used along with arrow buttons to represent variables whose values the trainee can change, whereas graphs and display boxes are used to represent the resulting effects of the particular set of variable values.

Direction, Commentary and Hints

Whilst manipulating aspects of the model, the trainee is presented with direction and commentary and can ask for hints.

Direction is given when the trainee progresses to the next stage of the model where typically more variables become active, ie the model becomes more complex. The direction will highlight those variables in the model that have become active and will encourage the trainee to manipulate their values. For example if the first stage of the model sets the cloud ceiling, flight visibility and zone type variables to be active and the time of day to be inactive, the direction presented to the trainee when moving to the second stage would be to highlight the fact that the time of day variable is now active and that it should be manipulated.

Commentary is presented when the variables have been set with particularly 'interesting' values. The commentary is intended to emphasise a particular point about the current state of the model. For example, if the time of day variable is set to 'night' the commentary might emphasise the fact that visual flight rules are not available at night.

If the trainee is unsure of what to do next he/she can ask for a *hint*. This is a suggestion generated by AVIATION2 prompting the trainee to change the value of one of the variables. The hint is generated on the basis of the training requirements that have been defined for the model. These specify which particular values of the variables the trainee is required to set and observe. Each stage in the model has its own training requirements and the trainee can move to the next stage only when the requirements of the current stage have been met.

Testing and Remediation

The AVIATION2 system provides testing and remediation facilities. Testing can occur either on the request of the trainee or on the initiative of the system. The system will typically test the trainee when progressing through the various stages in the model. The trainee can ask to be tested at any time when manipulating the model provided that the system believes that the trainee possesses the knowledge to answer the question correctly.

Four types of test are supported:

- single answer multiple choice
- multiple answer multiple choice
- key phrase
- model-based

The first two types of test require the trainee to recognise the correct answer(s) from a given list. For key phrase tests the trainee must recall and enter the answer via the keyboard. Model-based tests require the trainee to set the variables in the model to the correct answer, thereby demonstrating that the trainee can apply theoretical knowledge in a practical situation.

AVIATION2 can provide the trainee with remedial instruction if the answers given to tests are incorrect. This remediation is multi-media in that it can contain text, bit-map graphics, video stills and sequences. The content of the remedial material is determined by rules associated with each test which specify the misconceptions that the trainee is assumed to have if a wrong answer is given.

More Information

AVIATION2 provides the ability for trainees to ask for more information about a model. Typically this information will cover the theoretical issues that underlie the model.

Assumptions

Any assumptions that have been made when defining the model can be displayed on request. These assumptions will generally imply that the model is a simplified view of the real world, but sufficient for the educational task.

Progress Reports

When the trainee exits AVIATION2 a progress report is presented. This report details how much of the model the trainee has covered and how much remains to be done. The trainee's current test score is also displayed.

It is also possible for the trainee to obtain a more comprehensive report on his/her progress. For each model that the trainee has started there are details of which parts have been covered, test performance and any associated misconceptions.

Student Model

For each trainee registered with the AVIATION2 system there is a student model. This contains details of what the trainee has done with the system. The model is retained from session to session such that AVIATION2 can tailor the options and teaching material to the individual trainee. AVIATION2 knows which models have been started and how much the trainee has completed. It knows which tests have been attempted and remembers any unresolved misconceptions. The student model also contains details of individual pieces of teaching material that the trainee has seen either as remediation or when requesting more information. This allows AVIATION2 to present a different piece of material that teaches the same concept.

AVIATION2 Authoring System

The authoring system allows models, tests and remediation material to be specified and implemented. There are two parts to the system: a set of paper proformas and computer-based tools.

Paper-based proformas are used by the instructor to plan and specify the particular model. This is the more important stage of the authoring process; it is unlikely that quality training material will be created by using the tools without pre-planning.

The computer-based authoring tools are designed for use by non-computer experts. There is no programming involved; the approach is largely form filling. When a model is complete it is compiled by AVIATION2 into the code necessary for presentation to the trainee. The model can then be allocated to the relevant trainees using the administration system.

The authoring tools allow authors to build up libraries of teaching material which can be reused in various contexts. For example, a model which is similar to an existing one can be developed quickly by first creating a copy of the existing model.

In addition, AVIATION2 has an administration system which allows certain essential 'housekeeping' functions to be performed. Centralising the registration of trainees and authors in the administration system ensures that tight control can be maintained on the number of authorised users of AVIATION2. For similar reasons, the ability to allocate models to trainees is located within the administration system.

Evaluation of AVIATION2

AVIATION2 has significant benefits for the trainee. The system provides the means by which the trainee can consolidate his/her *understanding* of theory taught in the classroom by exploration of a simulated scenario. Because an individualised approach to the learning has been employed, the trainee can interact with the teaching material in a meaningful way, for as work progresses, the system builds up a unique model of the trainee. This model recognises progress and controls the content of the material presented and determines the appropriate teaching strategy, assessment and remediation in a way that is unique to each trainee. Thus the trainee feels supported by the system.

The problem which gave rise to the AVIATION2 project was first identified by students and instructors themselves during AVIATION1 trials. In the AVIATION2 trials it was gratifying for us to note that the students' responses to the system was favourable. The content and format of the simulation exercises were confirmed as pertinent to student needs and as satisfying the requirements. In addition, new buildings provided computer practical rooms and the necessary framework and logistics for integrating such a system into the CATC environment. Justification of the system was perhaps illustrated when two instructors observing the flight rules availability exercise had a difference of opinion about the accuracy of the answers that the system was presenting. Fortunately, the system had the correct answers.

As the AVIATION2 project progressed, staff at CATC took responsibility for the content of the coursework. This necessitated the instructors giving much thought to the content and current pædagogic style. Many basic assumptions were questioned which the instructors involved found revealing and beneficial. A working group was convened to address longer term issues such as the identification of areas of work at the College that would appear to be suited to delivery through the AVIATION2 system approach, and suitable methodologies to assess the efficacy of the innovation in the overall teaching strategy of the College. The group also reported an increasing confidence within its membership in dealing with computer-based systems.

DISCUSSION

In this section the major themes which are common to all three projects are discussed. The locus of intelligence in each training system is found to be rather different. The rôle of users in the development of the systems, and the sources of requirements are described. Finally we recommend a list of essential ingredients for developing intelligent training systems.

Locus of Intelligence

As a result of our experience in the three projects described, we believe their the intelligent aspects are useful features of training systems. They appear to promote flexibility, adaptability and individualisation of the training approach and allow the user a high degree of control over the dialogue.

In all of these systems an intelligent systems approach has been adopted. However, in each system the precise use or locus of the intelligence is different.

In CAI the intelligence resides in the representation of behaviour and the diagnosis of student errors, together with scenario generation facilities. In AVIATION1 the intelligent parts are the representation of modules of material and their flexible generation into lessons at a particular level, in a particular medium and using a particular teaching strategy on the basis of the subject, teaching objectives and student model. In addition, supporting material is made available for selection and offers the student initiative to follow an alternative course whilst still remaining instructionally sensible. In AVIATION2 the modelling is relatively smart but it is in the area of hints, directions and commentary together with backing material that the system is most 'intelligent'. Most of the intelligence in all of these systems concerns their behaviour.

The demand for intelligence and its locus is very sensitive to the training needs. Training needs influence the training approach adopted which in turn influences the kind of intelligence required. We believe that the use of conventional expert systems to provide the intelligence in training systems is not sufficient and sometimes not even appropriate.

An underlying objective of the early projects was the search for a truly generic ITS system. Given the experience of these projects we feel that it is the norm that each ITS project demands different forms of intelligence and that generic systems are not a sensible goal. It is probably more fruitful to try to develop ITS techniques to meet differing requirements, rather than off the shelf products which at best could provide no advantage over bespoke software and at worst could obstruct system development. Expectations of the level of genericism need to be scaled down. Genericism can exist at a certain level; AVIATION1 and AVIATION2 are genuine ITS shells, but they are heavily constrained by the underlying training approach.

User Involvement

At the start of these projects we appreciated the importance of having user involvement in the project. We needed it for three reasons. First, we are system developers and not instructors, and we need to work with instructors to provide them with the tools to produce ITS. Secondly, the instructors are application area experts, and thirdly, the evaluation of the system is best carried out by end users (students and instructors).

In CAI, instructors helped us to choose an application area and acted as domain experts. When the system was built they then returned to assess the suitability of the system. In AVIATION1, instructors were involved with the selection of topic areas and in the design and behaviour of the system. In addition, we held user trials with the AVIATION1 system using volunteer cadets. This was the point at which the critical components and rôle of the system within the CATC environment became clear. This meant that the AVIATION2 system was built based on realistic requirements. AVIATION2 was also the subject of user trials involving both students and instructors, and for the first time in these projects, instructors produced the bulk of the implemented training material.

User involvement extends to a variety of users. There are students who are the end-users of the system, the instructors who will design and encode the instructional material, and finally the managers and administrators who organise the training environment.

These users need to be involved in the:

- definition of the training problem
- selection and definition of the application area(s)
- definition of the constraints and scope of the system
- approval of the training approach
- evaluation of requirements, design and final system.

Requirements

In the CAI system, the focus of the work was to meet the technical challenges of building an ITS in the application area. This led to rather cursory examination of the training requirements. Evaluation came at the point when the system was complete and being demonstrated. During evaluation the preoccupation of the developers was whether the system behaved like the real world, the instructor evaluators were more concerned with how they would make use of the system if it was installed in the College. There was no consensus that the system was suitable for use by students. Indeed, such was the mismatch between the developers' and instructors' ideas that even the fundamental aspect of letting a single user operate a stand-alone training device was challenged. Some instructors considered it to be at best of dubious value; others voiced fears of it being positively dangerous!

As a result, the direction of work was steered away from training simulators towards more conventional CBT with intelligence for supporting background ATC knowledge. Again, the training requirement was self-evident, more students with less aviation experience having a stand-alone ITS to familiarise themselves in background areas of the theory of flight. However, during the user trials, students felt that a book would in some areas impart a similar level of information and would be much more portable. It was the modelling aspect of the system which most interested them, and this provided a list of ATC problem areas which could be usefully addressed using this technique. Instructors agreed that these new requirements would be genuinely useful to have incorporated in a stand-alone system.

AVIATION2 was designed to meet these requirements and provide a system which could be used in the ATC training syllabus. Evaluation with users established that the system met the requirements, and instructors confirmed that the tools for developing new models were satisfactory. The major requirement that remains to be satisfied in order for this system to enter use in the ATC cadet training course is sufficient time for instructors to generate models.

For all of these systems we were using a valid set of requirements to design the system. These requirements are comprehensive and exist at various levels. They are:

- user/student requirements (eg applicability, ease of use, effectiveness)
- instructor requirements (eg authoring)
- technical requirements (eg feasibility of approach, hardware, performance)
- programme requirements (eg how the system fits in to timetable)
- infrastructure requirements (eg accommodation)

- resources requirements (eg preparation time)

It is essential that these sources of requirements are captured during training system developments. Successive projects had an increased coverage of these requirements as they developed from demonstrators of feasibility to preproduction implementations.

CONCLUSIONS

On the basis of the experience we have gained on these and other ITS projects we conclude by listing the essential ingredients for building intelligent training systems:

User involvement
: Users (students, instructors and managers) need to contribute to the definition and design of the system

Requirements
: Training systems should be based on training *requirements* not training solutions.

Repertoire of techniques
: It is unlikely that the training needs will be satisfied using an existing off the shelf package. It is therefore important to have access to a comprehensive repertoire of techniques that can be used to formulate the requisite training approach.

Evaluation
: Evaluation should be conducted with users at regular intervals. Depending on the stage of development, the purpose of the evaluation may focus on requirements, design or final system implementation. The user interface in the form of static screen designs, prototype or full working system is a valuable artefact on which to base the evaluation.

ACKNOWLEDGEMENTS

We would like to acknowledge the support of the Chief Scientist's Division of the UK Civil Aviation Authority in the projects described in this paper. We thank the staff at CATC who over the years have contributed to this work. We also acknowledge the contribution of our colleagues within Logica.

SECTION IV
ADVANCES IN HARDWARE TECHNOLOGY

This section discusses the relative merits and cost effectiveness that are available when the most recent advances in hardware are used to deliver training to military personnel.

DVI AND SYSTEM INTEGRATION:
A FURTHER STEP IN ICAI/IMS TECHNOLOGY

Dr. Gian Paolo Noja

Gajon Institute of Technology
Genoa, Italy

ABSTRACT

Fault detection and isolation, analysis, repair and replacement procedures for complex systems have traditionally demanded massive quantities of technical data. Typically, a maintenance technician requires more time to locate a repair procedure than in actually making the repair. Training needs to be adequate with the complexity of modern sophisticated military equipment.

In order to provide rapid, convenient access to vital information and present it in an easily followed and highly comprehensive series of text, repair or replacement steps and training procedures, the Gajon Institute of Technology has developed a high technological approach based on Integrated Computer Assisted Instruction (ICAI) and Integrated Maintenance System (IMS) which includes a network of student/maintainer stations, teacher/supervisor stations, processors, videodiscs, CD ROMs, bidirectional audio/video/ computer lines for teacher/supervisor control and audio/video/ digital data handling, with extensive use of fiber optics and local area networks.

The impact of this technology has shown a reduction of 50% of time both for training and maintenance, a comprehensive recording of all data, an attentive supervision of the human brain, and a considerable reduction of gap between training and maintenance.

Advancements in technologies such as DVI, Virtual Reality, Speech Recognition and Communications have further enhanced the application of ICAI/IMS with some interesting projections for the nineties.

MILITARY PERSONNEL QUALIFICATION

Armed Forces have always applied the most advanced technologies in their time, and this

Advanced Technologies Applied to Training Design, Edited by
R.J. Seidel and P.R. Chatelier, Plenum Press, New York, 1993

is more evident today as warfare equipment is employing the most sophisticated electronic techniques.

This fact, beyond evident advantages, causes the necessity to provide operators and specialized technicians able to fix and keep in operating conditions very complex equipment and to train these people in the relatively short time of their service.

Normally suppliers of defense equipment organize courses for the personnel of the Armed Forces which bought the material, but this may happen only for the first group of technicians that are trained when the new device starts to be used.

Moreover, training in the factory usually gives bad results if trainees do not have a good basic knowledge and a certain experience in maintaining and repairing apparatuses, as the producer trains them only on that specific equipment.

For example, trainees studying a particular radar are supposed to have a general knowledge of radar principles. It must be realized that people who will later on operate and maintain this equipment need a specific practical training, completely different from the one needed by a university student who will only calculate and design the equipment.

What is really important for a good maintenance technician in the military field, is to be able to analyse a fault, to identify the damaged component and to replace it, to calibrate the apparatus making it operative in the shortest time.

To reach this target, one needs: a good knowledge of the apparatus and of its basic parts from the functional point of view, logic, capacity of analysis and a certain manual ability. Moreover, the technician should be able to understand the diagrams and the manuals of the equipment and to use the measuring instruments.

COMPUTER ASSISTED INSTRUCTION (CAI)

For a few years now, experiments have been conducted in various countries on the effectiveness and cost/effectiveness of Computer Assisted Instruction (CAI) versus conventional instruction.

In CAI, lessons are stored in the computer; each student interacts with the lesson at his own pace by means of a terminal. The computer responds by providing new material based on whether the student has answered an item correctly or incorrectly; the computer also maintains records of student progress, attendance and various details.

It is significant that CAI permits each student to procede at his own rate of learning, to correct errors or misunderstandings as soon as they occur and to reward him with knowledge of success when that occurs. This is difficult or impossible to achieve in conventional instruction where such knowledge is not immediately available to the instructor and where all students, regardless of their ability or style of learning, must procede at the same pace. The pace set in conventional instruction should be that needed to permit the slowest student to complete the course; in practice, the pace is set to meet the administratively acceptable failure rate and not the pace of the slowest student.

Anyway, few years ago people working with computers realized that the computer itself was not a sufficient media for instruction delivery, because of the lack of one important factor: real images.

Various users started new techniques, but no one could be satisfactorily implemented because of the limited capacity of the media. It has been necessary to wait for the development of the laser tecnique to have access to vast information, such as the one required for training and maintenance.

A good example of this necessity for using this technology was given by the US Army EIDS project, which in the background simply stated:

> *"Printed job aids and instructional material have proliferated to the point where the Army is drowning in a sea of paper"*
>
> *"Recent advances in integrated circuit technology, the development of low cost, high capacity computer memories, and the refinement of optical video disks will permit the development of new electronic delivery systems for instruction"*

Indeed, the newest and most promising storage device has been based on the laser technology. Its characteristics in terms of memory capacity, reliability, price and flexibility have put optical media in a class by itself, helping substantially the development of CAI.

The impact and application of this technology has already affected the way in which Armed Forces and related industries are planning the near future in terms of information delivery, both for training and maintenance.

Now, even if the superiority of CAI versus conventional instruction has been demonstrated, at least for time reduction, generally good results have been obtained mainly when limited subjects were involved.

This non complete success was mainly due to the unlimited trust in the computer as in a sort of magic box able to solve any problem, without taking into account that the

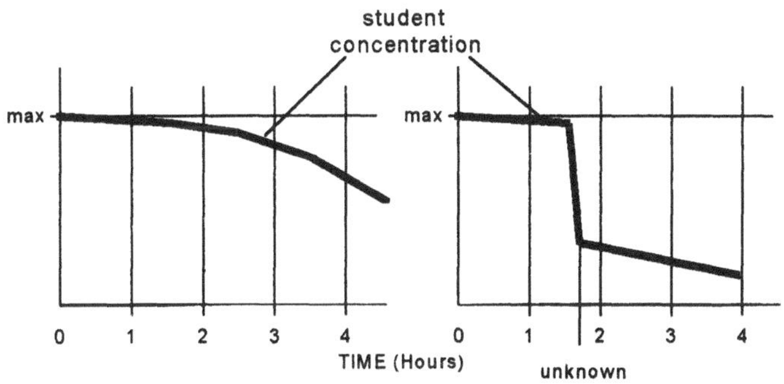

Fig. 1 Student concentration curve

learning process is the result of many elements, and it is not realistic to expect to reproduce this process simply by putting a student in front of a machine.

One should consider the curve of attention of the student (Fig. 1), which shows a loss of concentration during the time, and, most important, a deconcentration when the student is faced with an unknown element.

Nor must it be forgotten that a computer can simulate many things, but the complexity of problems produces an exponential cost, which must be contained in terms of cost/effectiveness. Consequently, one hour of lesson may take from 60 to 300 hours of development, with costs ranging from 3.000 to 15.000 Dollars. This means that a 100 hours course may cost from 300.000 to 1.5000.000 Dollars (Fig.2).

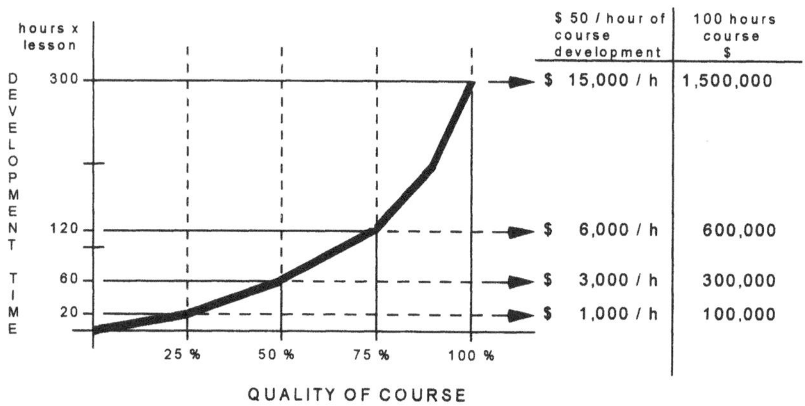

Fig. 2 CAI courseware development time

Another problem of preminent importance for a maintenance technician is the necessity of good practical experience, which can be achieved only with the ability of solving practical problems, which means, for example in the electronic field, to be able to use electronic instruments, to correlate different phoenomena and to use one's own hands.

To all this must be added the complexity and variety of military equipment, the difficult environment in which maintenance should be carried out and the usually short time allowed for training and maintenance.

The utilization of laser technology has helped in some way, but results with CAI and interactive courseware has maintained a number of limits.

INTEGRATED COMPUTER ASSISTED INSTRUCTION (ICAI)

To solve the limits imposed by the typical CAI, the Gajon Institute had to use a new approach. This approach considers an integration of different elements, all of which contribute to the learning process:

- student
- computer
- teacher
- instructional means
- software
- programmed learning

In the Gajon system, the student has an active behaviour, and consequently, on top of working according to his learning rate, is considered a teaching resource. In fact, often learning rate is slower than transmission of knowledge, and considering the usually scarce preparation of the teacher, it was compulsory to adopt a "learner centered" method.

The computer is considered not as "*the*" element but "*an*" element integrated in the learning system. Main characteristic of the computer is the facility given to the student to proceed at his own pace, the automatic control of audiovisual aids, the "standardization" of courses and the continuous checking and recording of student progress.

The aim of the teacher is mainly to overcome problems described in Fig. 1 and 2. Consequently, the teacher helps students who cannot proceed only with the computer. In fact, it is nearly impossible to prepare such a perfect and complex program able to take care of all possibilities. Moreover, after an initial curiosity, the student may lose his interest rapidly if he is not backed up by a real teacher. Another important point is that a teacher, helping students who are slow, keeps a better average in the class.

The instructional means comprise student stations, teacher station and communication lines. Fig. 3 represents the picture of a typical ICAI installation.

Fig. 3 Typical ICAI installation

Student stations, besides computer, printer, computer controlled videodisc and CD ROM, have all basic parts which may be necessary for the final target, referred to the maintenance task (Fig. 4).

Teacher station is a control point, where the instructor can check the progress of students and where data are continously collected. It also contains an integrated audiovisual section and a lesson composition section. Communication lines exist between each student and the teacher in such a way that a private communication can always be available in bidirectional audio, bidirectional video and bidirectional computer data transmission.

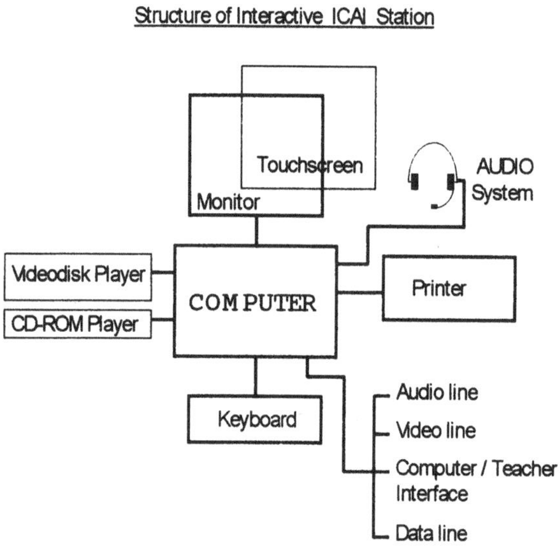

Fig. 4 Structure of student computer

Software has been designed in order to create a proper balance among computer data, audiovisual aids and practical experimentation.

The programmed learning utilized by the system is based on positioning the student as far as possible in the same environment of a scientist who is going to make experiments based on some reasoning process. With this system the student reaches his discoveries and feels like a scientist himself to some extent, thus remembering with great ease the material that he has studied.

To perform the activity of ICAI, a special evironment has been studied and experimented. This environment is composed by student stations, teacher station and communication lines. These are the characteristics of each device.

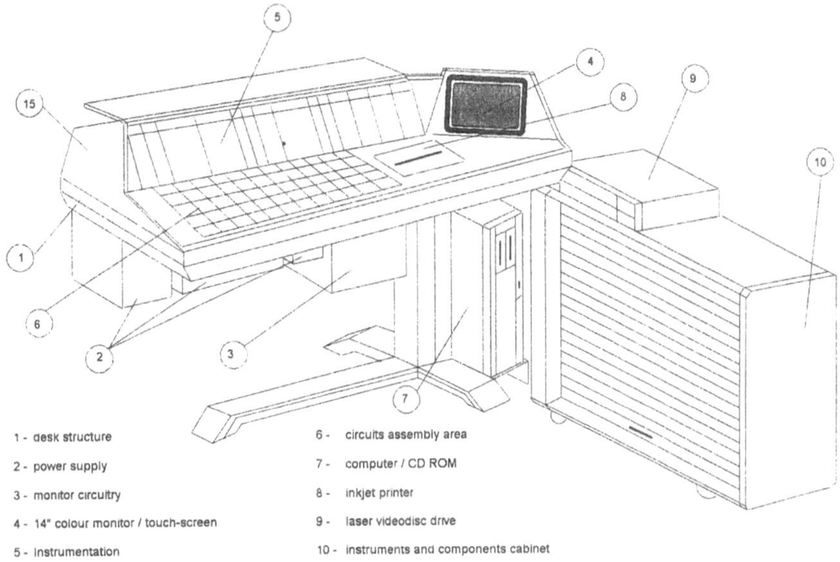

Fig. 5 Student station

It performs most subjects related to military training in fields such as electronics, mechanics, propulsion and weapon systems; as an example, these are the programs experimented in electronics:

- basic electricity
- basic electronics
- logic circuits
- basic computers
- advanced computers
- wire communications
- telecommunications
- pulse communications
- video techniques
- antennae
- microwaves
- radar
- sonar
- electrical circuits
- electrical machines and controls
- synchro
- automatic controls and servomechanisms
- gyrosystems

The student station is composed of the following sections:

Trainer section:
- circuits assembling open to any set-up
- circuits feeding integrated in the work surface
- instrumentation: 30 plug-in analog and digital modules

- electronic components: according to program module

Computer section:
- central unit: 80386 32 bit 33MHz CPU, 4MB memory, 120MB hard disk, 640MB CD ROM , VGA color controller, videodisk controller, touch screen interface, overlay controller, computer/ teacher interface, local area network
- keyboard
- monitor: 14" color, VGA resolution 640x480 pixels, touch screen 256 x256 resolution, 16 levels of pressure
- printer: ink-jet, 300 cps, extremely silent
- videodisc: 54.000 images per side, full motion and still control

Communications section:
- bidirectional audio line
- bidirectional video line
- bidirectional computer/teacher interface
- local area network

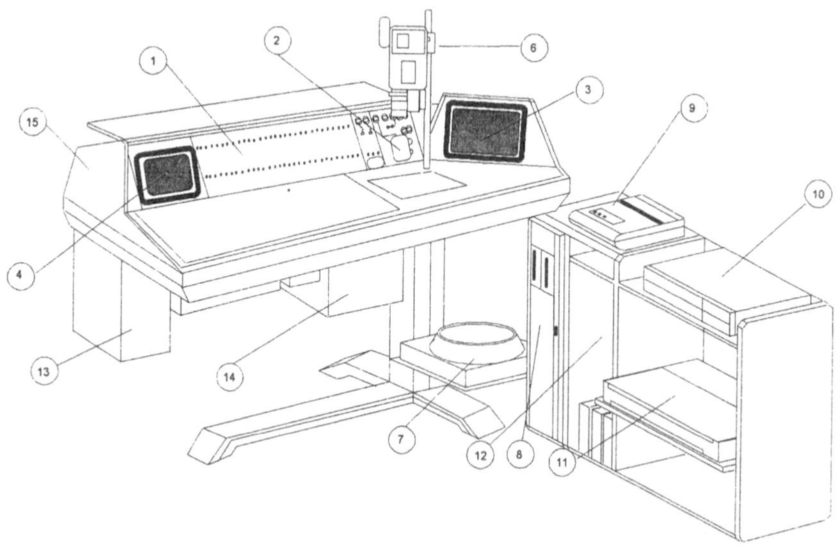

Fig. 6 Teacher station

Control section:
- bidirectional audio lines
- bidirectional video lines
- bidirectional computer/teacher interface
- local area network

Audiovisual section:
- CCTV
- videodisc player
- VCR

- CCTV slide projector
- CCTV overhead projector

Lesson production/computer section:
- computer (same as student, equipped with laser printer)
- student data collection
- direct video recording
- slides audio recording

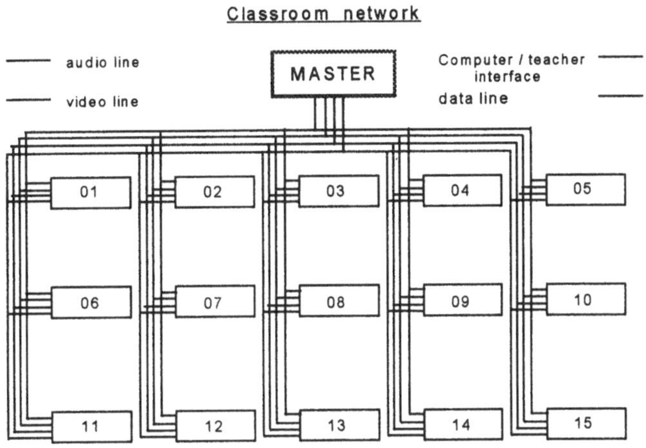

Fig. 7 Communication lines

Communication lines

Bidirectional audio line:
- direct and private conversation between student and teacher

Bidirectional video line:
- reception of video image from teacher
- pick-up by teacher of student image

Bidirectional computer/teacher interface
- request of help from student
- teacher permission at any step

Local area network:
- student data collection

Audio, video and computer/teacher ineterface can be single, group or general.

The utilization of ICAI technology has shown the following advantages:

- coherence: same approach is used through all the course
- cost: cost/effectiveness researches have shown an amortization in 4 - 5 years
- up-to-dating: computer data can be revised more easily than other courseware
- flexibility: the same system can be adapted for a variety of courses

- reduction of training time: researches have shown a 30 to 50% reduction
- better achievement of student
- instructor self training
- student self tailored training
- private student/teacher communications
- not as cold as normal CAI because of teacher role
- continuity of methodology when changing teacher
- teacher may dedicate more time to students in need of help
- continuous check of students
- student personalized manual printed during the lesson, thus avoiding pre-printed courseware
- automatic and continuous communications
- easy change of level of training

Large experiments with CAI and ICAI have mainly shown a significant convenience in terms of effectiveness and cost effectiveness. Various results of these experiments are reported in "ICAI findings" section .

TRANSFER OF TRAINING TO OPERATIONAL SYSTEMS

Fault detection and isolation, analysis, repair and replacement procedures for complex systems have traditionally demanded massive quantities of technical data. Typically, a maintenance technician requires more time to locate a repair procedure than in actually making the repair.

In order to provide rapid, convenient access to vital information and present it in an easily followed and highly comprehensive series of text, repair or replacement steps, it was necessary to develop an approach to the electronic delivery of technical information and a comprehensive Maintenance Management System.

IMS: Integrated Maintenance System

IMS is a natural evolution of PMS: Its main target is to make maintenance of a fighting Ship as reliable, automated and functional as possible, by giving maintenance technicians the necessary logical instructions at every single instance and step, through a network of equipment, computers and videodiscs, while controlling the whole operation and reducing down time substantially.

The second target is to maintain statistics of faults and spare parts of the Ship with the possibility of external communication.

Synthetically the system presents the following features:

- sequential operations assisted by computer

- sequential preventive maintenance assisted by computer

- sequential corrective maintenance assisted by computer

- certification of maintenance effected step by step for every maintenance technician

- registration of faults and maintenance of entire Ship

- registration of spare parts used and in stock

Basically the system is the same explained in previous chapter, substituting the teacher with the maintenance supervisor and the student with the maintenance technician.

The same high technological approach utilized in the classroom is made available on board of the Ship, with an integrated network which includes portable maintenance stations, supervisors stations, central processor with laser videodisc and CD ROM control, bidirectional audio/video/computer lines for supervisors control and audio/video/digital data handling. Consequently, the maintenance technician works in the same environment to which he has always been accustomed during previous training.

IMS combines videodisc imagery, high resolution color graphics, sound and animation with the data handling power of a computer and ease of operation of a touch activated display screen. Laser technology videodiscs provide nearly instant access to information in a format unmatched for completeness of content, clarity and detail, flexibility, speed and ease of use.

Starting from the rule that a system which is easy to use will result in faster and more accurate maintenance operations, IMS allows minor repairs to be carried out correctly and quickly at the first time even by inexperienced technician, and major repairs to be carried out precisely and in short time.

IMS network

Fig. 8 represents an initial example of IMS network for frigate prime electronic, electrical and propulsion systems, and should be extended to all other equipment on board of the Ship.

The system refers mainly to the possible existence of 25 locations of different equipment, which are supposed to be concentrated in 9 zones, where it is possible to transfer the basic maintenance equipment.

The system has, therefore, been considered according to the principal functions, payload and platform, with the respective grouping of the subfunctions Communications/ Information and Energy/Mobility. The result is a concentration of the function of the Ship in 3 services:

- operations

- weapons

- energy and propulsion

with 3 corresponding supervising officers.

The conformation of the system is therefore the following:

- 25 multiple outlets audio/video/computer distribuited on the Ship

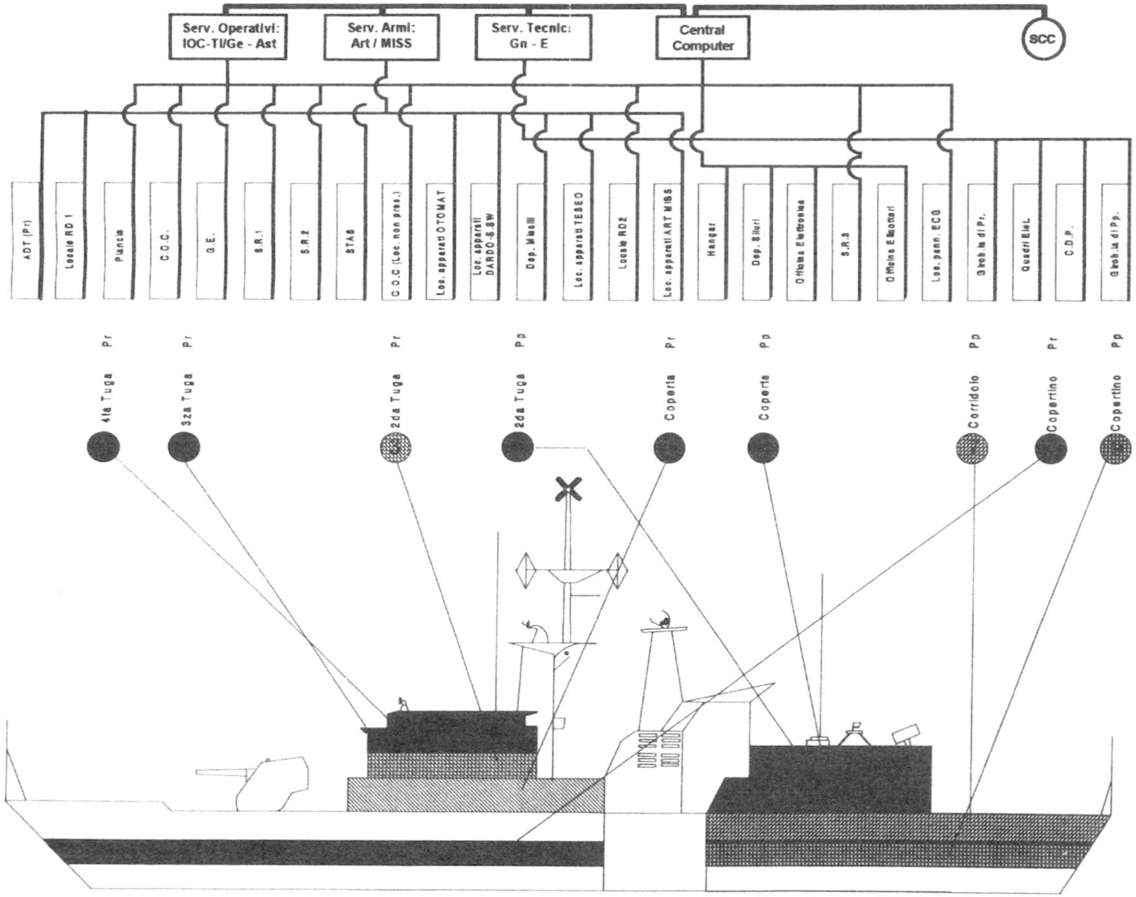

Fig. 8 IMS network

- 9 maintenance terminals (Fig. 9) distributed in 9 zones
- 3 supervision systems (distribuited in 3 supervising offices)
- 1 central processor
- 1 fiber optic communication network

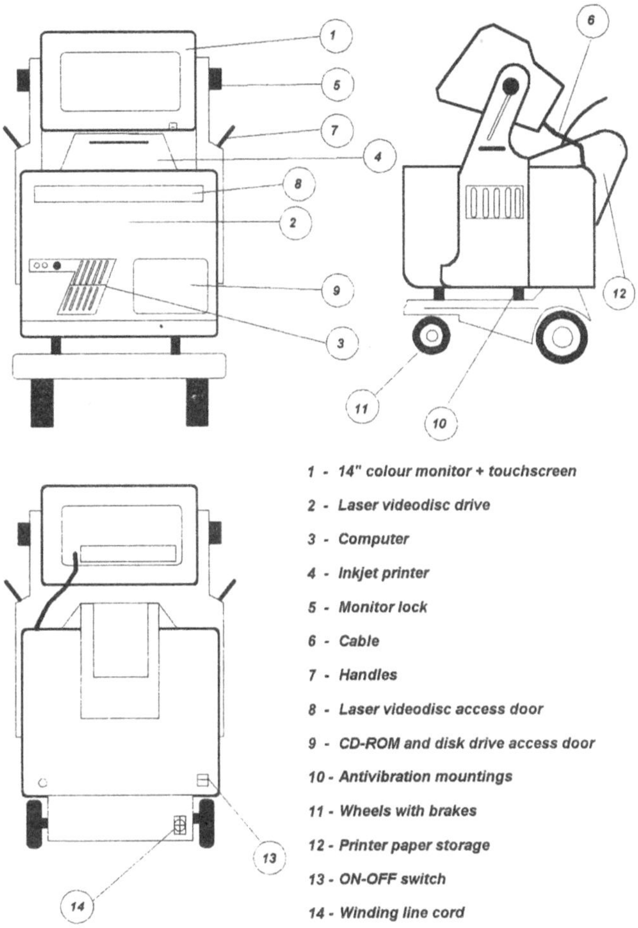

Fig. 9 Maintenance terminal

Equipment information contained in IMS

This is a set of information distribuited by IMS for electronic maintenance:

- theory of operation
- integrated working scheme
- equipment diagram

- assemby and disassembly procedures

- electronics assembly
  - assembly
  - disassembly
- electronics subassembly
  - installation
  - alignment
  - removal

- troubleshooting procedure
  - operational test
  - fault isolation
  - common problems

  - special tests

- supporting reference aids
  - auxiliary test equipment
  - schematic and wiring diagrams (general,blocks,flows)
  - tables
  - subassemblies and component location
  - part listing
  - safety summary

- automated maintenance procedures (in case equipment provision)

System at work

Operational procedure is the following:

- introduction of plug in nearest IMS outlet
- introduction of technician code through video keypad
- introduction of code of equipment subject to maintenance
- reception of all sequential instructions
- possible request for help through IMS intercom to supervisor
- during maintenance, recording of steps in central memory
- continuous supervisor control at a distance, even without technician knowledge

Operatively, all manuals and MRC of the Ship are contained in videodiscs, which distribute all information step by step and in logical sequence to each maintenance technician taking into consideration the type of maintenance, maintenance calendar and level of maintenance technician (classified through a code).

At each step, the technician can have information repeated, or call up additional, more detailed assistance if desired. Complex signal tracing functions are facilitated by easily identifying each part of test point on schematics and on photographs of actual circuit boards and components. With a single touch, the technician is able to scroll schematics to locate required sections. Selected areas of the schematics can be magnified, and each part called up individually, labelled and identified in close-up photography to eliminate confusion and mistakes.

Vital vawe forms are depicted in clear, dynamic illustrations, just as they appear on the technician's test instruments, so that quick and accurate comparison may be made directly.

Through the test sequences, the technician uses only a finger to control all command functions; there is no need for complex, multi sequence keyboarding. Each step, forward or back, is clearly and concisely labelled on the monitor screen.

IMS as on board training system

Although designed primarly to fully support real time maintenance, IMS is a powerful and versatile training resource on board of the Ship which teaches complex and delicate operations prior to being carried out on real maintenance jobs.

The level of technical information is fully under the control of the technician. Familiar sequences may be passed over when desired, or alternatively, more detailed procedures may be called up if needed.

Even support materials and test linkages are displayed with identifying information to avoid the possibility of confusion or error.

Realistic maintenance training can be given without tying up critical, prime mission equipment.

Advantages of IMS

It is obvious that IMS is not comparable to normal maintenance systems.

Equipment is becoming more and more complex and the time assigned to maintenance is not normally sufficient. It is necessary to bear in mind that a modern warship without efficient maintenance loses nearly all its value.

Here are some important advantages:

- rationalization of the maintenance system by proposing the exact sequence of periodic, daily, weekly, monthly, quarterly, semi-annual and annual preventive maintenance

- elimination of manuals and MRC (which may remain on board as reference)

- automatic display of exact maintenance procedures according to logical diagrams, including instruments and accessories, according to level of maintenance technician

- recording of operations carried out, name and code of technicians and utilization of spare parts

- certifying of substitution of spare parts

- remarkable decrease of down time, increasing operational readiness

- transmission to central ground system of preventive and corrective maintenance carried out, spare parts used and residual parts in stock

- low cost, because of the utilization of most advanced systems of data processing and image retrieval

- all parts of the system are interchangeable and can be substituted in minutes

- the system may be programmed to accept data deriving directly from the equipment to be maintained, if and when these are available, central processor may analyze data and suggest directly to maintenance technicians operations to be performed

- can easily be installed in the Ship in a very short time

- cost of IMS is recovered in a very short time through substantial reduction of manufacturers intervention

ICAI / IMS Technical problems solved

The following technical problems which have been solved represent a deep research in the interface between the student and the training system. Some of the problems have now been overcome by the latest technologies, but in the past they represented serious deficiencies found in other systems.

Communication lines

Taking around digital data in a networked system is not a critical problem nowadays for computers connected with a central unit. Adding a network of audio intercommunication was just an efficient educational idea to allow better and easier communications between student and teacher. But the best improvement was in the substitution of the typical blackboard with a completely private television closed circuit between each student and the teacher, and even better was the possibility of transferring student image to the teacher. Technically, every student has been divided in separate sections, with 3 commands: audio on/off; video sending/off/receiving; computer permission single pulse/off/continuous.

Modularity

Each single piece of the equipment has been designed so that all parts fit together with no connection cables in a format that allows quick substitution of each instrument, power supply or other components.

Maintenance

Maintenance has been conceived so that any teacher can make substitution of defective parts even during the execution of the lesson. To achieve this goal, acoustic and visual alarms have been provided to monitor most important components, and each instrument and power supply has been designed with many different plug-in boards, each corresponding to a function which can be easily detected in case of problem. Active components have been mounted on special sockets, to avoid dissoldering during the 3rd level maintenance, which is easily achieved with the substitution of the ICs, if necessary.

Display

It is not safe to use interlaced screen for a long period of time, because the "flickering" produced by these screens is very annoying, and with prolonged periods of time it can produce eye damage. However, until recently it was not possible to have in the same screen both the normal video signal coming out from the videodisk and high resolution RGB graphics. The problem has been solved unifying a 640 x 350 non interlaced RGB picture with the signal coming from the disk, obtaining a completely "non flickering" device.

Printing

It is not difficult to print the computer screen, but printing a high resolution graphic takes typically more than 3 minutes. Obviously, a student will become impatient if he should wait this long before having his paper written down for each step of the lesson. This is the reason why printing for each student is given as "optional" in CAI. But especially in the electronic field, printed material is essential for the same development of the lesson, because of the large quantity of material given. The problem has been solved printing both text and high resolution graphics at the speed of text printing, which takes a few seconds.

Motion transformation

This has been the most critical aspect, because a videodisk contains only 30 minutes of continuous motion playing. With an original approach, utilizing a laser CD ROM coupled with the videodisk, the length of the motion has been transformed into more than 20 hours of continuous playing with no noticeable difference from the standard playing (Fig. 10).

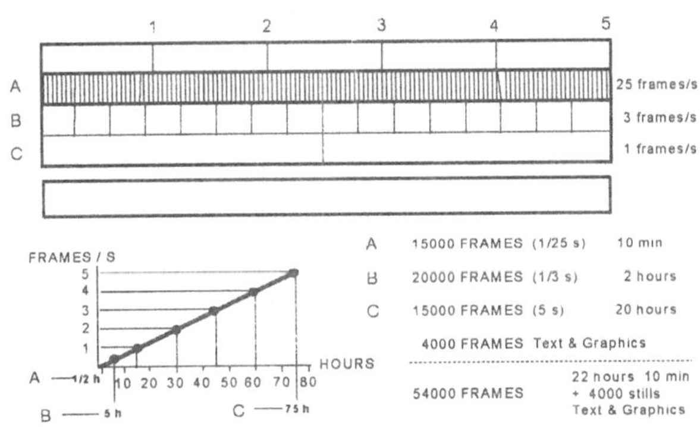

Fig. 10 Motion transformation

Authoring system and classroom management

Even if this cannot be considered a pure technical problem, it is indeed the most important aspect of the system. It is possible to have the most advanced technical system, but no real advantage can be found if the machine should be programmed by "experts". The solution proposed by the initial ICAI authoring system was treating each single piece of information and interaction as a scratch pad, where a teacher could write all type of information as he would do on a piece of paper, utilizing the "windowing" technique. When he had finished this job, he could give instructions to the computer on what to do with this information and how to interact with student answer. All these instructions were given in an extremely easy format, so that It took substantially less time to write a lesson.

Classroom management is made so that for each moment of the lesson a variety of data is collected automatically by the teacher station; at the end of each lesson and later at the end of each program segment, all data are shown in a 3D diagram (Fig. 11), so that the effectiveness of the instruction can be seen in real time and, in case of unsatisfactory result, corrections may be taken immediately.

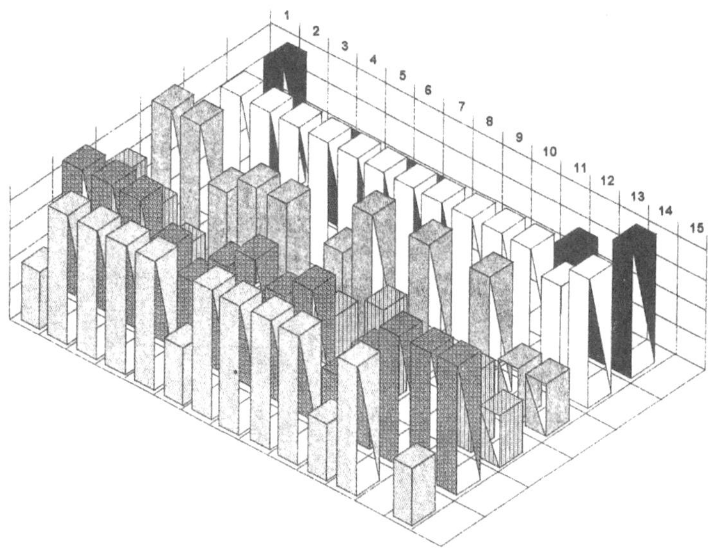

Fig. 11 Data collection

ICAI FINDINGS

Tests performed

The Gajon Institute of Technology training method has been operating since 1951, thus having a solid background of well proven tests. However, the ICAI approach was designed in 1978, developed in 79, 80 and 81 and was put into operation at the beginning of 82.

Since then, thousands of students have been trained and a number of courses have been executed as a means of comparison with conventional courses in a few different training centers (Fig. 12), with subjects ranging from basic electricity, 1st level of maintenance, to complex electronic techniques, 3rd level of maintenance.

Fig. 12 Geographical distribution of ICAI

The main result of these tests is that Armed Forces have decided to switch to ICAI for most of their courses.

Typical performance

The typical performance can be summarized as follows:

- Reduction of training time from 30 to 50%
- Positive increment of student practical ability during following maintenance courses on real equipment
- No need for separate theory and laboratory teachers
- Possibility of performing up to 4 turns per day in the same classroom with different subjects
- Possibility of performing different subjects during the same turn
- Better results obtained from low-grade students; consequent classroom uniformity.

Fig. 13 shows the superiority of CAI versus conventional instruction in terms of student achievement. Number of dots means courses which have been conducted and for each of these courses is shown when the student achievement was inferior, same or superior. On 38 courses, 1 was inferior, 22 same and 15 superior. For ICAI all 10 courses where superior.

METHOD OF INSTRUCTION	SYSTEM	SERVICE	LOCATION	STUDENT ACHIEVEMENT AT SCHOOL (compared to conventional instruction)			TYPE OF TRAINING
				INFERIOR	SAME	SUPERIOR	
CAI	IBM 1500	A	SIGNAL C&S		* * * *	* *	ELECTRONICS
		N	S. DIEGO			* * * * *	ELECTRICITY
	PLATO IV	A	ABERDEEN		* * *		MACHINIST
		N	S. DIEGO		* *	* * * *	ELECTRONICS
		N	N. ISLANDS			*	A/C PANEL OPERAT.
		AF	SHEPPARD		* *	* *	MEDICAL ASSISTANT
		AF	CHANUTE		* * * *		VEHICLE REPAIR
	LTS-3	AF	KEESLER		* *	*	ELECTRONICS
		AF	KEESLER		*		WEATHER
	TICCIT	N	N. ISLANDS		*		TACTICAL CO-ORD
	IDNOM	N	DAM NECK	*	*		FIRE CONTROL TEC.
	PLATO IV	N	DAM NECK		* *		FIRE CONTROL TEC.
				TOTAL 1	22	15	
CMI	NAVY CMI	N	MEMPHIS		* *		AVIATION FAMILIAR.
		N	MEMPHIS		* *		AV. MECH. FUNDAM.
	AIS	AF	LOWRY		*		INVENTORY MGMT.
		AF	LOWRY		*		MATERIEL FACILIT..
		AF	LOWRY		*		PREC. MEAS. EQUIP.
		AF	LOWRY		*		WEAPONS MECH.
				TOTAL 0	8	0	
ICAI	GIT	AF-N-A	VARIOUS			* * * * * * * * * *	ELECTRONICS MECHANICS
				TOTAL 0	0	10	

Fig. 13 Student achievement

Fig. 14 shows the superiority of CAI in terms of time reduction. Average reduction was 30% for CAI and 50% for ICAI.

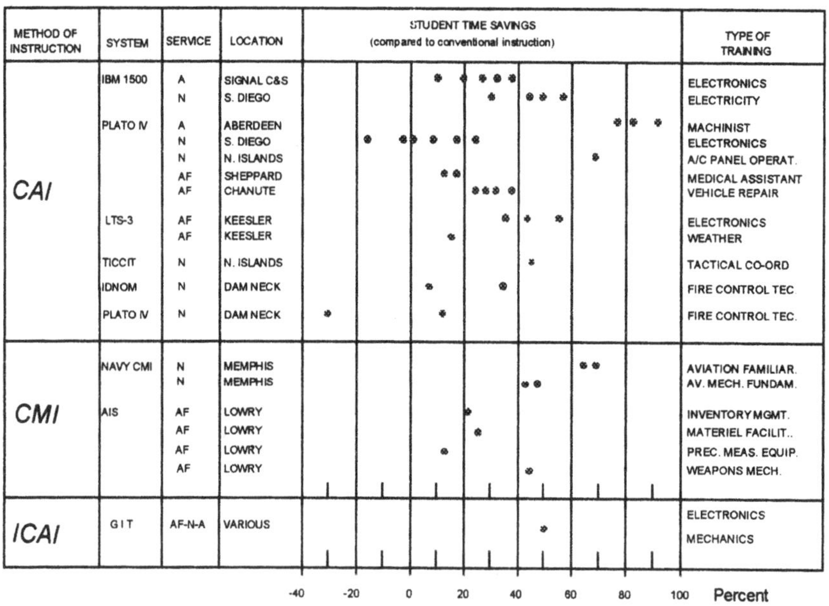

Fig. 14 Student saving time

If we should consider the relation between student achievement and time. It is evident that when we must reach a certain achievement, with CAI there is less time or, if the same time is available, a superior achievement is reached. With ICAI the ratio is even superior.

(CAI and CMI data contained in Fig. 13 and 14 by courtesy of Dr. Jesse Orlansky).

Student attitude

We must remember that if the software is not well developed, the student loses interest in the lesson in a very short time. Careful attention has been used in the ICAI method not to give students materials they cannot easily understand.

When the student is working experimentally and he is guided step by step by the computer, he is actually playing with it, and gets along very easily with the system.

It has become a fact that students have refused the 10 minute interval within the lessons, because they thought they were losing a 10 minute game.

The experience has also shown that it is even difficult to get the students out of the classroom when their turn is finished.

The last very attractive point from a military point of view is that students receive a much increased motivation in their job, which is normally scarce. This increased motivation strongly reflects into better discipline.

Teacher attitude

The first feeling of the teacher about the system is uncertainty and rejection, because of the apparent complexity of the teacher station and the difference with traditional systems.

But shortly the teacher gains confidence because he realizes that:

- students procede at own pace
- he can easily control students by video
- he has a much easier control of progress of students
- he has a much easier communication with the students

Practically the teacher starts feeling a better control over the classroom, and feels a substantial relief from the continuous interrogation process which is necessary in conventional instruction.

Teacher performance is also increased, because he must intervene at random points, taking away the routine of repeating always the same materials.

Lastly, the teacher feels he can easily get a preview of the lesson at fast speed, which helps him a lot in getting aquainted with the next materials.

Cost/effectiveness

No system can be declared useful if a good relation does not exist between cost and effectiveness. The ICAI system has shown a low 4,5 years amortization time, with extensive saving for the following years, as it appears from the following data:

- typical 15 student classroom cost, including customized software:
  - first classroom: $ 2.000K
  - additional classrooms: $ 1.000K (1)

- total cost for a typical 2 classroms installation: $ 3.000K

- typical life of equipment: 10 years

- typical turns per day in a 15 students classroom: 4

- total students per classroom per year: 60

- total students per 2 classrooms per year 120

- added cost per student per year: $ 300K/ 120 = $ 2.5K

- typical student cost per year with traditional training: $ 8K

- typical course duration with traditional systems: 2 years, 3rd level maintenance technicians, 3 hours technical training per day

- typical reduction with ICAI: 1 year, same course, 3 hours per day

- cost saving: $ 8K - 2.5K = $ 5.5K x 120 students = $ 660 K

- ACQUISITION COST: $ 3.000 K

- SAVING PER YEAR = $ 660 K

- AMORTIZATION = $\frac{\$\ 3.000\ K}{\$660K}$ = 4,5 years

- SAVING FOR REMAINING 5,5 YEARS: 660K x 5,5 = 3.630K

It is significant that the cost includes production and revision of software and all other related expenses, so that a firm point can be made on the positive and relevant reduction of cost.

(1) Additional classrooms cost is reduced to 1.000K because they utilize in common materials of the first classroom.

NEW TECHNOLOGIES

Computers are becoming more portable, with largely increased features. Architecture, memory management and VLSI are changing the way we were imagining the limits of small machines.

Analog instruments are slowly substituted with digital modules interfaced with computers, so that the reading is done directly on computer display.

The technologies we have been examining until now, even if perfectly working, will be substituted in the following years by some emerging technologies which are eliminating some of the problems that have been solved with a certain cost. Laser techniques have proliferated with the introduction of CDI and DVI, which have the clear advantage of working only with digital processing, avoiding thus the problems correlated to the analog signal processing of the videodisc system.

Voice recognition is advancing in researches, as it is virtual reality, which, although not completely well defined at this moment, promises an easier man/computer interface. Liquid cristals screens are step by step substituting ray cathode tubes, allowing easier, non fatiguing view of the images. New techniques are even changing the way in which liquid cristals were until now used. Pointing devices are greatly helping in the utilization of new breeds of software.

Software itself is more and more friendly user, and will allow in the future its complete utilization by people with no specific experience.

We can examine some interesting technologies that are available now, or are on advanced stage of research.

PROCESSING UNITS

Notebook computers are the leading effort of most researches. VLSI technology has allowed the reduction of a standard desk top computer into the size of 21 by 27 by 3 centimetres with the weight of 2.2 kilograms. Last notebook developments include 80486-25MHz CPU, VGA color screen, permanently installed trackball and other devices which help greately in the operation. Pen computers are appearing, avoiding the utilization of the keyboard. Last announcement has been the availability of a complete workstation in a notebook size. Sony has introduced the Data Discman, which includes computer, CD-ROM and screen in the size of a CD Player.

C D I

CDI utilizes a compact disk as storage means and has been long talked about. For most applications where real time motion is not a prerequisite, it seems it could comply with its promises. Surely, it reduces footprint and cost. Its main characteristics are:

- 34 hours of speech audio
- 8.000 high resolution images

Some drawbacks of this technology are: no real full motion video, no data compression, development usually done on Unix workstations, and necessity of professional development.

DVI

Digital Video Interactive (DVI) is the most recent technology developed by Intel, the manufacturer of the 80 series of CPU which has become the PC standard. It also utilizes a compact disk as storage and puts together all the benefits of the previous technologies, allowing better performance and reducing further the cost of both hardware and software implementation. Advantages are:

- 40 hours of speech audio
- 50.000 high resolution images
- 72 minutes of full motion video
- 160 : 1 video compression and 25 : 1 audio compression
- development on normal PC with some additional boards
- user level development

Many stations are appearing which utilize this technology because of its impact in several environments where the utilization of any other system is not really possible. The small dimension, large capacity and flexibility puts this system in a range of its own.

Fig. 15 shows a comparison between CDI and DVI.

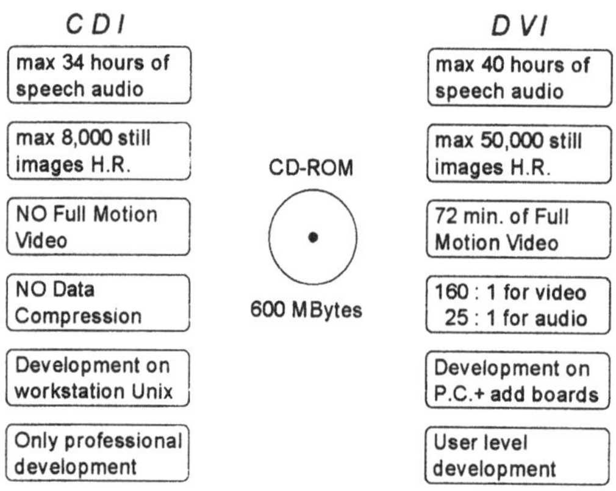

Fig. 15 CDI / DVI comparison

SPEECH RECOGNITION

Conceptually, it seems nearly obsolete to utilize the keyboard as means of interface with the computer. Many undergoing researches allow a more user friendly interface by means of direct voice recognition. A few low cost boards are available which enable the computer to translate any screen word and phrase into speech. A little more difficult for the computer is the interpretation of 100% of the complete sentences, but a great help is coming from new chips and the utilization of artificial intelligence. We can expect that in the near future most systems will be equipped with some sort of speech recognition and live answering.

VIRTUAL REALITY

Although there is not an agreement on the real broadness of this technology, examples which have been shown, even if at a beginning stage, show new possibilities of utilization of computers in a 3D environment, where different types of sensors put the operator in a virtual environment he can interact with. It will be extremely interesting to follow the development of the mathematical models and the input/output devices which will be especially necessary for the virtual world interface.

SCREEN TECHNOLOGY

LCD displays are slowly substituting ray cathode tubes. Particularly, with the possibility of having the full characteristics of a desk top computer in a notebook dimension, its seems compulsory to push this technology. At this moment, some models of portable VGA LCD color computers are available, i.e. Toshiba and Sharp, with the screen manufactured by Sharp. Their technology is based on TFT (Thin Film Transistors), where each one of the elements which constitute the screen (640 x 480 x3 RGB = 921.600 pixels) is made of a single transistor which is controlled individually. At this moment the screen allows full 256 colors in high resolution, and IBM, in cooperation with Toshiba, has shown a prototype of 1.100 x 900 pixels screen with 4.096 colors.

Another technology is based on the superimposing of three different planes, each one containing one basic color, which can produce a substantial variety of tones. Drawback of this technology is the slow speed of refreshing, which does not allow at the moment the vision of full motion.

Several researches are on the ground; just to mention one which is connected with virtual reality, a laser microscanner is expected to deliver high performance, low cost virtual display with the scanning of a color image directly into the retina, thus not addressing pixels, but rods and cones directly.

POINTING DEVICES

Mouse and trackball have become common devices since the introduction of the Apple Macintosh. Now, with the release of Microsoft windows, it appears that they are becoming an integrated piece of the computer environment. Some computers, as mentioned before, already permanently integrate this object in the body of the machine. Regarding the trackball, Microsoft has introduced an acceleration program, which makes movement of the pointer sensible and proportional to the velocity of the ball, consequently it is possible to precisely move the pointer at slow speed and rapidly cross the screen at high speed.

Another blend of pointing devices which are friendly user are touch screens. The latest models of touch screens include, on top of X and Y axis, also Z axis, alllowing the interpretation of the pressure on the screen up to sixteen levels. The result is a more human interface with much better interaction.

Also it should be mentioned that in one way or another, most virtual reality systems include or will include some pointing devices, some of which take advantage of human perceptions.

With the introduction of Microsoft Windows 3.0 and the expected 3.1 with multimedia extensions, most developers are fastly moving to this environment. CDI and especially DVI technologies are becoming a must for which an authoring system in the near future may be considered virtually out of the market if provision is not included for the utilization of these new breeds. This also means a more standard platform which can be implemented at least for a period of few years, while in the past such a standard platform did not really exist.

A reflection of this attitude will be found in the development of PC boards, which, being based on a common standard and consequently, suffering with more competition, will be obliged to increase power and decrease prices.

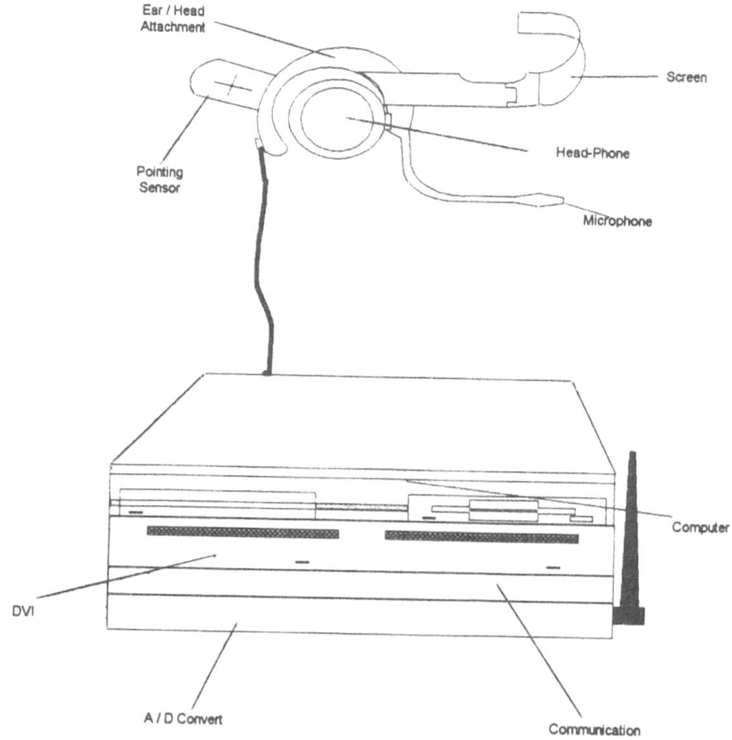

Fig. 16 ICAI / IMS integrated station

INTEGRATION OF TECHNOLOGIES

The efforts of Gajon Institute of Technology researches are directed towards the utilization of the previously mentioned new technologies, with the integration of any possible device which could be found on the shelves, trying to avoid the dispersion of resources.

Consequently, the actual development is a system which will be identically used both for training and maintenance, that utilizes the same ICAI/IMS approach used until now, but has no difference in model, size and weight among the training and field utilizations (fig. 16).

The system, very compact and integrated (based on 21 by 27 cm notebook computer footprint), is composed of the following separate but connected elements:

Computer module
- 80486 25MHz notebook computer, 8MB RAM, cash memory, 120 MB Hard Disk, 3,5" Disk Dive, compact keyboard, 256 colors VGA LCD touch and pressure sensitive display, bus and printer ports

DVI module
- DVI board, 2 x 640 MB CD ROM, bus port

Intercom module
- boards for audio with speech recognition, video, teacher interface, local area network, bus port, communication transfer both by air and fiber optics

A/D conversion module
- analog to digital conversion boards for oscilloscope, analog meters, digital meters and training/job dependent instruments, bus port

AVP (Audio/Video/Pointing device) head system
- headphone, microphone, virtual reality screen, pointing sensor

The system is capable of performing all the attributes of the actual ICAI and IMS systems, with the following advantages:

- completely digital processing of images
- 72 hours of video motion
- very small footprint (21 x 27cm) x 4 to 15cm maximum height, according to used modules
- very low weight (3 to 8 Kg, according to used modules)
- shoulder portable for difficult environment maintenance and field operation
- head motion and voice controlled pointer, voice recognition, when using AVP

The system allows the reading and utilization of the screen through the pressure sensitive touch screen, or the alternate utilization of the AVP when the conditions do not allow the utlization of the screen.

The system also allows the set up of any classroom in any condition and without any installation, because of its wireless communication system, or utilization in difficult environments (e.g. fighting ship) through its fiber optic link.

NEXT GENERATION

Portable CDI stations are now available at a reasonable cost. They have nearly the dimension of a Compact Disk player with cost decreasing in the measure in which the utilization is broadened to consumer market.

Again, the research effort of the Gajon Institute of Technology is directed towards a system which will have the dimension of a CD Player, equipped with a 3D AVP (Fig. 17).

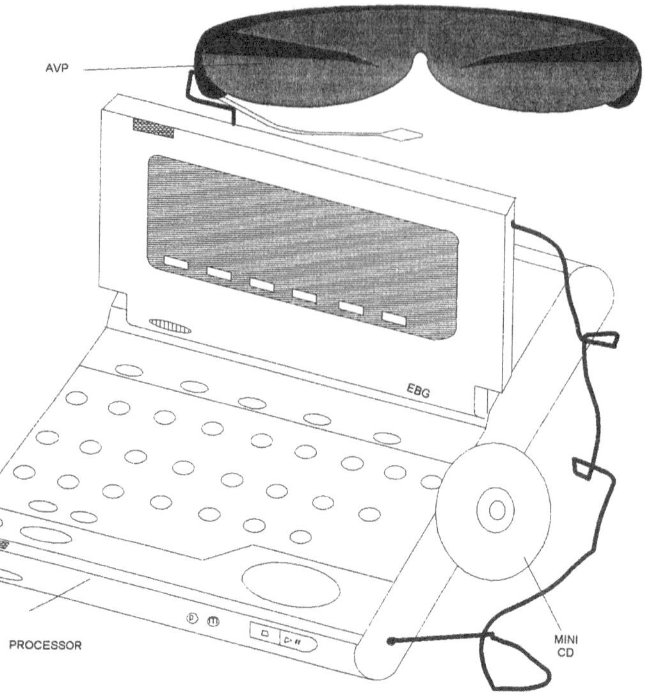

Fig. 17 Future ICAI / IMS integrated station

The system is going to use the following technologies:

- artificial intelligence
- virtual reality
- speech recognition
- audio and video digital recording
- cellular telephone and modem communications
- television
- Global Positioning System (GPS)
- output for fax, used also as printer

These technologies are already available, each one of them in compact size, having in common some devices as data processing, display and data access. Examples: Sony portable CDI, Motorola cellular phone, Canon digital camera, Sony GPS and a number of small LCD small television sets are already available in the consumer market.

As a result, in the near future it will be possible to use this small system attached to the belt, with interface through the AVP which will enable to give commands to the computer through head motion pointing and speech recognition. Available main menu will include, as an example, computer, communications, positioning, memo, video recording and playback, entertainment. Pull down subsets will include data base, spreadsheet, project management, graphics for computer area and geographic coordinates, maps, waipoints, navigation and information for positioning area, not to mention the other possibilities.

Data will be included in a 3 1/2 " CD ROM, small enough to slip in the pocket, with a capacity of over 200MB.

Training, maintenance and field operation will be major utilization of this device, while in the consumer environment its utilization will be even wider, as an integrated system for computing, communicating, navigating, image processing and entertainment, when the cost will be reduced to an acceptable standard.

As a closing remark, Fig. 18 shows a comparison of systems which have been available in the 70's, 80's, and at the beginning of the 90's; this enables us to forecast a projection for the near future. We can, therefore expect that the new device will be available in a reasonable lapse of time.

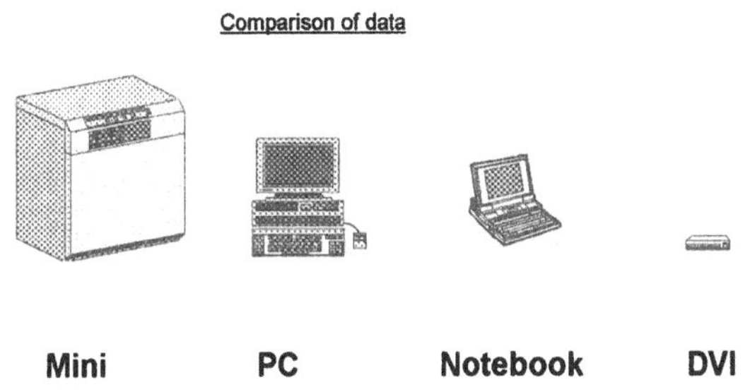

Years	70s	80s	90s	00s
Volume	0,5 m^3	0,03m^3	0,003m^3	0,0003m^3
Footprint	80 x 60 cm	42 x 36 cm	21 x 27 cm	10 x 10 cm
Weight	100 Kg	15 Kg	3 Kg	1 Kg
Power	8 bit	16 bit	32 bit	?
Cost	$ 50 K	$ 5 K	$ 2.5 K	?

Fig. 18 Systems comparison and projection

VIRTUAL REALITY: ITS POTENTIAL IMPACT ON EMBEDDED TRAINING

J. Michael Moshell
Ernest A. Smart
Richard Dunn-Roberts
Brian Blau
Curtis R. Lisle

Institute for Simulation and Training (IST)
University of Central Florida
Orlando, FL 32816

ABSTRACT[1]

Virtual Reality (VR) denotes a multi-sensory realtime simulation that immerses the user in 3d graphical space, allows freedom of movement within the space, and supports complex interactions including the modification of most features of the space itself.

Embedded Training (ET) involves the integration of training capabilities into operational devices. As such, ET seems to invoke the opposite spirit than VR. In ET, as little as possible is simulated; the actual device and its environment (e. g. the associated vehicle) are used. However, there are cases in which the experience of VR could meld very nicely with embedded training.

Consider the problem of integrating SIMNET-style networked simulations with real armor exercises. It has been suggested that tank commanders operating open-hatch or POP-hatch in field exercises could pick up a pair of special binoculars to see the virtual (simulated) components of the battlefield. Alternatively, the entire vehicle could wear display devices over its viewing blocks and gunsights, which would provide a mixture of real scenery and superimposed images of virtual participants.

---

[1] This work was sponsored by the Army's Project Manager for Training Devices (PM-TRADE). However, all opinions are those of the authors.

In order to explore these issues, the authors have developed and implemented the Virtual Environment Testbed (VETB) - the first laboratory for the realtime interconnection of multiple heterogenous VR systems in an object oriented environment.

This paper explores some of the issues in mixed (embedded and virtual) simulation and describe how the VETB can be used to advance our understanding of VR and our ability to create simulations which are robust and flexible enough to be used in embedded training.

Contents

1. Virtual Reality (VR) - an Overview

VR is a term which is widely used, with various definitions. Most workers agree that a Virtual Reality should provide the following experiences :

- Three dimensional interactive graphics
- Multiple sensory modalities (sound, manual manipulation, etc)
- Ability to interact with and modify objects in the space
- User presence (body image) in the graphics space
- Multiple interacting participants

Among users of the term "virtual reality", the most common technique for immersing the subject in graphical space is the head-mounted display. A small California company, VPL has marketed a low-cost head mounted display titled "EyePhones™" and an instrumented "DataGlove™" which are used in many experimental VR systems.

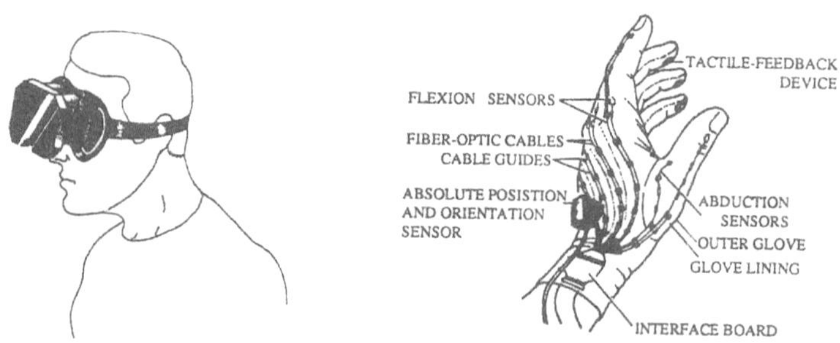

Figure 1: EyePhones™ and DataGlove™

It must be emphasized that VR is a *goal*, not yet a fully realized and deliverable technology. Fielded systems to date have coarse, slow displays and are very limited in the quality of the visual scenes. The intended civilian market cannot in general afford high performance image generators; instead, graphics workstations such as the Silicon Graphics Iris are used.

The contrasts between VR and traditional training technologies are not absolute; some existing high-cost flight simulators incorporate some of the same features - particularly head mounted displays. However, VR is intended to be a low-cost (under $200,000) generic reconfigurable interactive environment, with multiple uses. Some contrasts between VR and traditional training simulators are shown in the following table.

	Traditional Simulators	Virtual Reality Simulation
Versatility:	simulate a single system	reprogrammable; many "worlds"
Displays:	CRT screen, projectors	head mounted
Interaction:	interact via joysticks, controls	interact by touching, grasping
Sound:	some specialized sound	integral directional sound
Presence:	sometimes ownship in scene	personal body image in scene
Participation:	a few enemy aircraft (in SIMNET, many others)	other people present in scene
Feedback:	motion bases, pressure	force-feedback, touch sensors

One unfortunate characteristic of Virtual Reality is that it is presently being badly over-sold by some of its advocates, as well as the media. It will be several years before adequate graphical performance is available at moderate cost. Existing systems primarily demonstrate the feasibility of the concept.

2. Embedded Training - Some Terminology

Embedded Training (ET) involves the integration of training capabilities into operational devices. As such, ET seems to invoke the opposite spirit than VR: in ET, as little as possible is simulated, whereas in VR everything is simulated.

There is a variety of possibilities for how ET can be achieved. In ships and fixed onshore systems (e. g. radar stations) there is not a large penalty for weight, and so complete training systems can be built into weapons systems [Massey 86]. A mode switch simply transforms the actual display device into its own trainer. On some occasions an umbilical cord ties the ship at pierside to additional on-shore simulation equipment.

For smaller weapons platforms, the weight, size and power requirements of additional computing and display equipment have required that a *strap-on* approach be used. The GuardFist [Daedalean 91] system, for instance, consists of external image generation equipment and displays which are attached to the viewing blocks of the M1A1 tank. Overlays are strapped onto the tank's controls, and the crew is then made to believe that they are in motion by the provision of moving imagery and sound.

When special sensor systems are involved (e.g. night vision, infrared) a strap-on signal source is referred to as a *stimulator*. It is often easier to attach to the platform's electronics systems after the sensor (because of the difficulty of producing actual high resolution infrared or radar images). In this case the simulator must mathematically replicate the sensor, rather than stimulating a real device.

The Spectrum of Embedded Training Systems. In order to avoid controversy as to whether strap-on systems like GuardFist are in fact examples of ET, we introduce the concept of *Appended Training (AT).* An AT system consists of equipment added to operational weapons in order to temporarily transform them into trainers.

Now we can discuss a spectrum of choices:

- A weapons system requiring NO external augmentation equipment to provide training is said to provide pure Embedded Training.

- A weapons system equipped with strap-on systems which have no electronic connection to the operational weapon, but simply stimulates the operational sensors or viewports and senses control operation via overlays, is said to provide pure Appended Training.

- Most practical ET systems will fall somewhere between pure ET and pure AT.

We will now explore four aspects of VR, with respect to its potential impact on Embedded Training. These are respectively

- VR as an augmentation to embedded training
- VR as a potential prototyping tool for embedded training systems
- VR based weapons (such as tele-operation of robots)
- Embedded training for VR-based weapons.

3. VR Within ET Systems

Stereo Viewing. One salient feature of VR is its provision of stereo viewing. Within ten meters, the provision of separate images to the left and right eye significantly enhances the user's ability to understand a scene. In what domains would this capability enhance training?

Except for hand-to-hand combat, most modern warfighting takes place at ranges of hundreds to thousands of meters. However, many significant activities, such as mine emplacement or removal, vehicle driving and earthmoving, involve objects within stereo range. These skills are currently taught in the field with real equipment, and as stand-alone activities seem to have little need for advanced simulation technology.

However, in the combined-arms training scenario, there is a high degree of interest in enhancing the quality of participation by the engineers. Combined arms training in SIMNET and its possible successors will involve a mixture of full simulators, GuardFist-style strap-on devices and embedded trainers.

Driving today's tank simulators, whether low- or high-fidelity, quickly leads to the realization that it is very difficult to maneuver in close quarters. Parking a SIMNET tank in defilade in a prepared position is rendered much more difficult by the lack of a motion base. It is a reasonable hypothesis that the improved depth information available from stereo viewing could help to solve this problem.

Stereo views are naturally provided by a real M1A1's optical viewing blocks; but in a GuardFist Appended Trainer, only a single image is offered. Special glasses are available [StereoGraphics 90] which alternately block the view from one eye, then the other at 120 hz. If the display synchronously alternates between left and right-eye views, the user sees depth in stereo. Experiments are underway at General Electric (Daytona Beach, FL) to evaluate this concept, using GE's Tank Driver Trainer.

Instrumentation Overlays. Some of the original work on VR was done at the Air Force Human Factors Research Laboratory (Wright-Patterson AFB) by Tom Furness [Furness 88]. His *Super Cockpit* conceptual testbed advocated the idea of a high resolution head mounted display (HMD) which contained not only the out-the-window imagery, but also the complete simulated instrument panel.

Today's display technology cannot deliver the needed angular resolution in a HMD to accomplish this goal. However, this is an area of intense research, and we might speculate on how such a display, once built, could be used in an ET system.

A major difficulty with strap-on ET is the necessity to intervene in the weapon platform's instrumentation and controls, or overlay them. A Furness-style HMD could provide a complete visual scene which replicated the interior of a vehicle, gauges and all. DataGloves could also remove the necessity for overlaying sensors on the controls.

However, today's DataGloves and HMDs use magnetic sensing technology which would not work at all well within a metallic cockpit. New sensors on the market promise to overcome this problem by using a direct-current rather than an alternating-current signal. [Ascension 91]

It seems likely that by the time the VR technology progresses to the point where virtual instrumentation overlays are possible, full ET will also be possible and common. Strap-on systems are likely to fade away as ET is built directly into the electronics of weapons systems.

Dismounted Infantry. Another distinctive VR feature is the provision of images of the participant's body in the scene. Most prototype systems show only the hand, as transduced by the DataGlove. VPL has constructed a full body-suit, and many workers [Zeltzer 91] are developing algorithms to realistically animate the human form. Such animation systems would reduce the amount of detail and accuracy in data captured from a body-suit.

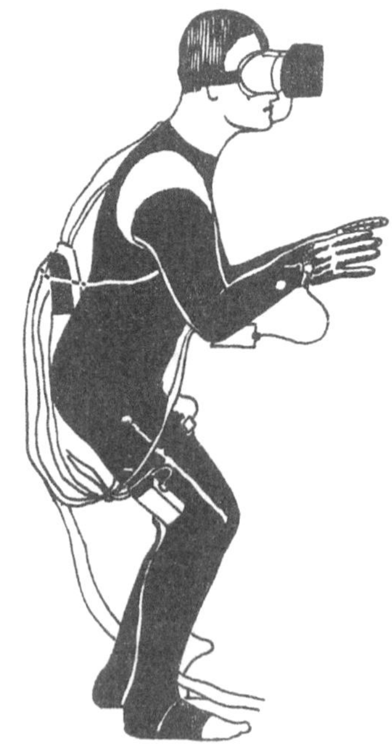

Figure 2. An Instrumented Body Suit

In fact, particularly in the simulation of infantry, the body suit may be a "red herring" - an unnecessary complication. Allowing a trainee to roam freely while wearing a head mounted display would be very difficult. The magnetic sensors presently used to determine head position and orientation have a range of only two meters. Myron Kreuger [Kreuger 89] is developing the concept of an *omni-directional treadmill* to overcome this problem.

The prototypical dismounted infantry system built by BBN for SIMNET uses simple icons to show the location of an infantry squad, and controls the squad's position via a spaceball. The squad leader views the world through eight panoramic monitors. According to the principle of selected fidelity, this may be enough detail to integrate infantry into the combined arms scenario without the complexities of body suits.

A longer-range view of the prospects for seamless integration of dismounted infantry into combined arms simulations is provided in [McBride 90.]

4. VR and the Development of Embedded Training Systems

Cockpit Design. In the weapons systems of the future, the view of the outside world is likely to be entirely through sensors, rather than via direct vision blocks. Sensors provide protection from energy weapons (e.g. lasers) and can mix visible light, infrared and radar imagery. Such viewing devices facilitate the addition of synthetic imagery in support of ET. Strap-ons and stimulators will no longer be necessary.

There are, of course, numerous options and design decisions to be made in constructing such a viewing system. How should symbology and actual images (visible, IR, radar) be mixed? How should the user select modes, when his hands are likely to be very busy? What automatic feature recognition systems and built-in warnings should be used to enhance the operator's cognition?

VR systems are likely to be ideal for prototyping and testing such choices. By their very nature, VR systems are software-intensive and rapidly reconfigurable. Instead of rebuilding a vehicle's cockpit to relocate display screens, one simply *re-draws* the cockpit in a CAD system, and the virtual displays are relocated.

Touchable Controls. Conventional *glass cockpit* development systems [AWST 87] already provide much of the desired reconfigurability for prototyping, in the form of touch sensitive screens upon which various versions of instruments and controls can be drawn. However, there are substantial psychophysical advantages to providing controls which can be actually grasped, rather than simply using colored patches on a CRT.

Advanced forms of VR DataGloves, incorporating force feedback, could perhaps provide the actual feel and kinematics of a simulated control handle or knob. However, this degree of accuracy in force feedback has not yet been achieved. This subject is discussed further in the following section.

Stereo Viewing. In the previous section we discussed the question of stereo viewing as a component of close range tasks such as tank parking and bulldozer operations. In the design of a sensor-based viewing system, the question of stereopsis will need investigation. Again, VR provides a ready testbed for such a decision. Both for sensors and for ET within such a system, it would be cheaper if stereo could be avoided. One rather than two image generator channels would be needed for ET without stereo.

5. VR-Based Weapons

(a) Personnel Amplification Systems

For several years, a team at Los Alamos National Laboratory conducted the PITMAN project [Gorman 90]. This engineering study examined the feasibility of powered body armor, with an integral viewing device/helmet/CBW protection system (a so-called *Personnel Amplification System.*) The project's concerns ranged across issues from power requirements, force feedback and servo design, to visual displays.

Force Feedback. A soldier wearing body armor would be bearing a total load (including power supply) of perhaps 400 kg. Clearly an individual's strength would not suffice; a powered *exo-skeleton* is required. Operating such a suit from the inside is equivalent to the remote operation of an anthropomorphic robot (from a very short distance away!).

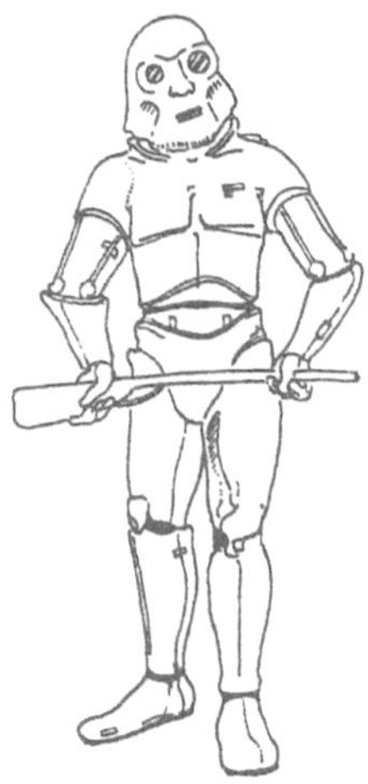

Figure 3 - Soldier in Body Armor

The engineering of such a capability is underway, but is many years from maturity. The computational requirements to support the predictive dynamics are large, probably in excess of 200 mFlops. The inertia and resistance of the sensors and exo-skeleton must be exactly compensated. Servo-based approaches have latency which can seriously affect the operator's balance, timing and fatigue.

Visual Display. Within the armor helmet, a high resolution display device is needed. Visual sensors, night vision and thermal information and perhaps other HUD symbology would be superimposed. Eye protection against lasers and fragments is provided (though the external sensors may be damaged).

Now - assuming the existence of such a "Ninja Turtle Warrior Suit" (a very large assumption), a number of interesting questions arise.

• Why wear it? Why not operate it at a safe distance, via a radio link? The danger of hostile electromagnetic interference (jamming) must be balanced against the reduced personal hazard of a remotely operated robot. It is likely that a man on the scene will be required for some time to come.

• What non-combat applications exist for full-body protection with augmented strength? Could firemen or divers use such help? Could some users afford to be tethered to a power source (greatly reducing the suit's weight?)

Is it Virtual Reality? Clearly the display technology is almost identical to the display requirements for VR. However, the soldier (or his tele-robotic *doppelgänger)*is in real space, not a simulated space. Personnel Amplification Systems (PAS) are thus not, *per se*, VR systems. However, the logical training system for a PAS does seem to be a VR. This is discussed in section 6 below.

(b) Stealth Platforms for Battlefield Visualization

An interesting component of the SIMNET system is the *Stealth Vehicle*. This system consists of three channels of imagery attached to a "silent invisible helicopter" - actually just a moving viewpoint, under the control of a SpaceBall™ (a six degree-of-freedom joystick). A plan view display is also provided on a separate monitor.

The Stealth vehicle allows a birds-eye view of the SIMNET battlefield. The viewpoint can be attached to any actual vehicle in the simulation, and can either see through the commander's eyes, the driver's eyes, or "over the shoulder", with the vehicle in the foreground of the view. Stealth is an invaluable tool during the development of Semi-Automated Force (SAF) systems; the programmed behavior of tanks and units can be visually observed and critiqued [Downs-Martin 89].

During the recent Persian Gulf conflict, this Stealth platform generated a great deal of interest as an actual tool for battle planning and visualization (as opposed to training). The desert terrain is uniquely easy to model by semi-automated means. In addition, the Defense Mapping Agency is moving steadily toward an all-digital approach to map production, which will greatly facilitate the rapid production of geospecific visual databases [DMA 90].

With the steady improvement of electronic sensors and their integration into C^3I systems, it will soon be feasible to locate one's own vehicles and many of the enemy's, in a continuous realtime display similar to the SIMNET Stealth system. One can then play out "what if" scenarios, by starting from a known configuration and playing forward, using SAF models of friendly and enemy tactics and doctrine.

In a Stealth system coupled with a head mounted display, a commander could move around on the ground as though he were walking up to a point of concealment; look through the bushes, and see what his forward observers would actually see (assuming that the database correctly represents the battlefield). Alternately, the commander could back off to a high-altitude "God's eye view" of the battle, with brigades and divisions represented by icons upon the landscape.

The distinction between VR as a pure simulation, and Stealth as a display of the real battlefield, is now almost completely blurred. A simulation based on good data is really just a new and more powerful kind of map - a planning tool.

6. Embedded Training for VR Weapons

For both of the previously described systems, embedded training is a natural extension. For a PAF armor suit, one could train the wearer by simply "hanging him up by the scruff of his neck" and coupling his suit's sensor circuits to a VR simulator rather than to live sensors. The trainee would run, climb, and react to fictitious forces imposed by the suit rather than to the forces of real terrain.

Live fire and deadly hazards could be presented in simulation with high accuracy, without actual risk to the trainee.

The main missing (or inaccurate) sensory cue is momentum. To simulate the impact of jumping off a wall, for instance, the PAF will have to deliver a force on the trainee's feet, which will accelerate the PAF and its wearer upward. A motion base will thus be required.

For the Stealth system as a visualization component of C^3I, the existing Semi-Automated Force systems already constitute an important component of an embedded training tool. If augumented with a rule-based after action review capability, commanders might find it useful to privately review their own battle management successes and failures, and that of their subordinates.

It should, however, be noted that in 1987, a NATO Workshop [NATO 87] explored questions of the appropriate use of ET, and concluded that ET is most likely to be useful for individual weapons or sensor systems. Its utility diminishes and risk increases as the size of the training exercise increases.

Because of the danger of confusion between real and simulated threats, and because of interference with operation of the actual theater level C^3I system, it was recommended that separate simulators be built for war-gaming at these scales.

7. The Virtual Environment Testbed

The Institute for Simulation and Training leads a consortium of major universities[2] which is developing the Virtual Environment Testbed (VETB). The VETB is built around the following principles:

(a) **Inclusion.** Whenever possible, include state-of-the-art CAD, GIS and visualization tools rather than building new ones. Find ways of using existing tools in new combinations.

(b) **Data Density.** Satisfactory virtual worlds will be very rich in both visual detail and interactive structure. Plan ahead to take advantage of rich datasets such as Project 2851 [PRC 90] and technologies such as Rapidly Reconfigurable Databases [Donovan, 90; GE 89].

(c) **Modularity.** Based upon the concepts of the DIS protocol [IST 91], the Virtual Environment Realtime Net (VERN) protocol allows the addition and removal of players of various kinds to a running simulation with minimal reconfiguration required. This provides a "mix and match" capability, which is essential for a testbed.

---

[2] including MIT's Media Lab, Georgia Tech's Visualization, Graphics and Usability Lab and the University of Iowa's Computer Science Department and Center for Simulation for Design Optimization.

(d) **Performance.** Use the best possible platforms: image generators, realtime computers, parallelism. Use million-dollar 1991 tools to simulate the performance of anticipated $10,000 tools in 1995.

David Zeltzer of MIT's Media Lab has created a descriptive coordinate system for describing the attributes of a particular Virtual Environment, as shown in the following figure.

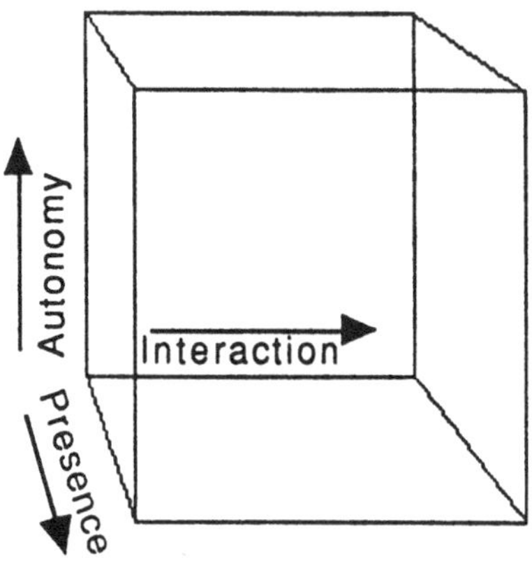

Figure 4: Zeltzer's Coordinates for Virtual Environments

• The *Interaction* axis measures the degree to which the user can influence the course of events in the simulation in realtime. Playing with a Nintendo is interactive; watching a movie is not.

• The *Autonomy* axis measures the amount of activity and responsiveness which the simulation itself generates - as opposed to activity which is pre-programmed. An animated, physically realistic human figure is highly autonomous; a puppet is not. The puppeteer must control every body joint.

• The *Presence* axis measures the immersion of the participant in the scenario. To the extent that the user can benefit from a lifetime of experience with three dimensional space, the "user interface problem" is banished[Bricken 90].

The Components. Each component in Zeltzer's VE space is addressed in the VETB Project by stating a formal goal, as follows:

<u>Goal 1: Presence.</u> Investigate the attributes of input and display systems which are critical to convincing the trainee that he is in a real world, not a simulation.

Display and Visualization development and research will address this goal. *Audio* devices enhance the trainee's sense of presence.

Goal 2: Autonomy. Investigate modeling techniques and simulation algorithms which support natural, unprogrammed behavior of objects in the virtual world.

Physical Modeling and Animation development and research address this goal.

Goal 3: Interaction. Investigate the kinds of input devices and modeling technologies which allow the trainee to modify, repair and construct parts of the virtual world and to interact with other participants at a distance.

User Interface and Object Oriented Software development and research, together with *Networking and Concurrent Programming*, will meet this goal.

Software and Hardware. IST has already constructed Version 1 of the VETB [Moshell 90] and is using this software for initial tests and conceptual development. VETB 1 consists of various software components running on Silicon Graphics, Sun and NeXT workstations.

These components represent experimental "test jigs" for work in stereopsis, physical modeling, object oriented user interface construction and realtime simulation networking. These components are built in C, C++ and SmallTalk.

In 1992, the Consortium will construct an integratedVersion 2 of VETB [Moshell 91] and embark on long-term testing activities to evaluate various VE components and strategies for training systems.

The following figure shows that each University has selected one area for primary emphasis, and that all will participate in the testing and evaluation activity under IST's lead.

● = lead
* = support

	Testing & Evaluation	Networking & Concurrency	Animation & Phys Modeling	Visual Representations	Obj Oriented Technology
UCF/IST & CS	●	●	*		*
MIT Media Lab	*	*	●		
Ga Tech VGU Lab	*			●	
U. Iowa CS & CSDO	*		*		●

The integrated, networkable software of the VETB will be made available to any laboratory working on applications of Virtual Environments to training, with the approval of the sponsor of this project.

VETB will also serve as the nucleus of IST's continuing research on the application of Virtual Reality to Embedded Training. Many of the issues raised in this paper will serve as starting points for new projects. Comment and involvement from the NATO training community is invited.

References

Ascension Technology Corp. *"Bird" Tracking System Product Description.* Burlington, VT. Feb. 1991.

AWST. "Crew Systems Lab Offers Rapid Feedback on Cockpit Concepts". Aviation Week & Space Technology, Dec. 7, 1987.

Bricken, Meredith. "Virtual Worlds: No Interface to Design". Submitted for publication to *CyberFace*, MIT Press, Cambridge, MA, May 1990.

Daedalean. *Guard Fist I Product Description.* Daedalean, Inc. Columbia, MD. 1991.

DMA. Tactical Terrain Data description, in *Digitizing the Future*, 3rd ed. Defense Mapping Agency, Fairfax, VA. May 1990.

Donovan, Kenneth B. "Mission Rehearsal Database Requirements and Technologies". *Proceedings of the 12th I/ITS Conference*, Orlando, FL., 6-8 November, 1990.

Downs-Martin, Steven G. *Replacing the Exercise Controller with the Enemy: The SIMNET Semi-Automated Forces Approach.* BBN Systems and Technologies Corp. Report 7211. Cambridge, MA. December 1989.

Furness, Thomas A. III."Harnessing Virtual Space." *Proceedings of the SID (Society for Information Display) 1988 International Symposium.* Playa del Ray, CA. May, 1988

GE. *Rapidly Reconfigurable Data Base Concept Demonstration Final Report.* General Electric Company. Daytona Beach, FL. July 17, 1989.

Gorman, Paul F. *SuperTroop via I-Port: Distributed Simulation Technology for Combat Development and Training Development.* IDA Paper P-2374. Institute for Defense Analysis, Alexandria VA. August 1990.

IST. *Entity Information and Entity Interaction in a Distributed Interactive Simulation.* Institute for Simulation and Training Report IST-PD-90-2 (revised). University of Central Florida, Orlando, FL. Jan. 25, 1991.

Kreuger, Myron M. *Artificial Reality II.* Addison-Wesley. Reading, MA. 1989.

Massey, D., Harris, M., Downes-Martin, S., and Kurland, L. *Embedded Training Technology Survey.* BBN Laboratories Report 6062. Cambridge, MA. 1986.

McBride, Dennis K. "Individual Portal (I-Port) in Advanced Distributed Simulation Research." in *Proceedings of the ADPA Conference on CLose Combat (Light) Simulation Technology.* Albuquerque, NM, 17-20 September 1990.

Moshell, J. Michael. *Work Plan for Dynamic Terrain Project: The Virtual Reality Testbed.* Institute for Simulation and Training, University of Central Florida. Orlando, FL. March 1990.

Moshell, J. Michael. *The Architecture of the Virtual Environment Testbed.* VSL Memo 91.6. Institute for Simulation and Training, University of Central Florida. Orlando, FL. April 1991.

NATO. Final Report of Research Study Group 12 on Computer-Human Interaction in Command and Control. Document AC/243 (Panel 8/RSG.12) D/7. NATO, 1110 Brussels, Belgium. Unclassified (Not for Public Release)

PRC. *ISWG Project 2851 Newsletter.* Planning Research Corp., McLean, VA. 1989.

StereoGraphics Corporation Product Directory. San Rafael, CA. October 1990.

Zeltzer, D. "Task Level Graphical Simulation: Abstraction, Representationand Control," in *Making them Move: Mechanics, Control and Animation of Articulated Figures."* N. Badler, B. Barsky and D. Zeltzer, eds. pp. 3-33. Morgan Kaufmann, San Mateo, CA, 1991.

ODIN

Dennis K. McBride

CDR, MSC, USN
Defense Advanced Research Projects Agency - ASTO
Arlington, Virginia
USA

ODIN LORE

Odin was a Norse God. A master of wisdom, magic and poetry, and protector of courageous noblemen in war, he could strike the enemy blind, deaf, and impotent--he was the god of the slain. Odin wore a golden breastplate and helmet, carried a magic spear, and rode an incredibly swift, eight-legged horse named Sleipnir. Odin sacrificed one eye to drink from the Well of Knowledge. On his shoulders perched two ravens, Hugin and Munin (thought and memory), who overflew the battlefield by night and reported to Odin by day. Little did Odin know that his paradigm for learning about the world around him would help stimulate a fundamental change in training technology and methodology in the late 20th century.

CENTCOM NEED

In response to an urgent and compelling need during Operation Desert Shield/Storm, the Defense Advanced Research Projects Agency (DARPA) formulated Project Odin. The effort was aimed at deriving and fielding innovative C2 capabilities to be utilized at various levels of command. Project Odin combines elements of several proven technologies, to provide an inventive network-based fusion of cutting-edge intelligence parsing, 2-, 3-, and 4-D terrain/order-of-battle map electronics, with computer-hosted force disposition and prediction capabilities. The system was designed for deployment in a rugged, dynamic, operational environment. The composite provides combat planners the capability to "fly" through a digitized rendering of tactically or strategically important areas of interest in order to observe terrain and the disposition of forces on terrain. The technology empowers the commander by providing a comprehensive picture of his geographical and battlefield geometry as it unfolds.

THE CONCEPT FROM THE USER'S PERSPECTIVE

Studies on the behavior of organisms as small and agoraphobic as rodents to those as large and claustrophobic as humans show clearly that an organism's ability to "represent" or "map" space and time is fundamental to its survival. The most common representation of space used by humans, of course, is the ordinary map.

Advanced Technologies Applied to Training Design, Edited by
R.J. Seidel and P.R. Chatelier, Plenum Press, New York, 1993

Cartography Development

Some archeologists identify the first use of cartography, the art and science of making maps and charts, as the markings associated with paintings found on cave walls. These markings depicted the game trails of animals. However, due to a lack of credible evidence, not all historians believe these to be the first maps. The oldest map specimen yet discovered, one which is positively a graphic representation of land features, dates back to 2300 B.C. This Babylonian clay tablet compares to Egyptian land drawings found on the walls of tombs dating circa the same period. Egyptian and Babylonian cultures developed mapping skills concurrently, showing the locations of fertile land and rivers, but neither represented the earth as a whole.

The credit for fundamental study of the nature of the earth and its formation is attributed to Greek philosophers and geographers. In pursuit of geographic development and knowledge, the Greek philosophers took to the sea to gain a greater perspective. They were preoccupied with the extent and formation of the earth which, at the time, was believed to be a flat disk. After years of maritime exploration, the theory of a spherical earth was gradually accepted. Thus ensued an exciting new challenge for geographers--creating a world map. The principal Greek astronomer and mathematician who is credited for the advancement of geography and cartography is Claudius Ptolemaeus (Ptolemy). His monumental accomplishment, the Guide to Geography, discussed the basic principles of map production and globe construction. It also provided a list of 8,000 names of places and approximate locations by latitude and longitude. Additionally, and most significantly, Ptolemy's work contained instructions for making maps using mathematical geography, as well as other elemental principles of cartography. This was a turning point in the reliability of existing maps.

During the 18th century, a renovation of cartography evolved. More factual content was represented including explanatory notes and there was an attempt to depict reliable proportion and perspective. This reformation was due to the fact that map making was now being done by scientists for independent means, and not for profitable output. Also, the cartography evolution has benefitted form more reliable and advanced equipment, i.e., the development of the chronometer which computes longitude less laboriously. With increasingly more dependable data, European power-conscious nations could accurately map their respective boundaries. Elaborate national surveys resulted prior to WWII and, in 1891, the International Geographical Congress proposed that participating countries jointly produce a 1:1,000,000 scale map of the world. The outbreaks of WWI and WWII hindered and facilitated progress in this endeavor. Although countries were not collaborating on the map making effort, their respective military forces were completing detailed surveys that would be used as viable input in the future.

At present, terrain data are available for most of the world, although some remain crude. Aerial photography has dominated as the current data gathering technology. But, as cartography continues to develop and exploration of the environment increases, old maps must be revised to account for the detail and granularity that the future will require.

The obvious limitation of maps, regardless of the quality of information which informs them, and independent of the granularity they indicate, is that they are two-dimensional (2-D). The third spatial dimension, Z, when represented on a map (i.e., elevation), is coded in some secondary way, as marked grid lines, color coded indications of depth, elevation, etc. This limitation is significant, especially in environments where Z is fundamentally important. Air traffic controllers, for example, must contend with a 2-D representation of aircraft locations and courses in order to manage the four dimensions of the real world under their control.

Military Application

Military officers have probably always used "sand tables" to represent the battlefield. Mental mastery of the very complex flow of forces over time and space has often been cited as a, or "the," criterion for success in operational planning. Commanders must be able to visualize military configurations in order to understand theatre dynamics. Moreover, they must use visual media, like sand tables, to convey their intent unambiguously to subordinates and detail planners.

The busy commander invariably "walks" over, around, and through his silicate rendition of the battlefield. By physically transporting himself, and by viewing the table from a variety of perspectives, he gains an ever greater mastery of the relative positions and apparent vectors of combat organizations. Hereagain, studies confirm what is obvious to automobile passengers--the proverbial "I can't remember how to get there, YOU were driving" phenomenon.

There is a wealth of anecdotage in which men in uniform use whatever means available to construct 3-D representations. For example, in SIMNET--the U.S. Army's Combined Arms training facilities developed by DARPA--there is typically a room set aside for real time monitoring of simulated battlefield training exercises. The room includes a large television screen, a computer interface into the "simulation network," and a manipulandum--typically a pressure-sensitive "Spaceball," which allows observers means to "space travel"--i.e., to fly around the battlefield without being detected by any of the trainees. This "Flying Carpet" is used extensively for After Action Reviews (AARs), as well. Because the debriefer can replay a recorded exercise, i.e., he can "time travel" and space travel, he is able to achieve the granularity and specificity he needs to identify tactical blunders, and to reinforce proper decision-making. AARs are done, by definition and by doctrine, after the action. More than a few clever commanders have tried to use the Flying Carpet to command during the action. Visual representation of the battlefield, they argue, affords them a decided edge for reasons outlined above. The doctrinal and correct response in training milieu like SIMNET sites is that the ambitious commander must return to his simulated Tactical Operational Center (TOC), or to his vehicle simulator, and be happy with the representation provided by "fielded" technology.

Reasoning at DARPA was clear and consistent with the ambitious training commander in August of 1991. Id est, if command needs a 3-D sand table; if command needs to travel in space and time; if command requires current and accurate representation of friendly and enemy forces on his 3-D map; and if field-rugged computer, communication, topological, and graphic technologies can support an "Electronic Sand Table," why not provide it? The technologies, although thought to be robust to the challenge, were thought to be dominated by a procedural or doctrinal one, however. The problems associated with getting timely force disposition into a "real time" electronic sand table had to be resolved, because invalid indications of enemy location and movement would have been less valuable than current sand table paradigms.

ODIN DEVELOPMENT

Three Installations

Three courses of development and field introduction for this project were managed. Each has succeeded in achieving fundamental changes in C2 planning and soon, training technology.

The first is being installed in the Commander's Briefing and Combat Assessment Centers CENTCOM HQ, MacDill AFB. The second is a completely self-sustained mobile ensemble, contained in a secure, 20-ft hardened truck. This mobile suite has been used extensively by the U.S. Army for evaluation and study by Aviation and Armor communities in support of a wide range of potential uses, including support to Combat Development, Training and Doctrine Development, and Plans and Operations. The third assembly is installed in secure facilities at the Army Topographic Engineering Center (TEC), FT Belvoir, Virginia.

Design Challenges

The early and principal design consideration was that of determining just who the user was going to be. Figure 1 is a notional portrayal of the design challenge. The combination of 2- and 3-D renderings of the battlefield seemed valuable; however, the value depended distinctly on the level of military seniority--and therefore the tactical or strategic responsibility--that was to be supported. At more junior levels, a comprehensive 2-D electronic map of an entire theatre would have been of little utility. Likewise, 3-D viewing of individual vehicles or even organizational symbols would have provided the Four Star little or no strategic advantages. The job was therefore one of selecting the service, and the level of command for which the sum of 2- and 3-D "travel" was maximized.

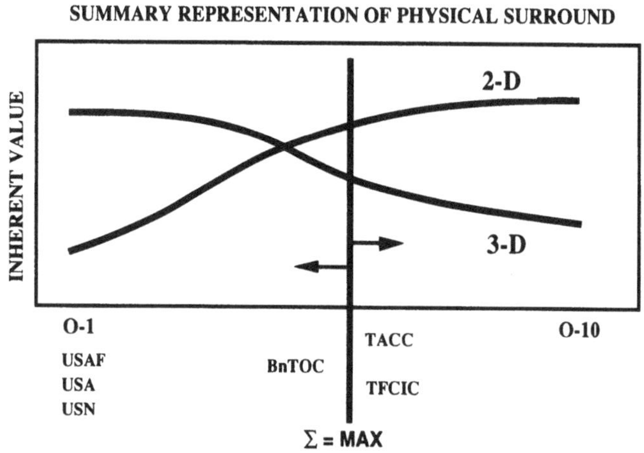

Figure 1. Notional Portrayal of Design Challenge

An additional constraint contributed to the market-selection process. Security requirements for an essential component technology (TACNAT, which will be described later) were such that the junior-most user of the complete Odin composite was Corps (assuming an Army application). The Air Force application was to support the Tactical Air Command Center (TACC). As will be described later, the application that was ultimately selected was US Army AH-64 Deep Operations Mission (and contingency) Planning.

THE COMPOSITE TECHNOLOGIES

The ensemble of technologies which were interwoven for Odin were, inter alia, RAPID TERRAIN GENERATION, TACNAT, FULCRUM, MACSAT, and FLYING CARPET. Each is described below. The component technologies were all successful, independently, prior to August, 1990. The risk associated with Project Odin was one of integrating known quantities to produce an operational suite. The electronic interfaces were intended to minimize or avert rework internal to the already proven components. The Odin informational flow was designed as indicated in Figure 2.

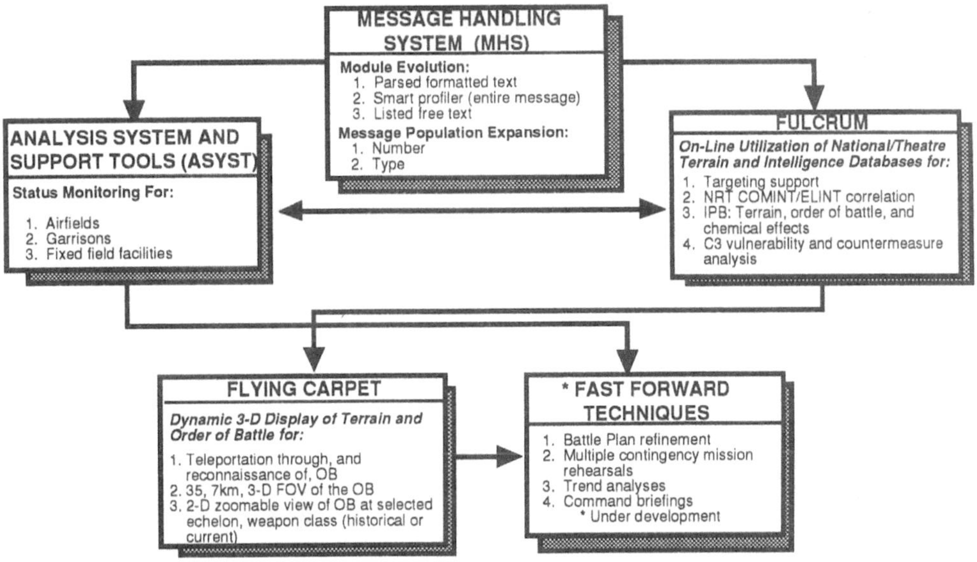

Figure 2. Odin Information Flow Architecture

Rapid Terrain Generation

The US Army Topographic Engineering Center (formerly the Engineering Topographic Laboratory), located at FT Belvoir, Virginia acted as the government agency for program execution. A fundamental function performed at TEC was the generation of the necessary terrain data base, and of the icons (visual images) endemic to the desert and its occupants (i.e., military vehicles, structures, etc.). Under direction from US CENTCOM science advisor, a 290 x 360 km terrain rectangle was specified. This terrain data base included (nearly) all of Kuwait, and significant portions of northern Saudi Arabia and southern Iraq. The data base became known as SAKI (for Saudi Arabia, Kuwait, and Iraq).

Under the brilliant direction of Mr. George Lukes at TEC, and the contractual support of Bolt Beranak and Newman, Inc., of Cambridge Massachussetts, what was feared to be a pacing variable--viz., terrain generation, became a pulling engineering variable. The TEC approach was to acquire Defense Mapping Agency data, SPOT Imagery, and any other information available in order to render the terrain and surface feature data base. Because of the aggressiveness and breadth of TEC methodology, Odin quickly became the most accurate and comprehensive representation of much of the desert theatre, and continued to be so throughout the campaign.

TACNAT

TACNAT was initiated in 1985 by Dr. Judith Daly, DARPA Program Manager. The system is an innovation which infers and apprises commanders about the status of enemy airfield activity or troop garrisons. The system monitors various message sources, and determines the intensity or vigor of specific enemy evolutions (like air sorties at an airport). If the metric (like number of sorties) exceeds a criterion set by an analyst, the system alerts the analyst. TACNAT provides a 2-D electronic map of the area of interest. Color codes are used to alert analysts. The analyst interacts with the system, and can inquire as deeply into the logic or data as is required in order to confirm the machine's inferences--as deep as the original source messages if necessary. TACNAT is even more valuable when coupled with FULCRUM.

FULCRUM

FULCRUM, also developed by Dr. Daly, is a video laser disc system which provides an interactive means of target development. The system is much more than an electronic "Jaynes," however. Terrain data bases are digitally stored in FULCRUM and can be analyzed for elevation, line-of-sight, masking potentials, precise distance calculations, and 2-D inspection of enemy force configurations. Because the FULCRUM data base is updated in the field with operationally required frequency, it is an extremely useful medium for command and control.

MACSAT

MACSAT was installed on the mobile Odin system to provide a means of communication from Continental U.S. or other locations to/from Odin, regardless of its location in theatre. The system is a 145 lb. SCOUT vehicle-launched satellite system. It operates via single channel as a long-haul store-and-forward data relay. The components assembled to support the ground Odin "subscriber" to this global message transfer/data base relay system are shown in Figure 3.

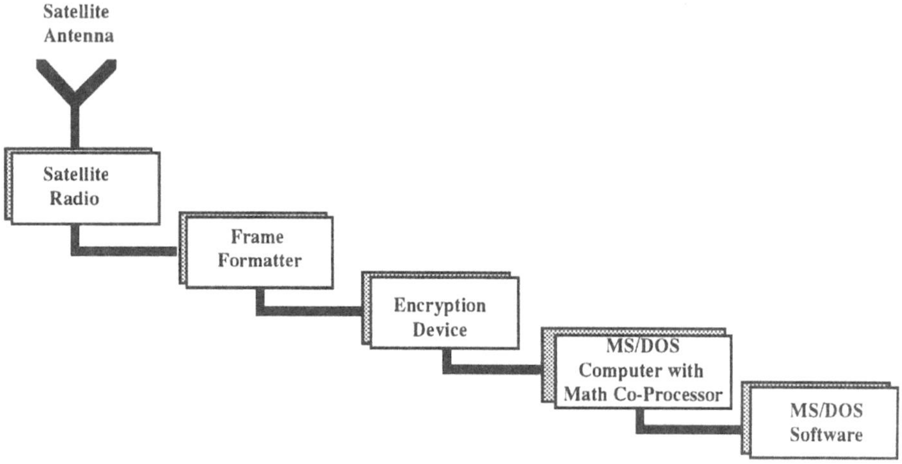

Figure 3. MACSAT Schematic

FLYING CARPET/Host Network

FLYING CARPET is a development of DARPA's SIMNET (Simulator Networking) program. Initiated in 1983 by Dr. Jack Thorpe, SIMNET effected a major paradigm shift in fundamental training technology. The SIMNET program implicitly held that the essence of a warfighting environment must be included if training is to succeed. The essence of warfighting is behavior--the behavior of many, many components on the battlefield. SIMNET succeeded in networking hundreds of such components (M-1 tank simulators, inter alia) and providing a capability for batallions of soldiers to practice and refine their warfighting skills.

As mentioned earlier, a key development of SIMNET derived technology is the FLYING CARPET. This sub-system can comprise any number of electronic components, although the function of FLYING CARPET remains as indicated above. For Odin, the terrain data base and force disposition are supported on a local area network of MIPS 3000 machines. Order of battle is interactively managed through an Order of Battle Generator (OBG) interface. The system is an object-oriented software environment. Terrain, architectural, and natural surface features are hosted on a hybrid data buffering Computer Image Generation scheme developed as a major technical/cost breakthrough specifically for the SIMNET program. The FLYING CARPET configuration is indicated in Figure 4.

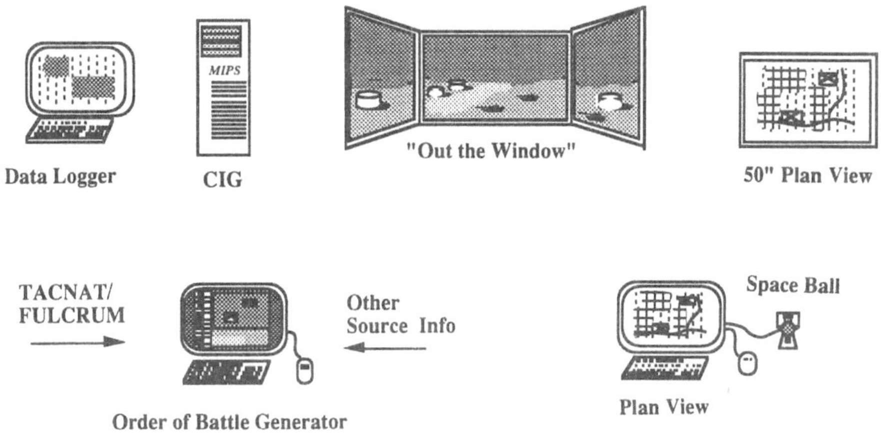

Figure 4. Flying Carpet Components

As mentioned in the previous paragraph, order of battle can be manipulated via the OBG. Most importantly, however, OB is also updatable via an electronic feed from the TACNAT/FULCRUM system. The configuration manager for Odin retains complete authority over the OB data base, and thus revises based on perceived validity of inputs, whether from human intelligence reports, or from a proposed TACNAT/FULCRUM update.

ODIN EVALUATION

DARPA engaged the Institute for Defense Analyses (IDA), an independent "Think Tank," to orchestrate a systematic evaluation of the Odin concept and its technology prior to transition to the customer. The evaluation plan derived by IDA included a field introduction and exercise at the UA Army Armor Center, FT Knox, Kentucky, and a follow-on evaluation at the US Army Aviation Center, FT Rucker, Alabama.

In addition to the field tests, Odin C2 capabilities were examined by US CENTCOM HQ officers. The Army Topographic Engineering Center was used as the test site for these evaluations.

Concept Evaluation--FT Knox

The mobile Odin system was deployed to the Armor Center at FT Knox in January 1991.

Four retired Army officers introduced the ensemble to combat developers and trainers at the Center. An exercise was developed and run in which the Odin gear was utilized to assist commanders as they conducted the mock evolution through the entirety of the Army decision cycle.

A detailed report was prepared on the basis of the FT Knox evaluation. In general, officers levied praise on the concept, and provided insightful suggestions for further refinement. Software immaturity led to system malfunctions on occasion, though the concept was summarily accepted by the Armor community as a significant addition to command and control. The request from the Armor Center to DARPA is for a permanent, residual Odin capability to support Intelligence Preparation of the Battlefield (IPB), situational monitoring, target development, collection management, Course of Action (COA) development, Logistics Sustainment Prediction (LSP), mission planning, mission rehearsal, communication of orders to subordinants, terrain analysis, and electronic sand table training.

Among the evaluators present at FT Knox were members of the Army aviation community. Perhaps the strongest suggestion provided by the wing-wearers was that Odin represented even greater potential to support Army Rotary Wing mission planning. This observation anticipated the follow-on implementation tests at FT Rucker.

<u>Implementation Evaluation--FT Rucker</u>

Mobile Odin was set up at the Aviation Center in Alabama in February 1991. The technology was arranged to support the development of mission plans for AH-64 Deep Operations (also known as Cross FLOT Operations). Deep Ops for the Apache is a very complex evolution, requiring precise timing of aircraft launch control, re-fueling, electronic coverage, etc. The Rucker tests utilized Odin to help plan routes, waypoints, contingencies, etc. The missions were then flown by helicopter pilots in the Apache simulators over the SAKI data base, modified and installed for this purpose. Results clearly supported the contention that Odin represented an ideal mission planning composite. The AH-64 environment was ideal with respect to the design constraints referred to earlier: (1) The milieu is one in which 2-D and 3-D representations of the battlefield are essential to combat planning, and (2), Deep Ops are planned at the Corps level, which meant that TACNAT could be exploited fully as a key component of Odin.

DEPLOYMENT TO DESERT STORM

The Operational Unit intended to accept Mobile Odin was identified by IDA. Logistical plans were refined, field kits, instructions and instructors were assembled, and deployment was prepared. Fortunately for mankind, but unfortunately for Odin, the Ground Campaign lasted fewer than 100 hours, and deployment was cancelled. The installation of the Headquarters suite met the same fate.

Based on requests from, primarily Army organizations too numerous to detail here, several courses were replanned for project Odin. Two strong training candidates are described briefly, below.

<u>National Training Center Scenario Capture</u>

The Army currently uses the National Training Center, FT Irwin, California as the graduation exercise for large combat organizations. Deployment to NTC is the culmination of months of garrison preparation wherein individual skills, crew and team skills, and mid-level command and control capabilities are finely honed. Deployment to NTC is "going to war" for soldiers. At NTC, entire Divisions are graded as they "fight" resident adversaries. The Division is tested for readiness--if unprepared, the entire Division will not be ordered to combat. The instrumented range on which the mock battles take place provides for the collection of vehicle position and weapon firing data, so that detailed debriefs can be supported after action. A current notion is to augment the Combat Instructional System "White Room" with the Odin technologies in order to broaden the debriefer's panorama of teaching tools. Details for injecting Odin technology are under consideration now.

Operational Marriage

A related plan calls for the "marriage" of Mobile Odin to an Operational Organization while it is in garrison at home station (FT Hood, Riley, for example), and to deploy with the organization when it deploys to NTC. The US Army Research Institute would analyze the apparent contribution of Odin C2 capabilities by comparing the advantaged organization's performance with a similar unit with only conventional planning media. The Army is currently identifying the appropriate organizations to support this evolution.

FUTURE CONSIDERATIONS

Several key features are being analyzed for future growth of Odin technology. These are described briefly, below.

Fast Future System

The Odin system currently provides 2-D, 3-D space travel, and historical time travel capabilities. A commander can "see" the battlefield of ten days ago, with whatever classes of weapons (armor, infantry, etc.), forces (enemy, friendly), or echelons (company to division) he desires. He can bring the battlefield forward and watch the configuration unfold over the past ten days and master the associated dynamics. A significant enhancement will enable the commander to interact with the machinery in order to explore the future. The order of battle will be ported into a war-gaming or newly-developed "fast-future" system, and with the infusion of a few indications of "commander's intent," the system will reveal tomorrow's potential battlefield, and what to look for if the enemy indeed tries to instantiate the predicted battlefield. The US Air Force Electronic Systems Directorate is helping determine the feasibility and utility of this enhancement.

Weather and Biological/Chemical Dispersion Modeling

Also under consideration are the addition of real-time weather feeds from various sources. The weather information is valuable to operational planning, and it is necessary in order to support the generation of physical dispersion models of hostile biological and chemical weapon effects. The Institute for Defense Analyses is examining this potential enhancement.

Commander's Intent--Remote Networking

There are two means identified which would support the "networking" of Odin into the field. One requires an ethernet land-line scheme for real-time networking. For non-real-time networking, an alternative motif is being studied. The latter would incorporate video laser disc television technology. Commander's intent would be organized at the Odin center, and down-loaded onto small laser discs and propogated throughout the command to soldiers down to very junior, or perhaps individual vehicle levels. Soldiers would be able to "space travel" and "time travel" through the virtual battlefield by interacting with a handheld television. This enhancement is under consideration currently, as well.

SECTION V
ADVANCES IN COGNITIVE THEORY

This section is the most theoretical of all the workshop sections. It sets the stage for the understanding of psychological aspects of technology and training. This section places much emphasis on the value and utility of cognitive theory and how it has affected tutoring models available for military training.

IMPLICATIONS OF ADVANCES IN COGNITIVE SCIENCE FOR INTELLIGENT TUTORING SYSTEMS

A. Samarapungavan, J. Beishuizen, F. T. M. Brazier,and A. F. Sanders

Department of Cognitive Psychology
Vrije Universiteit Amsterdam
Amsterdam
The Netherlands

SUMMARY

This paper reviews how recent developments in cognitive science can benefit those concerned with the design, development, and use of intelligent tutoring systems. The authors present an overview of the discipline of cognitive science describing its central concerns and methodology. In the following sections titled "Representing and Acquiring Knowledge" and "Knowledge Acquisition" theoretical and empirical advances in understanding the nature, acquisition, and use of knowledge in intelligent activity are described. The next section titled "Cognitive Science and Instruction" outlines some general principles for the design of instructional environments based on current theories of cognitive science and reviews some ITS systems in the light of the principles described. In conclusion, the authors discuss some strengths and weaknesses of existing ITS systems and highlight areas that require further research in order for progress to be made on these systems.

Introduction

Cognitive science may be described as the theoretical and empirical study of intelligent activity in all systems, human, nonhuman, and non-animate, that exhibit such activity. The central questions of cognitive science concern the origins, nature, and application of intelligent activity. Cognitive science may be characterized as a discipline in terms of certain core features which are as follows: a) Positing mental representations - the central idea is that the conduct of intelligent activity can best be described in terms of "mental representations" which are units of knowledge in the "mind" of the intelligent system. b) Metaphor of the mind as an intelligent computing system -there is an implicit assumption that all intelligent systems can be understood in terms of similar principles and that these principles can be formalized through the computational modelling of intelligent activity. Thus, computer simulations of intelligent "human" behavior are theoretical endeavors to formalize domain principles as well as practical or applied efforts to automatize components of information processing normally carried out by people. c) Interdisciplinary approach - Practitioners of cognitive science come from diverse fields such as philosophy, psychology, linguistics, computer science, anthropology, and neuroscience. The interdisciplinary nature of the field results from and facilitates a cross-fertilization of ideas across the above mentioned disciplines. This cross-fertilization includes the exchange of data, research

Advanced Technologies Applied to Training Design, Edited by
R.J. Seidel and P.R. Chatelier, Plenum Press, New York, 1993

methodology, and theoretical concepts among scientists from the different source disciplines that form a part of cognitive science. The perspectives of different source disciples are synthesized through the proposal that there are discrete though related levels of organization in terms of which a system may be described.

Cognitive science employs certain typical methodologies which include experimental methods drawn from traditional psychological research as well as newer methodologies that have evolved in attempts to address the central questions of cognitive science. These include protocol analysis, content analysis, computer simulation, meta-analysis, and logical analysis.

Representing and Acquiring Knowledge

The great contribution of cognitive science has been in providing a typology of different kinds of knowledge that are used in the conduct of intelligent activity. Important distinctions made between kinds of knowledge are as follows: (i) declarative knowledge - knowledge of facts typically embodied in statement of the form "I know that ....", (ii) procedural knowledge - knowledge of procedures as expressed in the ability to carry out the procedures and typically embodied in statements of the form - "I know how to ....", (iii) Content knowledge or knowledge specific to a certain subject-matter such as knowledge of Boyle's gas laws. (iv) Metacognitive or strategic knowledge which is knowledge about knowledge and thinking and therefore domain general.

One of the important findings of cognitive science has been with regard to the crucial role played by domain specific knowledge in the acquisition of new knowledge as well as in the conduct of problem solving activity. This domain specific knowledge has been described in cognitive science in terms of "knowledge representation." Knowledge representations (also called mental representations) are units of knowledge in the mind of the knower that can be characterized in terms of their structure, content, and interrelationships. Knowledge representations may be global and generic structures such as schemas and mental models, or they may be atomic such as "object concepts."

Knowledge Acquisition

One of the major concerns of cognitive science is the development of theories to elucidate the mechanisms by which conceptual change occurs as such mechanisms are the key to our understanding of learning. It has been proposed that there are several qualitatively different kinds of learning or conceptual change that occur. Accretion is learning by the gradual accumulation or addition of units of knowledge. In this kind of learning, the addition of new knowledge does not entail any reorganization of old knowledge or preexisting conceptual relations. Restructuring is learning where the acquisition of new knowledge entails major reorganization or replacement of existing conceptual relations or old knowledge. Examples of this may be seen often in the history of science as for example in the shift from the Ptolemaic to Copernican model in astronomy. The mechanisms whereby an individual must restructure knowledge are likely to be very different from those by which knowledge is simply added to an existing knowledge base. Accretion models of learning are based on processes such as discrimination and generalization of knowledge under the constraints of general principles of association and reinforcement. Restructuring models of learning are based on mechanisms that center on the creation of conceptual conflict and the resolution of this conflict. Such mechanisms include learning by anomaly or impasse drive learning. Additionally, mechanisms such as those involved in analogical learning can play a role in both accretion and restructuring. Restructuring models of conceptual change have most often been applied to declarative knowledge. Theories have also been developed to handle the learning of predominantly procedural knowledge of the kind required in computer programing. This type of learning is generally called "skill learning."

Cognitive Science And Instruction

Based on current theories of knowledge acquisition, some principles may be derived for the general design of instructional environments:

Domain analysis: Designers of instructional environments should begin with an analysis of knowledge in the to-be-taught domain. The different types of knowledge (i.e., declarative/procedural/strategic) required should be specified and the kinds of knowledge representations that the pupil is expected to acquire (e.g., rules/schemas/ mental models) should be determined.

Setting instructional goals: It is important to clearly establish the goals of instruction. Such goals are related to the domain analysis because they specify the kind and level of knowledge that a learner is expected to acquire following instruction. Where the instructional environment is designed to support the development of successive levels of expertise from low to high, care should be taken that what a pupil learns at the introductory levels provides an adequate basis for the building of advanced knowledge and skills at subsequent levels.

Diagnosing learner characteristics: As mentioned above, it has been shown that a pupil's prior knowledge and skills significantly determine his or her subsequent learning in a given domain. Therefore, diagnosing the prior knowledge and skills that a learner brings into the learning situation is crucial to the design of successful instructional environments.

Using principles of situated learning and cognitive apprenticeship: Recently, many researchers have noted that contemporary formal instruction in academic contexts results in the divorce of knowledge and skills from their applications in real world contexts. In a move to develop better instructional environments many researchers have called for methods of instruction that link knowledge and skills to their full range of potential uses by drawing upon traditional apprenticeship methods of learning a trade or craft. Important instructional techniques in apprenticeship training involve modelling of skills by an expert tutor, coaching, and the opportunity to learn skills in a realistic setting. Also important is the sequencing of feedback in coaching starting with a high degree of external support or scaffolding for beginning learners and moving to a gradual fading of support as learning progresses.

Evaluating instructional effectiveness: A program of evaluation to determine whether the instructional environment is meeting its objectives and to compare the performance of the environment against others designed to teach the same material should be undertaken. Factors to consider in addition to success in achieving the desired cognitive outcomes are: (i) the cost/effectiveness ratio of the instruction relative to other available instructions and, (ii) the effect of instruction on student motivation.

The Design of Intelligent Learning Environments and Tutoring Systems. The designers of Intelligent Tutoring Systems or ITS are constrained by specific considerations that have to do with problems of implementing the above principles not through a human instructor but through a computer tutor. ITS are computer programs which carry out educational tasks. The tasks carried out by the ITS require "intelligent activity" on the part of the system. Three components of an ITS are crucial in terms of the way in which its instructional environment will embody the general principles discussed above. These components are the expert component, the student component, and the tutor component.

The expert component: The expert component of an ITS contains a representation of all knowledge and skills necessary for the student in order to learn the target domain. In other words, it possesses the expert knowledge that the student will acquire from it. The accuracy and completeness of the expert component of an ITS is crucial to the success of instruction. Therefore the expert component must be designed based on domain analysis using standard cognitive science methods such as those of content analysis and protocol analysis of expert performance in the target domain. One problem that designers of ITS face is that for many domains it is currently difficult to specify the types and levels of knowledge required for successful performance. Especially difficult is the specification of strategic knowledge and the degree to which it should be considered domain independent. Even where the strategic or metacognitive knowledge can be theoretically specified, it is difficult to implement in the expert component of the ITS. This can be seen from the poor quality of most ITS explanation facilities which unlike human tutors can rarely explain or justify the path by which the system arrived at certain decisions or sequences of action. A second problem for designers of ITS sysytems is the potential instability of knowledge in some domains. Major revisions in current domain knowledge usually require that the expert component be redesigned, a process both time-consuming and expensive.

The student component: The task of the student component of an ITS is to maintain and update a model of the student's acquired knowledge and skills. This information is crucial as it will determine the choice of subsequent instructional strategies as well as the further sequencing of information presentation. Most ITS systems model student performance or learning in the very narrow context of the specific skill being learned. Thus for example, if a student makes a mistake in long division, the system can diagnose the error and provide feedback and additional practice. This kind of facility makes ITS very useful in providing individualized drill and practice instructions. However, unlike most human teachers, the student component of the ITS does not contain a rich composite picture of a student's general level of domain understanding as well as of other relevant factors such as motivation, and anxiety, or fear of failure. One important consequence of this is the inability of an ITS to draw out and relate a student's prior experiences or knowledge to the new material being taught.

The tutor component: The task of the tutor component of an ITS is to monitor the type and content of instruction and information presented in the learning environment. The tutor component must contain knowledge of when to react, to what, and how, requiring the coordination of domain knowledge and knowledge of the student with knowledge of instructional strategies. Several unresolved research issues with regard to optimal instructional strategy pose problems for the design of the tutor component. One such issue has to do with the provision, timing, and frequency of corrective feedback to the learner. Some researchers favor immediate, automatic corrective feedback to the learner when mistakes are diagnosed while others prefer systems where feedback is provided only on request from learners. Different strategies appear to be appropriate in different learning situations. If the pupil must acquire skills of self-diagnosis and correction as part of domain expertise or if the pupil must acquire the skills involved in exploration or discovery, it may be useful to delay the provision of corrective feedback rather than to provide it immediately. On the other hand, in skill learning of the kind involved in learning to program in LISP, immediate corrective feedback appears to enhance the acquisition of the skill considerably. One problem for the tutor component of the ITS is the lack of comprehensive knowledge about the student's knowledge, abilities, and plans. It is on the basis of such knowledge that a human tutor can make effective choices among alternate instructional strategies.

Conclusions

Based on recent advances in cognitive science several ITS systems (e. g., STEAMER, Sherlock, and QUEST) have been developed. A common domain of application has been the teaching of troubleshooting skills. Such ITS systems are valuable in providing instruction individualized to each pupil's current level of skill or knowledge and serve as convenient adjuncts to conventional instruction allowing for self-study in intelligent databases. Of particular utility, is the ability of the ITS to provide students with controlled discovery environments or "microworlds" in which the laws underlying domain phenomena, such as the laws of motion in physics, can be simulated. Such simulations, contained in the expert component of an ITS, simplify and make salient the theoretically important dimensions of domain phenomena in ways that are difficult to achieve in traditional classroom contexts. This feature of an ITS facilitates situated learning. However, current ITs systems also suffer from several limitations and constraints as discussed above. Therefore, more research is needed to determine the scope and effectiveness of current and future applications.

INTRODUCTION

This paper aims to provide a review of how recent advances in cognitive science can benefit those concerned with the design, development, and use of intelligent tutoring systems. In particular, we will focus on the implications of our current knowledge for the general development of effective environments for learning and instruction. As a starting point, it seems appropriate to broadly define the domain of study delimited by the term cognitive science. We will do this by presenting some of the central questions that motivate cognitive scientists and illustrating some basic methodological and conceptual aspects of their approach to answering these central questions.

The term "cognitive science" first came into use in the early seventies and several researchers (e.g. Bruner, 1983; Collins, 1977, Miller, 1979; Rumelhart, 1982) have since attempted to define the nature and range of this emergent field of study. In this paper we will adopt Gardner's (1987) definition of cognitive science as:

" a contemporary empirically based effort to answer long standing epistemological questions particularly those concerned with the nature of knowledge, its components, its sources, its development and its deployment." p. 6

The scope of cognitive science includes the study of all forms of knowledge, human, nonhuman, and non-animate. Cognitive scientists are interested in understanding all forms of intelligent activity whether they occur in people, other living organisms, or even nonliving systems such as computers. However, in this paper we will focus on the contribution of cognitive science toward understanding human cognition, drawing attention where appropriate to how the study of non-human intelligence has influenced our thinking about issues of human cognition.

Central questions of cognitive science

Cognitive science seeks to understand the relationship between what is known or the entities of the external world and the knower who cognizes these entities. Some of the concerns of the discipline thus correspond directly to both philosophical investigations in epistemology and to classical psychological research. For example, cognitive scientists are interested in the internal mechanisms or processes by which the individual comes to know or understand the external world such as the processes of perception, learning, memory, and reasoning. They are particularly interested in how knowledge is stored in the mind of the knower and how this knowledge is subsequently used in the conduct of intelligent activity.

Core features of cognitive science

1. **Positing mental representation**: Central to cognitive science approaches is the belief that in order to understand how people think, learn, and remember it is necessary to posit "mental representations." Mental representations may be conceived of as knowledge units inside the knower's mind. These units of knowledge can be characterized in terms of their content, structure, and relationship to other representations or units of knowledge. Thus, there is an essential level of analysis that is related to but distinct from purely biological/neurological levels of analysis and purely sociological or cultural levels of analysis. The necessity of positing mental representations becomes clear when one considers the metaphor of the computer as a model of mind. The capacity of an intelligent computing system to exhibit plasticity of behavior depends on its ability to interpret and manipulate internal symbols that designate external states.

2. **Metaphor of the mind as an intelligent computing system**: The metaphor of the mind as computer has been a core aspect of cognitive science since its inception. The metaphor highlights the implicit assumption of cognitive science that all intelligent systems living and non living can most likely be understood in terms of similar principles. The roots of this notion go back to Turing. Alan Turing (1937) demonstrated that a logical device called a "universal machine" could in principle simulate the activity of all intelligent

systems as long as such activity was describable in its formalisms. The metaphor of the mind as computer has impacted both on the development of human theories of cognition and on the development of intelligent electronic systems that are designed to model various aspects of intelligent human activity. On the one hand the problems encountered as researchers tried to develop intelligent computing systems helped elucidate the kinds of knowledge and reasoning that must occur in human beings as they routinely performed such activities. Examples can be seen in failure of the early attempts to create computers that could process natural language and perform automatic translation which lead to great advances in our knowledge of how people use natural language. On the other hand our understanding of how human beings use natural language has in turn furthered the development of intelligent computer systems.

3. **Interdisciplinary approach**: A third feature of cognitive science is its interdisciplinary nature. Its practitioners come from the fields of philosophy, psychology, linguistics, computer science, anthropology, and neuroscience. The goal of cognitive science is to create a synthesis of the different perspectives on intelligent behavior that these different disciplines bring.

One way of creating such a synthesis is for cognitive scientists to assume that an intelligent system has several discrete levels of organization which comprise its "cognitive architecture" (Posner, 1989; Newell, Rosenbloom & Laird, 1989). Some typical distinctions made between levels of organization in the cognitive architecture are given below:

(i) One level of organization is the semantic or knowledge level organization which embodies the content or meaning of what people know.

(ii) A second level of organization is the symbolic or syntactic level which embodies the codes in terms of which knowledge is expressed and the rules by which such codes are manipulated. The analogy can be made here with levels of organization for natural language. The "meaning" component of a natural language would correspond to a semantic level of organization. The grammar of a language along with the codes for representing meaning such as auditory and visual symbol strings would correspond to the symbolic or syntactic level of organization.

(iii) A third level of organization in the cognitive architecture of an intelligent system is the physical or biological level. This level of organization is the primary level of analysis for cognitive scientists interested in the biological or neurological basis of intelligent behavior as well as for computer scientists concerned with the hardware used for the development of artificial intelligence systems. It is of less interest to cognitive psychologists or linguists who study primarily functional aspects of intelligent activity.

Methodologies in cognitive science

Reflecting in part the interdisciplinary nature of the field, cognitive scientists can choose from several eclectic research methodologies that range from computer simulation, to recording electrical activity in the brain, to naturalistic observation. Many of the methods used are a fairly direct incorporation of methods used in the source disciplines from which cognitive science emerged. For example, the use of classical linguistic analyses as developed in the field of linguistics are also part of the standard methods of contemporary cognitive science. Likewise, traditional **experimental methods** from psychology used in research on psychophysical judgements, memory, decision making, and concept acquisition are incorporated in the general body of cognitive science methodologies. Experimental methods are used in exploratory research, the results of which guide subsequent theory construction, as well as in the testing of hypotheses derived from an articulated theory. Exploratory research often employs correlational techniques to explore potential relationships among theoretically interesting variables. Correlational analyses are also important when variables of theoretical interest are not amenable to experimenter manipulation and control such as sex, socio-economic status, and cultural background of subjects. Experiments are also designed to test specific hypotheses that derive from a given theory. Where at least some of the variables studied can be controlled and manipulated by the experimenter, analysis of variance and regression techniques are often used to analyze results and test hypotheses. There are some important new methodologies that have become associated with the practice of cognitive science. These methods are protocol analysis, content analysis, artificial intelligence simulations, meta-analysis, and logical analysis. These will be reviewed below.

Protocol analysis. Protocol analysis uses the verbal reports of subjects as data and has become a very important tool in the study of human intelligence. As Simon and Kaplan (1989) say, "... it is one of the few methods in cognitive science that gathers data with sufficient temporal density to test models that account for behavior nearly second by second." Protocol analysis differs from introspectionist use of verbal report in many important ways. The introspectionist psychologists treated subjects' self-reports about their cognitive performance as valid theories of cognition. However, protocol analysis treats the verbal reports of subjects not as theories about thinking in and of themselves, but simply as verbal data to be analyzed in terms of an experimenter generated theory. Sometimes, this experimenter generated theory may be modelled as a computer simulation of some intelligent behavior.

Protocols can be obtained from subjects either by asking them to "think aloud" about something they are doing or have done or by asking them a series of directed questions. The "think aloud" method is less intrusive and there is less chance of biasing the subject than in systematic questioning. However, directed questioning has the advantage of getting subjects' verbal reports about some aspect of the task that they might omit in their spontaneous report but that is important from a theoretical viewpoint. **Concurrent protocols** are obtained from subjects while they are engaged in task performance. Most researchers use a conjunction of concurrent and retrospective reports to obtain complete data about performance. The verbal reports of subjects as they perform the task are first taped and then transcribed and coded on the basis of prior theoretical analysis. This coded data is then analyzed from a theoretical perspective. **Retrospective protocols** are obtained from subjects after the performance of the task has been completed. The term **log analysis** is used when the process of obtaining concurrent protocols from subjects as they interact with a system and the process of analyzing these protocols is automatized.

Content analysis. Protocol analysis usually provides data about human cognitive performance over a relatively short period of time. Usually the collection of data contained in a single protocol can take from a few minutes to a few hours. Content analysis on the other hand allows the researcher to examine cognitive behavior over extended periods of time such as months or years. As the name implies, content analysis comprises of the theoretical analysis of written text materials (Stone et al, 1966; Weber, 1985) such as articles or laboratory notes.

There have been various applications of content analysis. For example, Kulkarni and Simon, (1988) developed a theory of scientific discovery processes which they implemented as a computer program and tested against a historic record of an actual scientific discovery process as recorded in a series of laboratory logs.

Content analysis has also been used to understand how people perform complex cognitive tasks such as those involved in reading comprehension. An example of this can be seen in the work of Van Dijk (1980) and Van Dijk and Kintsch (1983). The starting point in content analysis is to decide upon the units in terms of which one wants to code or analyze a text. These units can be fairly small such as a word or a sentence, or they can be large such as a paragraph. Further, one can analyze various aspects of text (Mandl, Stein, & Trabasso, 1984; Meyer, 1979). For example, in the case of reading comprehension, a text may be analyzed in terms of syntactical aspects, semantic aspects such as meaning of words and sentences, rhetorical aspects such as the manipulation of text structure to achieve different kinds of rhetorical effects, or at the level of "idea units" expressed in the text ranging from main ideas to subordinate arguments and ideas. This kind of content analysis allows a researcher to specify what kinds of cognitive operations a person must possess in order to successfully comprehend a written text.

Artificial intelligence methods and computer simulation. Artificial intelligence broadly defined is the study of how computers can be programmed to perform activities that in human beings would be regarded as "intelligent." Examples of such activities are solving arithmetic problems, summarizing the main ideas of a piece of text, and answering questions from a database. When computers are programmed with the intent that they simulate human cognitive processing or that they perform an intelligent activity in the same way that people perform it, the simulation becomes an implementation or formalization of a theory of human cognition. In such cases the implication is that the computer simulation will perform successfully and unsuccessfully in the same conditions that human beings would. Formalizing a cognitive theory by way of computer simulation thus enhances the ability of the researcher to make testable predictions about human behavior.

An example of the applications outlined above is a computer simulation called READER (Just & Carpenter, 1987). READER is a production system that models certain processes supposed to occur in reading and comprehending text. The system can model certain real time aspects of human performance in the domain of reading. It takes more time to process complex words in the same way as people do. The system constructs an internal representation of the text as it reads and can answer text related questions or summarize textual information on the basis of this internal representation.

The READER system was build as an implementation of Just and Carpenter's theory of reading processes. For example, the researchers could compare the time taken for processing each word in the text (indicated by duration of a gaze on a given word as measured by tracking eye movements in reading) by human beings and READER. This was accomplished by comparing the number of processing cycles that occurred in READER per unit of text with the time taken by human subjects to read that unit. The system also revealed differences in the way it and human subjects processed information in reading. READER encoded each word in the text as it read while people tended to skip as many as 60% of short function words while reading text. The researchers speculated that this difference arose from the ability of people to encode short words in peripheral vision while they fixated on the words preceding them. This lead to a further revision of the READER system.

One advantage of computer simulation is that it can reveal hidden inconsistencies or fuzziness in cognitive theories that are verbally expressed. Certain constructs may prove impossible to operationalize within the simulation and may need to be refined. However, many complex cognitive activities that people are capable of cannot as yet be implemented as computer simulations. This is partly because we still do not have completely specified cognitive theories for many complex human activities. Another reason is that such complex activities often require vast amounts of highly structured knowledge of the world which cannot as yet be technically implemented in computer simulations.

Meta-analysis. Meta-analysis is a quantitative reviewing procedure that allows one to synthesize the results of groups of studies that investigate a certain phenomenon (Kaplan, 1987; Wolf, 1987). The procedure involves the formulation of basic questions pertaining to the phenomena of interest. The studies reviewed are then coded on the basis of how they answer the fundamental questions posed. Statistical procedures for combining the results of different experiments addressing the same question are used to draw general conclusions across a variety of studies.

Logical analysis. Philosophical and logical analyses with regard to questions of an epistemological nature have had a profound influence on cognitive science at certain stages of its development. For example, modern symbolic logic underlies the very foundation of artificial intelligence as a branch of cognitive science. Important contributions in this regard have been those of Turing (1937) and Godel (1962).

A second meeting point for logical/philosophical analysis and cognitive science has been in modelling human rationality under conditions of uncertainty (Kahneman, Slovic, & Tversky, 1982). The interesting development in this area has been the demonstration that people in general do not perform in accordance with normative statistical and logical models of rational judgement under probabalistic or uncertain conditions.

A third meeting point has been the study of notions of causality and of mechanisms or processes of scientific discovery in the philosophy of science and in cognitive science. In general, logical and philosophical analyses of basic epistemological questions have often served as a source of hypotheses for cognitive science and stimulated the development of theoretical discussion.

The section above described some key features of cognitive science as an emergent discipline. Our purpose was to provide a disciplinary framework in the context of which we could discuss recent advances in the field and their implications for the science of instruction. In the following chapter, we will discuss current views of knowledge representation and knowledge acquisition.

REPRESENTING AND ACQUIRING KNOWLEDGE

One of the great advances of cognitive science has been in providing a typology of knowledge. It is clear from current research that in order to understand common psychological phenomena of perception, memory, comprehension, and reasoning one must posit a number of different types of knowledge. In this chapter we will discuss some of the fundamental distinctions that have been made between categories of knowledge. We will also discuss how these types of knowledge might be represented, what processes might operate upon them, and how they might be acquired or learned.

Types of Knowledge

The declarative/procedural distinction

Many researchers (Anderson, 1983; Bransford, Franks, Vye, & Sherwood, 1989; Winograd, 1975) have made the distinction between declarative and procedural knowledge.

Declarative knowledge may be conceived of as a static collection of stored facts or declarations. Declarative knowledge is typically expressed in statements of "knowing that." An example of declarative knowledge is the statement, "I know that one can compose, store, and edit documents on a personal computer."

Procedural knowledge is dynamic and can be regarded as a collection of actions or procedures that an intelligent system can carry out. Procedural knowledge is typically expressed in statements of "knowing how." The procedural knowledge that corresponds to the declarative knowledge that computers can be used for word processing is the ability to do word processing on computers. Procedural knowledge need not involve motor action and can be purely cognitive in nature. The ability to mentally multiply two digit numbers is procedural knowledge that is purely cognitive.

The current consensus is that most complex cognitive tasks require both declarative and procedural knowledge. Indeed many forms of mental representation posited by researchers such as "schemata", (Rumelhart, 1980) "scripts", (Schank & Abelson, 1977) and "mental models" (Johnson-Laird, 1980; Norman, 1983; deKleer and Brown, 1983) combine procedural and declarative information within a single representational structure. In general, there has been a growing trend to emphasize the importance of linking declarative knowledge to procedural knowledge or skills under varied conditions of use. Examples of this can be seen in the recent work of Bransford, Sherwood, Vye, and Reiser (1986) and in the work of Collins, Brown, and Newman (in press).

The main point of recent research has been to show that learning subject matter such as the laws of physics or units of political geography in declarative mode is insufficient for successful performance in the domain being learned. Learning only the factual or declarative aspects of domain knowledge typically results in the problem of inert knowledge or knowledge that the pupil cannot subsequently apply to solving problems in the domain. What is required is to proceduralize the declarative knowledge or to teach pupils the procedures by which the facts they have learned can be applied in a variety of problem solving contexts. This aspect of current theories of mental representation will be revisited later when we discuss the knowledge-based or domain specific approach to understanding cognition.

Strategic or metacognitive knowledge

In addition to the declarative-procedural distinction, many researchers (Brown & Kane, 1988; Nickerson, Perkins & Smith, 1985; Schoenfeld, 1982) distinguish between content knowledge or domain specific knowledge and strategic or metacognitive knowledge. Strategic or metacognitive knowledge is knowledge about general priniciples of thinking and reasoning that is applicable across specific content domains. Historically, theories of cognition have tended to emphasize the importance of strategic and domain-independent knowledge. In contrast, the early proponents of the contemporary knowledge-based view have tended to de-emphasize the importance of strategic knowledge.

In the following section we will discuss the issue of whether the critical knowledge and cognitive skills needed for intelligent functioning are domain specific and context-bound or general and applicable across a variety of contexts.

Intelligent performance: Domain specific knowledge or general reasoning skills?

In the 1950's and 1960's psychologists tended to believe that successful intellectual performance, such as good problem solving, was a manifestation of general reasoning strategies supported by a high level of "general" ability or intelligence (The "g" of intelligence testing). It was assumed that people who possessed such general cognitive skills or reasoning strategies could apply these successfully to any "database" or any specific domain of information, be it chess or military strategy.

There were several sources of support for this approach to understanding cognition. For example, the mathematician Polya (1954, 1957) argued that successful problem solving in mathematics depended not on the formalities of mathematical proof and derivation (i.e., domain specific knowledge) but on the application of heuristics or general strategies for problem solving. Such heuristics were to break the problem into sub-problems, solve simpler problems that were related to the main problem, use diagrams to explore alternate representations of a given problem etc. Many of these heuristics were clearly applicable to problem solving in domains other than mathematics.

Early artificial intelligence programs such as Newell and Simon's (1972) "General Problem Solver" which used a general heuristic called means-end analysis again reflected the belief that the key to modelling successful intellectual performance was in identifying general strategies that could be applied to any base of knowledge. Many general skills and strategies for problem solving, creative reasoning, decision making, learning, and efficient mental management were identified. In contrast there was little interest in the role that domain specific knowledge such as knowledge about the rules of chess or knowledge of particular axioms of mathematics played in cognition.

In the late sixties and early seventies however, studies of expert performance in a variety of domains led researchers to question the assumption that general reasoning strategies rather than domain specific knowledge were central to understanding cognition. The studies of de Groot (1965) and Chase and Simon (1973) of memory for chess configurations in novice and expert chess players showed that the advantage that the experts had over the novices was in their domain specific knowledge. Expert chess players remembered as many as 50,000 chess configurations that emerged in actual games of chess played while novices had relatively few representations of chess configurations or "chess schemata" stored in their memory. There were no differences between the novices and experts on general cognitive skills such as the ability to recall random layouts of chess pieces which did not corresponded to likely configurations in real games. However, the experts were much better than novices at recalling chess configurations that were drawn from actual games of chess played.

A number of studies of expertise in domains such as physics, biomedical diagnosis etc. (Glaser, 1984) demonstrated that expert performance depended on the possession of knowledge about a vast number of domain specific patterns of information (eg. typical chess configurations, typical use of conservation laws in physics). It also rested upon the ability of the expert to quickly recognize situations where such patterns, often called "schemata", applied. Further, experts showed a pattern of "forward reasoning" where they moved from the recognition that a pattern applies to a problem situation toward a solution of the problem by working with the pattern. In contrast novices who lacked the knowledge of such domain organizing patterns or could not recognize them rapidly were forced to use backward reasoning or going from the unknowns to try and find rules to get to the givens. In all the studies of expert and novice differences, general heuristics did not seem to play a very important role in explaining differences in expert and novice performance.

Simultaneous to studies of expertise in people came the recognition from within the artificial intelligence community (Feigenbaum & McCorduck, 1983) that in order to enable computers to solve real world problems rather than "toy problems" such as the Tower of Hanoi" problem, it was necessary to build "expert systems" into the computer rather than mere general heuristics of the kind embodied in Newell & Simon's General Problem Solver. An example of the new approach to the building of expert systems was Feigenbaum's DENDRAL. The system contained large amounts of domain specific knowledge such as knowledge about organic compounds.

Finally, a variety of studies have failed to show the transfer of general cognitive skills from the contexts in which they were originally learned to other applicable contexts. As early as in 1901, Thorndike and Woodworth reported experiments showing that training in Latin and mathematics did not improve performance on other cognitive functions contrary to the prevalent belief that they trained general mental skills. More recently, it has been shown, (Hayes & Simon, 1977) that training on one type of logical problem did not improve the performance of subjects on solving isomorphic versions of that problem presented in a different way. Nor did teaching children general cognitive strategies improve their performance in domains other than the ones in the context of which such strategies were originally taught. These general developments have lead to a novel approach to issues of conceptual development such as the acquisition and representation of knowledge. Broadly called the knowledge-based approach, (Feldman, 1980) it assumes that knowledge is modular, or that mental representations and the cognitive processes that operate upon them differ across different content domains such as those of mathematics, language, physics and history. Although, the knowledge-based approach emphasizes the importance of domain specific knowledge, there has been a recent interest in exploring how domain specific knowledge and strategic or domain-independent knowledge interact to enable cross-domain application or transfer of knowledge. An example of this attempt to specify different types of knowledge and their interaction is the FERMI expert system developed by Larkin, Reif, Carbonell, and Gugliotta (1988).

Domain Specific or Knowledge-based Approaches in Cognitive Science

Knowledge-based approaches assume that human cognition can be best explained by examining the content of knowledge representations in specific domains rather than the form of representations and of the reasoning procedures that operate upon them (Johnson-Laird, 1983; Feldman, 1980; Wason & Johnson-Laird, 1972). This approach has been applied to cognitive activities as diverse as reading comprehension (Schank and Abelson, 1977) and physics problem solving (McCloskey & Kargon, 1988; Wiser, 1988). One consequence of the knowledge based approach is that there has been a shift in interest from studying logical faculties of the mind such as the formal operations posited by Piaget (1930) to studying the structure and content of people's "world knowledge" or their mental representations of the external world.

Several different kinds of mental representations have been posited in the cognitive science literature and they will be reviewed below. A fundamental distiction we will draw is between levels of representation. Our claim is that there are at least two levels at which representational structures are posited. The first is a global or molar level and the second is an atomic level.

Knowledge representation

Global and generic representations. One of the important findings of cognitive science has been that in order to carry out everyday cognitive activities such as devising plans of action, comprehending conversations, and navigating through space, human beings process their environment in terms of generic and global knowledge structures that involve the coordination of several different concepts such as concepts of space, time, human intentionality, and physical causation. These global and generic knowledge representations, often referred to as schemata (Rumelhart & Ortony, 1977) or mental models (Johnson-Laird, 1983; Norman, 1983) provide people with informal theories of classes of events in the world such as "dining at fancy restaurants" or "terrorist hijacking." These global and generic knowledge structures contain variables that can be bound to different aspects of the environment or to different "instances" but the relationships between variables specified in the schemata are constant across contexts. To illustrate how such representations function, we will provide an example from Rumelhart (1980).

Readers in the US were given the following fragment of text to read:

Business had been slow since the oil crisis. Nobody seemed to want anything really elegant anymore. Suddenly the door opened and a well dressed man entered the showroom floor. John put on his friendliest and most sincere expression and walked towards the man.

Most readers generated an interpretation of this text wherein "John" was a car salesman who was currently unable to sell many cars because of the oil crisis. He probably sold rather large "elegant" cars like Cadillacs. John regarded the well dressed man who entered the show room as a potential customer and tried to make a good impression on him in the hope of making a sale.

Rumelhart suggests that in arriving at such an interpretaton of the text, people must use their generic knowledge or their schemata about buying and selling. The first sentence is likely to activate such generic knowledge. Included in the schema are variables of a buyer, seller, sale object, and conditions of sale that are related in systematic ways. For example, people might use their generic knowledge about conditions of sale which indicates that people are less likely to buy a product when the cost of using it goes up. They could then infer that since the oil crisis, increases in the price of gasoline had decreased the sale of automobiles, particularly of large automobiles. The word "showroom" was also likely to make subjects assign the value of "automobile" or "large automobile" to the variable of sale object in the buying and selling schema.

People also used their generic knowledge about human intentionality and motivation to interpret the last sentence in the fragment. For example, they might have activated their generic knowledge that in order to persuade other people to do what you want, you must make a good impression on them or get them to trust you. This generic knowledge is combined with the generic knowledge that the motive or intention of a "seller" is to persuade a "buyer" to purchase the "sale object" from him or her.

The crucial thing to note in this example is that there is no explicit information in the text itself that assigns specific values to the variables of "buyer", "seller", "sale object", or "conditions of sale" from the entities contained in the text. The text does not explicitly state that John is a salesman, or that he sells automobiles, or that he is trying to sell a car to a man he regards as a possible client. The fact that people can routinely comprehend such text indicates that they do use generic or schematic knowledge to interpret phenomena.

The early research on molar knowledge representations tended to focus on peoples' knowledge of routine interpersonal situations or events. Recently, researchers have investigated peoples' molar and generic knowledge representations of physical systems (White & Horowitz, 1987; deKleer & Brown, 1983) such as electrical circuits. Research in this domain often refers to peoples' mental models of physical sytems as "qualitative" or "causal models."

Atomic representations. Global and generic mental representations of the kind discussed above have been useful in modelling people's knowledge of molar or "large" phenomena" such as "taking a vacation abroad" or "understanding how a chemical factory functions." However, there are other kinds of knowledge that are modelled by smaller "atomic" representations such as object concepts, propositions, or rules.

For example, our knowledge of "natural kinds" such as "dog", "tulip", or "tree" as well as of "objects" such as "chair" can be modelled in terms of a representational structure traditionally called a "concept" (Rosch, 1975; Smith and Medin, 1981). Concepts of the kind just discussed, are believed to be abstractions of the entities they represent. The abstractions contain both core or definitional features as well as non-definitional but characteristic features (i.e., features frequently though not necessarily present in exemplars of the concept). Thus, a person's concept of "mammal" might include in its representation the definitional features that all mammals give birth to their young directly and suckle their young. However, also included in the concept of mammal might be such non-necessary but characteristic features as "lives on land." Other atomic representational structures are rules or propositions that have been used to represent knowledge of arithmetic (Siegler and Richards, 1983) and of the working of a physical apparatus such as a balance beam (Siegler & Klahr, 1982).

Qualitative Aspects of Knowledge Representation: Issues of Cohesiveness, Internal Consistency, and Generativity

The theory-like nature of knowledge representations

There has been a growing trend in several different streams of research such as developmental psychology, science education, and cognitive science to focus on theory-like properties of people's mental representations. When referring to the cognition of lay people or non-experts, such mental representations have been called "naive theories" (McCloskey, 1983).

There are several different kinds of mental representations that have been discussed within the "naive theory" framework ranging from such molar mental representations as schemata, scripts, and mental models to fairly atomic representations such as object concepts, rules, and propositional representations.

The common theme that underlies the application of the term "theory" across this fairly disparate set of representational constructs appears to be that:

(a) Knowledge representations do not merely describe common categories of experience but also explain them.

(b) Knowledge representations have predictive power and allow people to generate and test expectations about novel circumstances.

(c) Individual elements of our overall conceptual structure such as concepts, rules, schemata, and mental models interrelate and mutually constrain each other as they participate in more global organizations and reorganizations of knowledge.

Schemata/scripts/mental models etc. Elements of the view of mental representation outlined above are implicit in much of the work on schemata, (Rumelhart, 1980) mental models (Johnson-Laird, 1983), and similar classes of knowledge representations that have been proposed in the domain of psychology and artificial intelligence. These structures are molar and represent global knowledge packets within which many individual concepts are interrelated. There has been considerable emphasis on the theory-like character of schemata and the similarity between theory change and schema restructuring in the course of knowledge acquisition (Rumelhart & Norman, 1981). An example of this position can be seen in the following quotation from Schank (1986):

> People are, in essence, theory creators. The theories may not be very elaborate or especially scientific. People create theories about how certain events will turn out and if they are wrong, they attempt to modify those theories by incorporating the new data into the old theory. (p. 32)

Concepts. Traditional research on conceptual structure has focused on atomic representational structures such as "object concepts" or "natural kind" concepts (Rosch, 1975; Smith & Medin, 1981). Recently however, there has been interest in how peoples' generic knowledge about the world and their "domain organizing theories" (Medin & Wattenmaker 1987; Murphy & Medin, 1985) influence the construction of more atomic representations such as object concepts and how they influence inter-conceptual relations as well. Murphy and Medin (1985) argue that the structure of individual concepts as well as the processes of conceptual change can only be explained if one takes into account the influence of domain organizing theories.

A similar line of reasoning is reflected in the science education literature on a novice's understanding of the concepts of light (Stead & Osborne, 1980), animal (Bell, 1981), and gravity (Stead & Osborne, 1981). It appears that the content of novice's science concepts can undergo dramatic changes as a result of more global changes in knowledge at the level of domain organizing principles.

Additionally, researchers such as Gelman and Gallistel (1978), Siegler and Richards (1983), and Klahr and Wallace (1976) working within a rule learning paradigm, have emphasized the importance of content knowledge in children's abilities to master problems in the domain of number or with a physical apparatus such as the balance beam. From a rule learning perspective, the knowledge base is best characterized in terms of the sets of rules or propositions on the basis of which a person reasons in any given domain. Such rules or sets of rules that a person comes to generate have also been called the person's theory for the domain by some researchers (Holland, Holyoak, Nisbett, & Thagard, 1986; Karmiloff-Smith, 1986). From within a rule learning framework it is possible to model characteristics of performance in domains such as addition and subtraction skills (Beishuizen, Beishuizen, & Felix, 1990) at successive levels of expertise.

Atomic concepts can be shown to participate in and be constrained by more global knowledge structures that are used to predict and explain the physical world, and it is in this context that they become an appropriate level of analysis for the naive theory approach.

Naive theory. The term "naive theory" constitutes the most direct analogy between the novice knowledge and scientific theory. Across many domains researchers have documented theory-like changes in the knowledge representations of novices as they gain expertise in a domain. Researchers in this tradition emphasize the global cohesiveness of novice concepts for a domain. They suggest that novices construct interpretive frameworks that provide them with categories of explanation and with ways of predicting events in the natural world.

For example, McCloskey (1983) has proposed that novices often appear to approach problems of force and motion in physics from a conceptual framework that strongly resembles medieval or Aristotelian theories for these domains rather than a Newtonian one. He says, "... people develop on the basis of their everyday experience remarkably well-articulated naive theories of motion" (p. 299). The intuitive theories assume that motion must have an initial cause such as the application of a force in a given direction. Once in motion, a body remains in motion only as long as it has "impetus" or an internal motive force. When the impetus is exhausted or dissipates, the objects returns to a state of rest. In contrast, in Newtonian theory there is no force or "impetus" needed to keep a body in motion once it has been set into motion. In the absence of an applied force, a body in motion will stay in motion just as a body at rest remains at rest. Novice beliefs in an impetus theory of motion were discovered through standard techniques of protocol analysis during problem solving.

Driver and Easley, (1978) in a major review of the conceptual change research in science education conclude that pupils have "alternative frameworks" which interact with formal instruction in interesting ways. This view is echoed by Osborne and Wittrock (1983) who also find that pupils often approach problems of classical mechanics from an Aristotelian framework. The term "alternative framework" has strong parallels to Kuhn's (1962, 1977) notion of paradigm in the philosophy of science literature. The implication is that novice conceptions of a domain are like alternative paradigms to the currently dominant "scientific" view presented in instruction. Further, there is the suggestion that the shift from novice to expert view involves radical theory change. The process of conceptual change is likened to a Kuhnian revolution whereby the novice's old paradigm, beset by overwhelming anomalies that it cannot resolve, is abandoned in favor of an alternate paradigm.

As can be seen from the research reviewed above, the notion that novice knowledge representations have theory-like properties has been extensively invoked in many different streams of inquiry. While introducing the knowledge-based approach to cognition we emphasized that it was a modular, decentralized view of cognition with a corresponding shift in emphasis from general, context-independent cognitive skills to domain specific knowledge and context-dependent skills. However, from within the knowledge-based approach outlined above, there has been a resurgence of interest in certain general cognitive skills or strategies, many of which comprise an area of research called the study of "metacognition." This interest has been rekindled in part by dissatisfaction within the knowledge-based approach with a strong form of the "naive theory" view. Some important criticisms of the naive theory approach to cognition will therefore be presented.

Criticisms of the naive theory approach

Not all proponents of a domain specific or knowledge-based approach to cognition agree with the view that novices' intuitive knowledge or beliefs about a domain should be viewed as naive theories which are qualitatively similar to scientific theories for a domain although they rest on different assumptions or concepts about the world. For example, diSessa (1988) altogether rejects the notion that novice knowledge representations in the domain of physics are "theories" of the physical world. Although, diSessa agrees with the notion that novices have intuitive beliefs or conceptions about physical phenomena such as the phenomena of "motion", he claims that these naive conceptions differ in important ways from scientific theories. In particular, diSessa (1988) observes that novice conceptions of the physical world are "a fragmented collection of ideas, loosely connected and reinforcing, having none of the commitment or systematicity that one attributes to theories" (p. 50).

A similar argument is made by Norman (1983) with regard to the mental models of physical systems that people construct when interacting with devices such as computers, cameras, and watches. Norman rejects the notion that mental models are coherent and internally consistent knowledge representations that act as an "inchoate theory of things" (Johnson-Laird, 1983). Norman suggests that mental models are incomplete, unstable over time, and unscientific because of their unstable, fragmentary, and inconsistent nature.

Critics of the naive theory approach also point out that differences between novices and experts are not confined to differences in the content or organization of their beliefs about the world such as differences between an Aristotelian and a Newtonian view of motion (McCloskey, 1983). There are also important differences in the ability of experts and novices to apply, regulate, and modify their knowledge representations. These differences are differences in metacognitive knowledge rather than content knowledge.

For example, Chaiklin (1985) and Solomon (1983) claim that novices unlike scientists are not sensitive to information that disconfirms their beliefs about the physical world. Kuhn, Amstel, and O'Loughlin (1988) suggest that this is because novices unlike experts do not make distinctions between the conceptual entities that constitute their theories of the world and the empirical evidence for these entities. Consequently, novices are poor at coordinating their knowledge or "theories" with evidence while scientists are not. There is growing evidence that differences in cognitive performance arise not only because of differences in content knowledge but also because of differences in the strategic or regulatory processes involved in the monitoring of one's own cognitive activity. The study of how cognitive monitoring and strategic control over one's own cognitive activities affect performance constitutes the domain of metacognition.

While critics of the naive theory approach agree that differences in domain specific knowledge of experts and novices are important for understanding cognitive performance, they also stress that one should study differences between experts and novices in metacognitive or strategic knowledge.

Metacognitive knowledge and strategies

As the term metacognition implies, the domain of study is people's knowledge about their different cognitive activities such as remembering, learning, reasoning, problem solving, and understanding. As is the case with content knowledge or knowledge of subject matter such as arithmetic, metacognitive knowledge can also be declarative as well as procedural.

Knowing that you do not know the phone number of president Bush, or knowing that you cannot remember all items on a list of 95 items after a single reading of that list constitutes metacognitive knowledge of the declarative kind. Knowing how to learn a 95 item list by using appropriate strategies such as subdividing the list into smaller sections, grouping similar items together, and rehearsing each subsection to automaticity constitutes metacognitive knowledge that is procedural in nature.

As discussed above, researchers from within the knowledge-based approach have become interested in questions of how the maintenance, use, and transformation of knowledge is effected by the cognitive system. Initially, researchers such as Rumelhart and Norman (1981) proposed that generic knowledge structures such as schemata that embodied our theories of the world were dynamic. They contained within themselves the regulatory procedures that monitored their "goodness of fit" to information from the external world and initiated restructuring of knowledge when the fit was judged to be inadequate. In this scenario the strategic or regulatory functions of the kind involved in comprehension monitoring for example, were directly tied to the modular or domain specific representations in terms of which an event was being comprehended. This approach could work for the comprehension of highly prototypical or routinized patterns of events for which there were easily applicable generic representations available. However, matters became problematic when one had to model more complex cognitive activities of the kind that required the coordination of several schemata in novel or atypical problem solving contexts where no pre-packaged generic knowledge was directly applicable.

Recently, Clement (in press) studied how physics experts solved atypical problems in their domain. He found that although they made use of their extremely rich domain specific knowledge in their attempts to solve such problems, the experts also used many general strategies. For example, the experts often tried to solve an atypical problem by analogy with other systems that they understood better. They also deliberately searched for misanalogies or counter-examples to their initial analogy. Further, experts tried to qualitatively model the target system in their minds, using mental models based on their physics intuitions. They also tried to establish how their solution would work if they assumed extreme values for variables in the problem. Some of the strategies such as breaking down the problem into simpler component problems were not unlike the strategies described by Polya. The important insight from these studies seems to be that experts did possess a variety of general executive or control strategies that directed their problem solving efforts. However, these general strategies were closely linked to and coordinated with the expert's domain specific knowledge of such principles of physics as the conservation of energy and conservation of momentum.

A confirmation of the need to link general cognitive strategies and skills to domain specific knowledge and skills comes from recent developments in artificial intelligence. FERMI (Larkin, Reif, Carbonell, & Gugliotta, 1988), an expert system that reasons in natural science domains, encodes domain specific knowledge of scientific principles and strategic problem solving knowledge in separate but related semantic hierarchies. The system thus provides for the specification of knowledge at different levels of generality in a hierarchical organization and allows FERMI to apply common problem solving principles of invariance and decomposition to a variety of problem domains such as those of fluid statistics, DC-circuits, and centroid location. Similarly, GUIDON 2 (Clancey, 1987) a medical diagnosis expert system, combines domain specific medical knowledge with general reasoning strategies to teach strategies for the management of diagnostic hypotheses.

Further evidence for the importance of strategic skills or metacognitive skills of the kind that govern access to and use of knowledge comes from the examination of a variety of training studies to determine the conditions under which knowledge acquired is transferred to different contexts. Researchers (Brown & Campione, 1981; Sternberg, 1984) have shown that people differ considerably in their ability to spontaneously access and use knowledge that they have previously acquired. Brown and Reeve, (1987) have argued that learning environments which taught students domain specific or task specific skills alone were less successful in producing transfer of what was learned than environments which also taught students why the skills were useful and when they should be used. This latter kind of knowledge is called "executive" or "metacognitive." Recently, a number of researchers from within the knowledge-based approach to the study of cognition such as diSessa (1988) and Glaser (1988), have emphasized the importance of studying how metacognitive skills interact with domain specific knowledge in understanding cognition.

Thus far, the discussion has been confined to the nature of knowledge representation. However, there is a second component of the knowledge-based approach to cognition that has to do with the study of knowledge acquisition and the mechanisms or processes of conceptual change. The following section will review some historical approaches to the issue and then discuss the knowledge-based approach.

Knowledge Acquisition

As mentioned in the introduction, some of the central questions of cognitive science have to do with issues of knowledge acquisition and conceptual change. We will now discuss some major approaches that attempt to explain how people acquire knowledge and how conceptual change occurs. We will begin by discussing approaches to conceptual change that have been elaborated with regarded to the kinds of knowledge that have a large declarative component. We will then elaborate approaches that have been developed to account for the acquisition of knowledge that is largely procedural or "skill learning."

Research on the Mechanisms of Conceptual Change

Accretion models of conceptual change. In psychology, much of the traditional research on concept attainment was conducted within an behaviorist context and emphasized an accretion model of learning. The growth of individual knowledge was characterized as the gradual accumulation of facts in accordance with the basic principles of reinforcement. This process resulted in the enrichment of initial concepts and their differentiation or integration to form new ones. What such a view implied for complex domains of knowledge such as physics and history was that complex knowledge was acquired through the concatenation of simple facts. Accretion continues to be an important way of describing much of what we learn. For example, the kind of learning that occurs as a result of the Voyager discovery that Uranus has 15 moons rather than 5 as previously supposed is best described as accretion.

The introduction of more molar forms of representation such as schemata and mental models necessitated a search for new mechanisms of knowledge acquisition since molar reorganizations of the knowledge base did not lend themselves easily to accretion-based accounts. In the section on knowledge representation discussed above, we stated that an important approach to understanding expert-novice differences was the "naive theory" approach. Proponents of the naive theory approach claim that the process by which conceptual change occurs in people is akin to a process of theory change.

This conception of knowledge acquisition, generally runs along the following lines:

The initial representations of the world constructed by novices have some theory-like qualities and these representations may be called initial or naive theories. As a result of direct experience of the phenomena themselves and cultural experience (through instruction) with expert interpretations of the same phenomena, novices come to restructure their initial representations.

Knowledge based-approaches and models of conceptual change. The search for candidate mechanisms underlying molar changes in the individual's knowledge base yielded a new set of constructs for characterizing the differences in an initial knowledge state and a subsequent state presumed to have occurred as a consequence of learning. In an important paper, Rumelhart and Norman (1981) proposed that knowledge acquisition often required the restructuring of prior knowledge rather than the mere accretion of new facts. Restructuring refers to changes in the organization of knowledge that result in the creation of new knowledge structures.

The notion of restructuring has been adopted by researchers such as Wiser and Carey (1983) as a means of accounting for the way in which the initial knowledge structures or theories of the child/novice are replaced by expert ones. The strong implication behind the notion of restructuring as used by them is that knowledge acquisition in the individual is like paradigmatic theory change in the history of science as described by Kuhn (1962). This is also a view adopted by researchers such as Driver and Easley (1978) in the field of science education.

Recent studies in the domain of astronomy (Vosniadou and Brewer, 1987; Samarapungavan, 1989) indicate for example, that high school students typically believe that seasons on earth are caused by the differences in the earth's distance from the sun as the earth revolves in elliptical orbit around the sun. The students do not use their knowledge about the earth's tilt to explain the seasons. They also believe that the solar system is in the geometric center of an expanding universe. As they acquire expertise in the domain of astronomy, the students do not merely add on new facts but must restructure the relations between known facts and conceptual entities. Thus, they must understand that it is the tilt of the earth rather than its distance from the sun along its orbit that causes the seasons. They must also understand that the solar system is not at the geometric center of an expanding universe but on the expanding surface of a balloon or bubble-like universe whose geometric center is unknown. Such restructuring requires the acquisition of new conceptual relations and the elimination of old ones within the conceptual system. It also requires the comprehension of new explanatory mechanisms for known phenomena (as in the case of the seasons) as well as an understanding of the problems with the old explanatory mechanism that was used.

Some researchers have emphasized the creation of conceptual conflict and the recognition of anomalies as the key to knowledge restructuring novices (Anderson, 1977; Collins, 1986). Others (Clement, in press) have focused on analogical reasoning as a way of generating new knowledge structures. Brown (1989) and Vosniadou (1989) cite considerable empirical evidence that even young children are capable of comprehending analogies and using them to understand new domains. They propose that analogical learning is a mechanism by which novices can generate new knowledge structures with which to comprehend aspects of the physical world.

More recent approaches to issues of conceptual change (White & Frederiksen, 1987) emphasize that models of knowledge acquisition must encompass not only changes in the content of concepts but also changes in strategic or metacognitive knowledge about a domain such as problem solving skills. Such approaches will be discussed in more detail in the second chapter which describes some areas of knowledge and skill acquisition in which cognitive science research has been fruitfully applied in the design of teaching and learning environments. The research on mechanisms of conceptual change discussed above deals largely with the acquisition of declarative knowledge. We will now examine research with regard to procedural or skill learning.

Skill learning

As noted before there has been a considerable interest within cognitive science in the acquisition of skills. One of the most comprehensive attempts to account for the acquisition of skills is the work of Anderson (1987) and his colleagues. They have developed a model of skill acquisition in domains such as generating geometry proofs, computer programming, and text editing.

The knowledge that underlies skilled performance is represented in Anderson's ACT* theory in terms of productions. These productions are pairs of conditions and actions. They specify that if a given state "X" occurs in working memory, then a specific action "Y" should be carried out. The action to be carried out may be mental or physical. An example of a production in the system is as follows:

IF the goal is to write a solution to a problem and there is an example of a solution to a similar problem

THEN set the goal to map that template to the current case

This production is a way of embodying a specific problem solving skill which is the search for a structural analogy to the current problem. If such an exemplary problem that can serve as an analogue to the current problem exists in the system's memory, it will be retrieved and used in the solution of the current problem. The underlying assumption of the theory is that skilled performance in people operates on the basis of the kinds of rules embodied in the ACT* productions. The productions are units of knowledge and can be used to represent steps in problem solving as well as stages of knowledge acquisition.

Although as a model of skill learning Anderson's ACT* theory emphasizes procedural knowledge, the interesting aspect of the theory is that all knowledge in the theory starts out in declarative form and must be converted into procedural form. According to the theory, knowledge initially comes in declarative form and then this knowledge is used by some very general or "weak" problem solving methods, such as problem solving by analogy, to generate solutions to novel problems. In this process of problem solving, a solution is generated and then stored in memory. Eventually these stored solutions are used as models in the solution of new problems in the domain resulting in the acquisition of a domain specific skill. The process of proceduralizing declarative knowledge and storing it is called knowledge compilation. Additionally, processes of generalization (extending the scope of application of procedures) and discrimination (restricting the scope of application of overgeneralized procedures) also play a role in skill acquisition.

Several general conclusions may be drawn from Anderson's model of skill learning:
a) The only way to acquire procedural knowledge or skill is by "doing" or practice. Learning occurs in the process of problem solving whereby declarative knowledge is interpreted and proceduralized.

b) A second conclusion (Lewis and Anderson, 1985) is that immediate feedback about performance is very important to skill acquisition in people. The researchers found that subjects learning to play computer games learned faster and performed better when they were informed of errors as soon as the errors occurred rather than when they were left to discover the errors on their own.

c) A third conclusion that can be made is with regard to the importance of "model tracing" whereby the system that learns builds up a record or history both of successful performance or application of skill as well as of the learning process by which the skill was acquired.

d) Since in the course of knowledge compilation prior successful solutions to problems are stored and used as models for new applications, it is very important for the learner to have experience with a wide range of problems and to be exposed to different types of problems.

e) The acquisition of new knowledge places burdens on working memory. Therefore, it is very important that there is a minimization of cognitive load or working memory load for aspects not directly related to the acquisition of the target skill in the learning environment.

The general principles embodied in Anderson's approach to skill learning are applied in the development of an intelligent computing system that is used to teach novices programming skills in LISP.

One important aspect of expert skilled performance is the speed of knowledge application. Experts apply knowledge fast even though human ability is limited in the process of performing competing, attention demanding tasks (Shiffrin & Schneider, 1977; Schneider, 1985). To achieve such fast skilled performance it is very important to integrate the basic skills necessary in the performance of complex cognitive tasks with the advanced skills and strategies required. In complex cognitive tasks such as reading, attention must shift between basic skills such as letter and word recognition to higher level skills of comprehension such as the activation of prior knowledge relevant to the subject read and the monitoring of comprehension. Although the component processes involved may each be well executed by novices in isolation, novices have trouble coordinating and using the different processes together in complex cognitive tasks (Perfetti & Lesgold, 1979). In the development of expertise therefore, certain component skills basic to cognitive activity have to be automatized so that they can be carried out without conscious processing. The conscious processing capacity is then freed for higher level, strategic skill components of the task (Glaser & Bassok, 1989).

The focus of this section has been on a particular model of skill learning developed for domains such as LISP programming. However, the principles of skill acquisition that hold for these domains can also be extended to higher order strategic skills of the kind involved in monitoring reading comprehension (Brown, Bransford, Ferrara, & Campione, 1983).

Having reviewed basic issues in the representation and acquisition of knowledge we will now turn our attention to the design of instructional environments. In the next chapter we will review some principles for the design of instructional environments based on current advances in cognitive science. We will then describe some examples of instructional environments that have been designed
to illustrate these principles.

COGNITIVE SCIENCE AND INSTRUCTION

In this chapter we will propose some general principles and procedures for the design of instructional environments based on current theories of cognitive science. We will then provide some examples of instructional environments based on such principles.

Pointers for the Design of Instructional Environments

Domain analysis

The starting point for the design of an instructional environment should be an analysis of knowledge in the target domain (Glaser & Bassok, 1989). Important points to consider are whether the to-be-learned knowledge is largely declarative, largely procedural, or a combination of both. Domain analysis is accomplished by methods of content analysis and protocol analysis described earlier. FERMI, an expert system that reasons in an variety of physical science domains such as fluid statistics and DC-circuits (Larkin, Reif, Carbonell, & Gugliotta, 1988) provides a good example of the importance of distinguishing between different types of knowledge needed for expert performance. Further, there should be a hierarchical specification of the kinds of units in terms of which the knowledge is to be acquired by the pupil such as rules, schemata, or mental models. This specification should be both in terms of levels of representation (such as global and molar versus local and atomic) and in terms of the content of representations at each level.

Failures to perform careful domain analyses are evident in a variety of instructional studies in academic as well as technical domains (Anderson, Boyle, Farrell, & Reiser, 1984; Greeno, 1978; Kieras, 1987). For example, it is well established that strategic control knowledge is important in planning and goal setting components of expertise in domains such as geometry (Greeno, 1978), computer programming (Anderson, Boyle, Farrell, & Reiser, 1984), and simple device operation (Kieras, 1987). However, where instructional materials in these domains have been evaluated against cognitive models that are based on careful domain analysis, they have been found lacking. Greeno, (1978) found that the principles used in setting appropriate goals for geometry proofs were never made explicit in texts. Anderson and his associates (1984) found that in some geometry and computer programming texts, goal structure knowledge was incorrectly expressed!

Setting instructional goals

An important point in the design of instructional environments is to clearly establish instructional goals. Such goals are related to the domain analysis because they specify the kind and level of expertise that a learner is expected to achieve in the instructional environment. Some instructional environments may be designed to teach only limited aspects of domain knowledge and skills. Instructional environments may also be limited to teaching knowledge and skills at a certain level such as an elementary level or an advanced level. Decisions about the ultimate level of expertise that the learner will acquire in the instructional environment will influence the way in which information must be presented by the tutor. For example, complete novices must often be taught basic, declarative domain knowledge before they can achieve complex procedural and problem solving skills in a domain. For intermediate and advanced learners, instructional environments that provide for the acquisition of domain knowledge in a variety of problem solving contexts tend to be more effective. Similarly, providing immediate corrective feedback and a great deal of external support appears to facilitate basic knowledge acquisition in beginning learners. However, discovery processes of the kind created in inquiry teaching (Collins, 1986) that require more independent organizing activity on the part of the learner appear to facilitate the acquisition of complex, dynamic knowledge and problem solving skills. Where the instructional environment is designed to support the development of successive levels of expertise from low to high, care should be taken that what pupils learn at introductory levels provides an adequate basis for the building of advanced knowledge and skills at subsequent levels.

Diagnosing learner characteristics

One of the important findings in cognitive science has been that pupils' prior knowledge and skills significantly determine their subsequent learning in a given domain. Therefore, diagnosing the prior knowledge and skills that a learner brings into the learning situation is crucial to the design of successful instructional environments. An effective instructional environment is one that activates all relevant prior knowledge in the learner and relates it in appropriate ways to the new information being provided. If for example, the learner has correct but incomplete intuitions of knowledge about a domain prior to instruction such knowledge must be systematically integrated in the general framework of knowledge being taught. Where the learner has significant misconceptions these need to be explicitly addressed and disconfirmed in instruction.

Using principles of "Apprenticeship Training"

In recent times, many researchers have noted that contemporary formal instruction in academic contexts results in the divorce of knowledge and skills from their applications in the real world (Collins, Brown, & Newman, 1987; Papert, 1980). In a move to develop better instructional environments many researchers have called for methods of instruction that link knowledge and skills to their full range potential uses. The general name given to such methods is **situated learning** or **cognitive apprenticeship**. Much of this thinking draws upon observations of traditional apprenticeship methods of learning a trade or craft. What is new is the ability to analyze the target knowledge and skills to be acquired based on recent advances in the study of cognition.

Important instructional techniques in apprenticeship training involve modelling and coaching (Palinscar & Brown, 1984). Modelling is simply the execution of skilled performance by an expert tutor for the benefit of the learner. The learner can observe how knowledge is actually used in some context of application. Modelling also allows the tutor to make tacit knowledge, such as strategic knowledge involved in planning a task, overt for the learner. Coaching refers to the tutor's function in providing immediate feedback and correcting errors or highlighting potential sources of error. It also covers such functions as providing explanations and missing pieces of knowledge where necessary. Modelling and coaching provide external support or scaffolding from the tutor for the learner as the learner tries to develop expertise in the target domain.

The important aspect of apprenticeship training methods is that learning activities are carefully sequenced to match the students' changing levels of competence and instructional needs. There is a gradual fading of external support as the pupil acquires expertise.

Evaluating instructional effectiveness

So far we have confined our discusssion to principles for the design of instructional environments. However, the evaluation of such environments is vital for the development of a science of instruction. Instructional environments must be evaluated from a variety of standpoints. Minimally, there should be evidence that pupils who learn in the environment achieve the specific cognitive outcomes that the environment was designed to produce. This can be done using standard pre- and posttest comparisons of student performance in the target domain.

A stronger test for an instructional environment is one of relative efficiency which involves a comparison of a given system of instruction with other systems designed to teach the same material. Such tests require the comparison of groups of students who are similar in terms of prior knowledge and ability but who are assigned to different instructional conditions. The results of such evaluations should show whether a given instructional environment is as effective or more effective than other instructional environments available for the domain. Instructional environments should also be evaluated for certain non-cognitive outcomes such as their effects on student motivation and their cost-effectiveness. Having discussed some general issues that need to be considered in instructional practice, we now turn our attention to the design of intelligent tutoring systems.

Implications of Current Cognitive Theories for the Design of Intelligent Learning Environments and Tutoring Systems

Computers are currently in used as aids in formal and non-formal education as well as in vocational training. Three broad categories of applications can be distinguished:

1. **Instrumental applications** in which the computer serves primarily as an information presentation device such as a word processor, database, spreadsheet, testbank, or drawing facility. Such applications do not contain any specific instructional component. For this purpose general, commercial software is often employed.

2. **Instructional applications** in which explicit educational objectives, such as the acquisition of a list of prespecified skills, are achieved with the help of computers. Included in this category are the traditional drill-and-practice applications as well as the more sophisticated intelligent tutoring systems.

3. **Learning and discovery environments**. Such environments, often called "microworlds", are artificially constructed worlds designed to simulate theoretically important aspects of a "real" world. For example, the microworld may be a simulation of a complex ecological system or of a programming language. The microworld provides

students with a universe of discovery in which they can systematically explore and manipulate phenomena to discover their underlying laws or systemic principles.

In this section we will focus on the second and third categories of application discussed above, paying particular attention to their implementation in intelligent tutoring systems.

Intelligent Tutoring Systems

Intelligent tutoring systems (ITS) are computer programs which carry out educational tasks. An ITS replaces or complements the function of the teacher in instruction. Since Carbonell's (1970) pioneering work in this domain, several intelligent tutoring systems have been developed, mostly as prototypes for demonstration purposes (Sleeman & Brown, 1982; Tennyson & Park, 1987; Kearsley, 1987; Mandl & Lesgold, 1988). The tasks taken over by the ITS require "intelligent" activity on the part of the system. Five components can be distinguished in an ITS: (i) the expert component containing a representation of the subject matter, (ii) the student component in which all student-related information is stored and updated, (iii) the tutor component comprising the pedagogical knowledge of the system, (iv) the interaction component which monitors the communication between student and system, and (v) the executive component which coordinates the collaboration between the components (Clancey, 1982).

In this section we will focus on the first three components of an ITS: the expert component, the student component, and the tutor component. Problems with respect to the executive control of an ITS will be discussed in relation to the three main components.

The expert component. The expert component of an ITS contains a representation of all knowledge and skills necessary for the student in order to learn the target domain. An ITS needs complete knowledge of the target domain in order to diagnose the student's performance, to provide appropriate instruction, and to record the achievements of the student. The architecture of expert components is identical with the structure of expert systems (Hayes-Roth, Waterman, & Lenat, 1983). As discussed in the preceding sections, knowledge can be represented in various forms (Bruin & Wielinga, 1984) such as declarative knowledge, procedural knowledge, and heuristic or strategic knowledge.

Apart from the knowledge base, expert systems are equipped with an explanation facility which elucidates the way in which the system carries out tasks. The system may do this by demonstration, or by actually solving a problem or conducting a skill. The explanation facility is important in educational settings because it is the basis for feedback and instruction. The problem with many expert systems is the poor quality of the explanation facility. The very narrow domain-orientation and task-orientation of most expert systems results in the exclusion of more general declarative and procedural knowledge necessary for an effective explanation facility.

An example of the problems encountered when an existing expert system is used as a component in an ITS comes from Clancey (1987). The original MYCIN expert system (Shortliffe, 1976) provides advice on antibiotic therapies for infectious diseases. It contains 450 rules or heuristics used in a backward chaining reasoning strategy in which a question is decomposed into subquestions until an answer is found. The explanation facility answers "why" questions through a backward goal regression: the answer is given by stating the previous goal in the inference stack which was created by decomposing the original problem.

MYCIN does not contain an adequate explanation facility because it cannot explain, (i) why an applied rule is correct, and (ii) why the reasoning strategy is chosen (Clancey, 1987). In order to achieve these goals GUIDON (Buchanan & Shortliffe, 1984) was built as an ITS. Although GUIDON could discuss how a rule was achieved, and could also evaluate a student's response to the system, it lacked the necessary medical knowledge and knowledge of the world to give a complete account of the arguments leading to a conclusion by the system. Therefore, a third system, NEOMYCIN, was developed in which both metacognitive and domain specific knowledge was embedded.

In NEOMYCIN domain-dependent disease knowledge was organized in terms of four different knowledge "sources" intended to represent different epistemic categories: (i) a hierarchical taxonomy of etiological information, (ii) causal relationships between the entries in the hierarchy, (iii) knowledge of the world, and (iv) disease process knowledge (Clancey, 1987, p. 208). General procedures for making diagnoses were represented in forward-chaining strategies for interpreting incoming data and a hierarchical set of domain-indepenent meta-rules determining the selection and testing of hypotheses. These procedures diverged from the original, purely backward-oriented reasoning strategy in MYCIN. In NEOMYCIN the original black-box knowledge structure of MYCIN was replaced by a "glass box" structure (Goldstein & Papert, 1977) to be used in instructional explanations (Mandl & Lesgold, 1988).

The implicit conclusion to be drawn from current implementations might be that ITS consider expert performance to be the basis for instruction. However, novice students are often unable to understand instructions based on expert performance because they lack the prerequisite skills to put the prescriptions into practice. Many authors assume that the acquisition of cognitive skills can be described as a progression of stages. Each stage can be characterized in terms of a relatively stable complex of rules which are used to solve some important subset of domain problems (Siegler & Klahr, 1982; Case, 1982). According to this point of view, instruction should depart from the current level of expertise of the student and strive for two consecutive goals: (i) a student's mastery of skills at the current level of expertise, and (ii) movement to the next level of expertise.

Goldstein (1977) developed an intelligent tutor for a computer game, WUSOR. The game requires the student to learn several probabilitistic rules concerning imminent dangers in a network of caves to be explored. Goldstein specified several levels of expertise in this domain and proposed a variety of mechanisms for transition from one level of expertise to the next. Some of these mechanisms such as refinement/simplification and generalization/discrimination resemble conventional accretion models of conceptual change while others like analogy resemble restructuring models of conceptual change.

The number of discrete levels of expertise to be distinguished for instructional purposes, depends on the nature of the domain being taught. As mentioned before, in order to properly specify levels of expertise and kinds of knowledge needed at each level, designers of instructional environments should conduct a domain analysis (see above). The efforts to create an instructionally valid tutor on the basis of MYCIN, show that several types of knowledge play different roles in an explanatory facility with real instructional power.

The student component. The task of the student component is to maintain and update a model of the student's acquired knowledge and skills during the student's interaction with the ITS. In Norman's (1983) distinction between kinds of mental models, the content of a student component represents the conceptual model and can be defined as the system's model of the student's mental model of the target domain. An ITS needs this information in order to provide appropriate instruction tailored to the individual student's specific deficits.

Several kinds of data may be used to derive a representation of the student's knowledge and skills. Such data include: (i) the problem solving behavior of the student showing the rules not yet acquired or mastered, (ii) answers to questions raised by the system to check the state of the student's knowledge, and (iii) assumptions based on the learning history of the student (Clancey, 1982).

Two widely known examples demonstrate at least one dimension on which students' models may differ. Using a "glassbox" representation (see above) of the rules to be mastered in his WUSOR game, Goldstein (1982) derived an overlay model of the actual status of students' knowledge of the individual rules (i.e., passed, failed, not yet encountered). Brown & Burton (1978) applied a deviation modelling technique in their BUGGY system, which provides training in subtraction problems. By comparing students' answers with an empirically collected inventory of subtraction errors, the system identified students' misconceptions and provided them with remedial instruction.

The way in which the student's knowledge of the domain is modelled by the system depends on the means of instructional intervention available to the system. Systems with tutor components that foster learning by accretion of knowledge (e.g., in a domain with many topics to be mastered) tend to use an overlay model which represents what the student already knows about the domain. Systems with tutor components that use intervention strategies for restructuring students' prior knowledge through the creation of conceptual conflict (e.g., impasse-driven learning, VanLehn, 1988) use a deviation model to represent students' misconceptions about the domain.

A second dimension on which student components can be distinguished has to do with the criteria used to assess the student's gains from instruction. There are two types of comparisons that can be used in assessment: (i) comparison of the student's performance with the ultimate expert's level of performance, and (ii) identification of the current performance level of the student by comparison with performance along a series of successive levels of expertise. Beishuizen, Beishuizen, and Felix (1990) developed an ITS in which a genetic model describing the development of mental addition and subtraction skills was used as a basis for establishing a representation of the student's progress.

Some authors (Tennyson & Pask, 1987) suggest that the student component should diagnose and represent a number of student characteristics such as: (i) measures of intelligence, aptitude, cognitive style, and prior knowledge, and (ii) measures of motivation and self-esteem. However, as long as a clearcut theory about the relationship between a given student characteristic and the student's behavior in an educational setting is lacking, the inclusion of that characteristic in the student model appears unwarranted. As Tobias (1987) pointed out, the only student characteristic which has produced substantial interaction with treatment variables in instructional settings is prior knowledge of the subject-matter taught. Further research is necessary to identify important student variables to be included in a student model of an ITS. Some candidates worth considering are, fear of failure or test anxiety, serialist versus holist learning styles (Van der Veer & Beishuizen, 1986), and intrinsic motivation or internal versus external locus of control.

The tutor component. The most important task of the tutor component is to monitor the type and content of instruction and information presented or made available within the learning environment. The knowledge that the system needs to ask the right questions and to focus on the appropriate issues in the course of instruction is stored in this component. Decisions on responsiveness and adaptivity are based on this knowledge. The criteria on which the decision to increase the complexity of a microworld are based (Fischer, 1988) for example, are formulated within this component. The problems with which the designer of the tutor component of a learning environment is faced are multifold. Knowing when to react, to what, and how, mandates use of knowledge of the domain, of the student, and of instructional strategies. Within the tutor component, knowledge of instructional strategies is of utmost importance.

The issue of choosing appropriate instructional strategies for the tutor component has been addressed by Merril (1988). In his opinion, intelligent systems should be used to implement instructional strategies that would be difficult or impossible to implement with live teachers. Tutorials are useful to focus student's attentions or to help overcome misconceptions. Experiental instructional models allow for the use of intelligent systems for the creation of controllable microworlds. Issues such as learner versus computer control and intervention, timing, and content are thus of importance. In a review of current literature on cognition and learner control, Steinberg (1989) points out the limitations of full learner control within a learning environment. Students with a limited knowledge of a domain (who do not possess the necessary declarative, procedural, and strategic knowledge) do not benefit from open learning environments. There are differences between students in their ability to take control of their learning in a non-directive environment. The needs of the student with regard to external structuring or direction for learning also change as learning progresses. The availability of adaptive advice within a system is seen as a way of increasing the effectiveness of learner control. The ideal solution for timing and content of intervention has yet to be found. As mentioned above, Anderson (1987) favoured immediate feedback for instructional purposes. VanLehn (1988) supports this view within the context of impasse-driven learning, recommending that the tutor should intervene if the solution found by the student is incorrect. In contrast, cognitive apprenticeship (Brown, Collins, & Duguid, 1988) stresses the importance of fading of feedback as learning progresses. Brown and his associates suggest that an interfering tutor can be annoying. They recommend that students should be taught to devise and execute plans without interference.

One major difficulty in the design of cognitive apprenticeship systems is the recognition of plans both at the level of task performance and at the meta level of learning strategies. Whether advice should be provided only when solicited by the learner is another issue which remains unclear. Individuals differ in the degree to which they initiate help-seeking activity. Such differences in help-seeking may reflect differences in personality as well as differences in expectations. Although timing is no longer of importance in systems where advice is given only when solicited, the effect of the interaction may be questioned. A "good" teacher does not always wait to be asked for help before intervening in a learning situation.

In the following section we will review some computer-based instructional environments that have been constructed to teach troubleshooting in domains that involve knowledge of physical systems. As we review these systems we will discuss them in terms of the components (expert, student, and tutor) described above.

Learning Physical Systems and Troubleshooting

Broadly speaking troubleshooting refers to procedural skill in the diagnosis of complex systems. Troubleshooting can be analyzed within a general framework for understanding human performance with devices or systems. In this framework, performance is the selection and execution of procedures to interact with the device or system to achieve certain goals (Gott, 1989, Gott & Pokorny, 1987). Consequently, there are three types of knowledge (Gott, 1987; Gott 1989) that comprise expertise in troubleshooting:

i) Knowledge of the system or device with which one must interact referred to in the literature as device knowledge. In the literature on troubleshooting, the term device knowledge refers to declarative knowledge about the system being studied.

ii) Knowledge of the procedures with which one investigates the device to make diagnoses about its functioning or procedural knowledge.

iii) In complex devices there are several options for investigating the device which require strategic decisions about which goals to pursue and which procedures to implement in which sequence. This constitutes strategic knowledge.

There have been several different approaches to the development of instructional environments for teaching troubleshooting skills. Three of these, STEAMER, Sherlock, and QUEST, will be discussed below.

STEAMER: Some systems such as STEAMER (Hollan, Hutchins, & Weitzman, 1984) emphasize the development of device knowledge as the key to effective troubleshooting. The central idea in STEAMER is that a computer-based "interactive inspectable simulation" of a steam propulsion plant enables a pupil to acquire a qualitative process understanding of the system. This knowledge enables effective troubleshooting with the system.

Implemented in STEAMER is an expert tutor. The goal of the tutor is to train pupils in the procedures for operating and maintaining a steam propulsion plant. However, the instruction emphasizes the learning of rich qualitative models of the plant's functioning or dynamic device knowledge rather than the learning of specific diagnostic procedures. What the tutor tries to teach is not a physical model of the system but rather the conceptual model of the steam plant that experts use when they provide qualitative explanations of the plant's functioning (Hollan, Hutchins. & Weitzman, 1984). This conceptual model of the steam plant and its functioning comprises the expert component in STEAMER.

The expert component contains 100 hierarchically organized views or graphical representations of the plant ranging from a global overview of major steam plant parameters to detailed of views of its subsystems. Underlying this series of graphical representations is a mathematical simulation model of the plant. Qualitative interpretations or images of this underlying quantitative model are made manifest in the graphic interface. These qualitative interpretations make the graphic views "runnable models" that make the dynamic operation of the system and the interrelations among its components explicit. The models allow pupils to observe aspects that they could not observe in a real steam plant such as the flow rate in pipes. STEAMER teaches procedural representations that are abstract rather than situation specific with regard to the functioning of the steam plant. For example, one such procedure is, "Before admitting steam into a closed chamber, you should first align the drain valves." A more specific procedure could have been, "Before opening valve 13, align the drain valves FWD-E245 and FWD-E239." Because the procedures are represented at an abstract rather than situation specific level a much smaller number of procedures need be represented. The system also contains a set of ordering relations that constrain the sequence in which the procedures are to be used. This knowledge, combined with the understanding of why the procedures work based on device knowledge, are the basis for acquiring expertise in troubleshooting in the STEAMER environment.

The pedagogical principle embodied in the tutor component of STEAMER is that system understanding evolves from observation and exploration of the device simulations provided. The tutor requires students to compose, execute, and explain procedures as they interact with it in the course of instruction. If the procedures composed by the student are wrong (e.g., the sequencing of steps is inappropriate) STEAMER corrects the student and explains what the mistake was. For example, feedback concerning aligning drain valves would be as follows:

"According to the principle which requires that whenever you admit steam into a closed chamber you should first align the drain valves, before opening valve 13 you should align the drain valves FWD-E245 and FWD-E239."

As evident from the description above STEAMER embodies many of the principles we discussed in the beginning of this chapter as part of its instructional environment. The presentation of device knowledge and procedures is based on cognitive analysis of expert performance in steam propulsion plants. The tutor provides for an active, situated learning context in which the pupil learns general system procedures by carrying them out. The tutor also teaches in a manner consistent with the principles of apprenticeship training by providing external support, modelling, and coaching for the pupil.

STEAMER does not provide explicit instruction in strategic skills. Indeed, the tutor lacks an "expert" model of troubleshooting procedures. The assumption seems to be that strategic knowledge about choice of appropriate procedures will come from the qualitative and causal understanding of how the steam propulsion plant functions. In this sense, STEAMER seems a good instructional environment for teaching qualitative causal reasoning about complex devices but probably needs to make strategic knowledge necessary for troubleshooting procedures more overt and situation specific. There is also no attempt to sequence the presentation of material based on the level of prior knowledge or expertise of the learner.

Sherlock: Sherlock is a computer-based avionics troubleshooting training system developed by Lesgold and his associates. The system was field tested in American Air Force training centers in 1987. The goal of Sherlock is to enhance a technician's procedural investigation of airborne electronic devices and other related test equipment. The tutor teaches difficult diagnostic procedures needed in investigating the test equipment used to maintain F-15 fighter aircraft.

The curriculum of Sherlock was based on two sources. The first source was a cognitive analysis comparing the performance of highly skilled and less skilled apprentice technicians on realistic fault isolation tasks (Gitomer, 1984, Glaser et al., 1985; Gott, Bennet, & Gillet, 1986). The second source was a series of knowledge engineering studies of air force avionics experts (Gott, 1987). The performance of each of the groups studied was represented in terms of a problem space-based formalism (Glaser et al., 1985; Newell & Simon, 1972). From this data it was possible for each class of problems (e.g., equipment faults) to abstract the hierarchies of plans and procedures used by most technicians in order to create a prototypical solution structure. It is this expertise, embodied in prototypical expert troubleshooting strategies which are represented as structured steps through a problem space, that Sherlock aims to teach.

Sherlock is designed to teach trainees a "fundamental mental model" (Lesgold, Lajoie, Bunzo, & Egan, 1988) of an electronic test. Additionally, the system is designed to develop the trainee's goal structure or plans for investigating the equipment, to teach procedural knowledge in form of specific fault isolation actions, and to teach strategic knowledge to help direct and monitor the course of problem solving.

The simulation of the system in Sherlock models the external control of equipment rather than the internal behavior of that equipment. For example, the components of the test station such as control panels of station drawers are only functionally simulated, allowing users to take voltage and ohm measures from them at certain test points. The graphics interface with which a trainee works is designed to be physically similar to those that will be encountered in the real work environment. Unlike STEAMER, Sherlock does not directly present the pupil with an abstract device model. However, the problem solving procedures or plans that are taught are based upon the abstract device model.

The tutor component in Sherlock provides a problem oriented practice environment. Students are expected to acquire the necessary device, procedural, and strategic knowledge as they solve instructional problems designed to make manifest the characteristics and cognitive demands of the work environment. These instructional problems take the form of 34 trouble shooting situations. Each situation carries with it a list of hierarchically organized menus that teach relevant procedures and provide hints for the progression of problem solving.

A troubleshooting scenario is activated when a series of electronic tests run by the test station through a set-up procedure results in a failure. When this happens the pupil can access a top level plan menu to begin the fault activation process. This provides the pupil with the plan options at the top of the hierarchy of problem solving procedures. For example, at the first level of the problem solving process the user can subdivide the equipment space into Unit, Test Package, and Test Station and then select one of them for examination.

Once the pupil has adopted a plan or some steps in the execution of a plan, he or she can access a hint menu to help evaluate the choice and its outcome. The hint menu also provides a direction for moving into the next level of procedures in the problem solving hierarchy. For example, if a technician began a fault isolation attempt by choosing to examine a component of called the "Unit" and proceeds to have it tested, he or she could obtain the following hint from Sherlock:

Option (Level 5): Your initial thinking was the unit was bad. That was a good choice because the unit was sent to the shop for repair. The TO (technical order) states that the UUT Shutdown Command Output may be bad. Because you swapped the unit, we learned that the unit is good for this test, so the problem must be in the Test Package or Test Station.

Hints such as the one above are usually given by Sherlock only in response to requests from the trainee. However, Sherlock does provide automatic feedback and help if the trainee has made a clearly wrong decision about procedures to be carried out or when the trainee plans to carry out redundant procedures that will not add information toward fault isolation. Tutorial principles in Sherlock's tutor component include situated learning, and intelligent coaching from the expert tutor. The emphasis of the instructional environment is on the teaching of specific troubleshooting strategies and procedures based on a model of expert troubleshooting performance.

Sherlock's effectiveness was evaluated at two separate Air Force F-15 flying wings. The study used 32 trainees pretested on several indices of technical proficiency. These indices were used to divide the trainees into matched experimental and control groups at each of the two sites. The basic instrument used to measure the effect of instruction on troubleshooting performance was a structured "think aloud" verbal protocol instrument called "Verbal Troubleshooting Tasks." A pre- and posttest paradigm was used. The experimental group spent 20 hours working in the Sherlock environment. The control group continued to participate in its usual on-the-job training program during that time. The results of statistical analyses showed that there were no significant differences in troubleshooting performance between experimental and control group prior to instruction. However, the experimental group that trained with the Sherlock tutor was significantly better in performance on the " Verbal Troubleshooting Tasks" than the control group at the end of the intervention period. Trainees who had used Sherlock were asked to evaluate it at the end of their training. Trainees rated Sherlock highly on teaching procedural skills and on providing a functional understanding of the equipment and circuitry used. They rated it low on providing flexibility in selecting troubleshooting plans and on providing explanations or hints that were meaningful.

The system provides the trainee with a clear task structure to model. However, the expert component of the system fails to provide any overt conceptual model of the underlying device with which the trainee must interact while troubleshooting. It is assumed that such device understanding will be abstracted by the trainee in the course of learning to carry out the full range of troubleshooting procedures that Sherlock teaches. Thus the focus of the expert component is quite narrow. The explanation facility within Sherlock is also limited. It is possible that the emphasis on canned or deterministic troubleshooting models might intefere with the transfer of procedural knowledge to atypical problem contexts.

One problem that the evaluation of Sherlock revealed was that its lack of flexibility was particularly limiting to users with more advanced troubleshooting skills. In this sense Sherlock shares the limitations of STEAMER in not accommodating to different levels of learner knowledge and skill as such information is absent from its student component.

QUEST: The name of the system is an acronym for Qualitative Understanding of Electrical System Troubleshooting. This is a computer-based instructional environment that teaches qualitative reasoning about electrical circuits. One of the distinctive features of QUEST is its developmental approach to instruction. The target domain is presented to the learner in a hierarchical succession of mental models designed to correspond to increasingly advanced levels of expert understanding. Thus, unlike the previous systems examined, QUEST is explicitly designed to accommodate the changing needs of learners with different levels of knowledge and skill.

The construction of the expert component which comprises of a hierarchy of models that contain the declarative, procedural, and strategic domain knowledge was based on exhaustive domain analysis. This included a study of the structured goals that experts pursue in troubleshooting and the kinds of knowledge they use to achieve the goals.

The content of the expert component is a series of mental models corresponding to different levels of conceptualization of circuits. The mental model for each level of circuit understanding incorporates declarative or device knowledge of electrical circuits, procedural knowledge for exploring such circuits, and a control structure that guides strategic decisions about the use of declarative and procedural knowledge. The system can thus model both the qualitative and causal aspects of the device being studied as well as expert reasoning about that device at varying degrees of abstraction and elaboration.

The key to the sequencing of instruction through a "progression of causal mental models" (White and Frederiksen, 1987) is that models at each level in the hierarchy are designed to be compatible with the more advanced models that will be taught at subsequent levels. Further, each level of model is designed to handle some part of domain problems that is not handled by models at the lower levels of understanding. The models vary in perspective, and degree of elaboration, as well as in level, or order in which they are intended to be acquired. The system models reasoning processes in troubleshooting and provides explanations of device behavior.

The tutor component of QUEST supports a variety of learning activities such as the exploration of electronic circuit "microworlds" and problem-driven learning. The student can choose learning conditions, such as open-ended exploration or problem solving, as well as areas to explore and types of problems to pursue. For example, the student may solve problems involving the prediction of device behavior in variable conditions. The student may be given a faulty circuit with a test light inserted and asked to list all possible faults that are compatible with the behavior of the test light. The student may also be asked to determine the location and type of fault in the circuit. The student may be asked to design new circuits to achieve certain goals or functions. Information about the student's performance is stored in the student component and used by the tutor for further sequencing of information.

The QUEST system is an excellent example of an instructional environment based on principles drawn from cognitive science research. Its presentation of domain knowledge is based on an analysis of the components of expertise but also on an analysis of stages of progression in learning. Thus, the system decomposes expert knowledge in ways that allow a novice to acquire a progressive understanding of the expert model of the domain. Another advantage of the system is its flexibility. QUEST not only adapts to the current expertise of learners but also allows them to select different conditions of learning such as domain exploration versus problem solving. Further, the explicit modelling of strategic or metacognitive knowledge in conjunction with domain specific procedural and declarative knowledge increases the likelihood that what is learned will be applied even in novel contexts.

The possible limitation of the general approach of QUEST may lie in the complexity of the target domain to be taught. QUEST deals with a fairly simple electrical system yet requires the implementation of a large number of mental models that are hierarchical conceptualizations of the system at different levels of complexity and detail. Whether such implementations would be feasible with more complex systems or whether increasing domain complexity would result in an explosion of models remains to be seen.

In this chapter we have attempted to give a flavor of how current advances in cognitive science have influenced the design of instructional environments. In the final chapter we will summarize the main points of our review and discuss unresolved problems as well as some directions for future exploration.

CONCLUSIONS

In this paper we have attempted to review some of the major advances of cognitive science and to highlight their application in the design of computer-assisted instructional environments. As we have reviewed, advances in theories of knowledge representation have enabled educators, including designers of ITS applications, to specify educational objectives with greater precision than before. Similarly, advances in theories of knowledge acquistion have helped to guide the development and evaluation of various instructional strategies.

In addition to the increasing specificity of theories of knowledge acquisition, advances in cognitive science have led to increasing concerns with the ecological validity of such theories. There is consequently a great deal of emphasis placed on describing knowledge acquisition and application as it occurs in the performance of complex, "real world" activities such as reading comprehension, computer programming or biomedical diagnosis. This concern with the domain specificity of knowledge and the ability to describe "situated cognition" or real world cognitive performance is reflected in the development of ITS systems as reviewed above. There are however, several issues with regard to the design of instructional environments in general, and ITS systems in particular, that need further research. We will discuss these issues in terms of their implications for the design of expert, student, and tutor components in its ITS applications below.

Expert Component

Changes in domain knowledge

As noted before, current theories of knowledge acquisition are based on a specification of the nature of knowledge in specific content domains such as chemistry or physics. Instructional design is closely linked to such domain analysis. Consequently, when there are radical changes in the nature of domain knowledge of the kind embodied in major scientific revolutions (e.g., the Copernican revolution in astronomy, Einstein's relatvity theory,) models specifying knowledge representation as well as knowledge acquisition for the entire domain have to be revised or restructured. Such major restructuring of domain knowledge has consequences for conventional instruction requiring the revision of standard textbooks and the retraining of teachers. However, it presents even bigger problems for the expert component of the domain in an ITS which must be rebuilt from scratch. This process is both time consuming and expensive for ITS designers. Therefore, in considering the cost/effectiveness ratio of ITS, one should consider the prospective stability of current knowledge in the to-be-taught domain. For example, one can be fairly sure that an ITS designed to teach knowledge about the operation of internal combustion engines will not change radically over time and can be used for as long as the device itself is in use. On the other hand, an ITS designed to teach military strategy in the South Pacific might have to be rebuilt in a matter of years if the expert conceptual models or knowledge of the domain undergo radical change.

Specifying types of knowledge

Specifying the different types of knowledge (i.e., declarative, procedural, and strategic) that comprise expert performance is still problematic for many domains. Especially difficult is the specification of strategic knowledge and the degree to which it should be considered domain independent rather than context specific. To the extent that current cognitive theory has problems with the specification of types of knowledge necessary for expert performance in a given domain, these problems will be reflected in the design of expert components as well.

Distinguishing levels of expertise

Many expert components fail to distinguish between levels of expertise and contain only one model of expert performance representing a single level of expertise. Thus, more research is needed to help specify different levels of expertise in the target domain.

Student Component

Individual differences

Dealing with individual differences in students along dimensions such as motivation, prior knowledge, aptitude, and learning style is problematic both for conventional class instruction and for ITS systems. Part of the difficulty in dealing with individual differences is pragmatic and has to do with the cost/effectiveness ratio of individualizing instruction. However as mentioned before, part of the problem is a lack of theory relating individual differences on dimensions other than prior knowledge to learning outcomes in current theories of knowledge acquisition. If there are theoretical advances with regard to the problem of individual differences it should be possible to match groups of students who have similar status on a specific dimension or dimensions to appropriate instructional environments.

Tutor Component

Selecting instructional strategies

Several research issues with regard to the tutor component have been discussed in the previous section. As noted (Fischer, 1988), it is currently difficult to specify the knowledge that an ITS tutor component needs in order to obtain the right information about pupil performance for the appropriate sequencing of instruction. An effective tutor component must be based on knowledge of the domain as well as on knowledge of student characteristics and of appropriate instructional strategies. One problem with the development of the tutor component is a lack of scientific consensus with regard to what constitutes an appropriate instructional strategy in a given situation. For example, there is considerable debate about the degree of learner control that should be allowed in an instructional environment and the degree and timing of help and feedback that should be provided to the student. Much research is needed to empirically establish the relative advantages and disadvantages of different instructional strategies in different kinds of domains and with different kinds of student populations.

Need for human tutor

Group-based conventional instructions of the kind encountered in traditional classroom situations have pragmatic restrictions on individualization of instruction. Because of the need to teach an aggregate of students it is often impossible to tailor the pace, sequence, and type of instruction to the needs of each student in the group. On the other hand, expert human instructors are very good at diagnosing a variety of individual needs of students (e.g., motivational needs). This diagnostic knowledge allows them to experiment with ways of individualizing instruction within a group context. For example, a human instructor who is aware of a pupil's shyness or hesitance in help-seeking may be able to elicit help-seeking behavior in the pupil through special interaction techniques. Tutor components in current ITS systems do not possess such knowledge about the motivational and emotional needs and characteristics of students that affect instructional outcomes.

As noted above, expert human tutors have certain advantages with regard to diagnosing and catering to individual needs of students on dimensions that are not handled well by ITS systems. Further, there are likely to be some advantages of conventional instruction that stem from the motivating effects of human interaction. However, the lack of interpersonal-interaction may not always be demotivating to students. There is some indication that students regard evaluations of performance from an ITS tutor as fairer and less biased than those of a human instructor (Papert, 1980). ITS systems appear very useful in providing drill or practice essential to the mastery of various cognitive skills such as those required in LISP programming. They are valuable in providing instruction that is individualized to each pupil's current level of skill or knowledge. ITS systems also have great utility as adjuncts to conventional instruction that allow for self-study from an intelligent database.

Peer interaction and cooperative learning

Finally, it should be possible for ITS designers to provide for peer interaction and interaction of students with a human tutor electronically. This will increase the flexibility of such systems.

To conclude, although current advances in cognitive science have been of help in identifying both the potential and the limitations of computer-assisted instruction, more research is needed to determine the scope and effectiveness of current and future applications.

REFERENCES

Anderson, J. R. (1983). The architecture of cognition. Cambridge, MA: MIT Press.

Anderson, J. R. (1987). Skill Acquisition: Compilation of weak method solutions. Psychological Review, 94, 192-210.

Anderson, J. R., Farrell, R., & Sauers, R. (1984). Learning to program in LISP. Conitive Science, 8, 87-129.

Anderson, R, C. (1977). The notion of schemata and the educational enterprise: General discussion of the conference. In R. C. Anderson, R. J. Spiro, & W. E. Montague (Eds.), Schooling and the Acquisition of Knowledge (pp. 415-431). Hillsdale NJ: Erlbaum.

Beishuizen, J. J. Beishuizen, M., & Felix, E. J. M. (1990). The emergence of addition and subtraction skills. In L. de Leeuw, J. M. Pieters, & P. R. J. Simons (Eds.), Computers and Instruction. Lisse: Swets & Zeitlinger.

Bell, B. F. (1981). When is an animal not an animal? Biological Education, 15, 213-218.

Bransford, J. D., Franks, J. J., Vye, N. J., & Sherwood, R. D. (1989). New approaches to instruction: Because wisdom can't be told. In S. Vosniadou & A. Ortony (Eds.), Similarity and Analogical Reasoning (pp. 470-497). Cambridge, England: Cambridge University Press.

Brewer, W. F., & Samarapungavan (1991). Child theories versus scientific theories: Differences in reasoning or differences in knowledge? In R. R. Hoffman & D. S. Palermo (Eds.), Cognition and the Symbolic Processes: Vol. 3. Applied and Ecological Perspectives (pp. 209-232). Hillsdale, NJ: Erlbaum.

Brown, A. L. (1989). Analogical learning and transfer: What develops? In S. Vosniadou & A. Ortony (Eds.), Similarity and Analogical Reasoning (pp. 369-412). Cambridge, England: Cambridge University Press.

Brown, A. L., Bransford, J. D., Ferrara, R. A., & Campione, J. C. (1983). Learning, remembering, and understanding. In J. H. Flavell, & E. M. Markman (Eds.), Handbook of child psychology: Cognitive development (Vol. 3, pp. 77-166). New York: Wiley.

Brown, A. L., & Campione, J. D. (1981). Inducing flexible thinking: A problem of access. In M. Friedman, J. P. Das, & N. O'Connor (Eds.), Intelligence and Learning (pp. 515-530). New York: Plenum Press.

Brown, A. L., & Reeve, R. (1987). Bandwiths of competence: The role of supportive context in learning and development. In S. Liben (ed.), Development and Learning: Conflict or Congruence? (pp.173-223).

Brown, J. S. & Burton, R. R. (1978). A paradigmatic example of artificial intelligent instructional system. International Journal of Man-Machine Studies, 10, 323-339.

Bruin, A. J. & Wilenga, B. J. (1984). Expertsystemen. In G. A. M. Kempen & C. Sprangers (Eds.), Kennis, Mens en Computers (pp.31-40). VNU: Amsterdam.

Bruner, J. S. (1983). In Search of Mind. New York: Harper & Row.

Buchanan, B. G., & Shortcliffe, E. H. (1984). Rule-based Expert Systems. Reading, MA: Addison Wesley.

Carbonell. J. (1970). AI in CAI: An artificial intelligence approach to computer assisted instruction. IEEE Transactions on Man- Machine Systems, 11, 180-202.

Case. R. (1982). Intellectual development: A systematic reinterpretation. New York: Academic Press.

Chaiklin, S. (1985, April). The Stability of Conceptions in Novice Physical Science Reasoning. Paper presented at the annual meeting of the American Educational Research Association, Chicago.

Chase, W. C. & Simon, H. A. (1973). Perception in chess. Cognitive Psychology, 4, 55-81.

Clancey, W. J. (1982). Tutoring rules for guiding a case method dialogue. In D. Sleeman & J. S. Brown (Eds.), Intelligent Tutoring Systems (pp. 51-78). London: Academic Press.

Clement, J. (in press). Nonformal reasoning in physics: The user of analogies and extreme cases. In J. Voss, D. N. Perkins, & J. Segal (Eds.), Informal Reasoning. Hillsdale: NJ: Erlbaum.

Collins, A. (1977). "Why Cognitive Science?" Cognitive Science, 1, 3-4.

Collins, A. (1986). A Sample Dialogue Based on a Theory of Inquiry Teaching (Tech. Rep. No.367). Urbana: University of Illinois, Center for the Study of Reading.

Collins, A., Brown. J. S., & Newman, S. E. (1987). Cognitive apprenticeship: Teaching the craft of reading, writing and mathematics. In L. B. Resnick (Ed.), Cognition and Instruction: Issues and Agendas. Hillsdale, NJ: Erlbaum.

Collins, A. Brown. J. S., & Newman, S. E. (in press). Inference in text understanding. In L. B. Resnick (Ed.), Knowing, Learning, and Instruction: Essays in Honor of Robert Glaser. Hillsdale, NJ: Erlbaum.

deKleer, J. & Brown, J. S. (1983). Assumptions and ambiguities in mechanistic mental models. In D. Gentner & A. Stevens (Eds.), Mental Models (pp.1550190). Hillsdale, NJ: Erlbaum.

diSessa, A. (1988). Knowledge in pieces. In G. Forman, & P. B. Purfall (Eds.), Constructivism in the Computer Age (pp. 49-70). Hillsdale, NJ: Erlbaum.

Driver, R., 7 Easley, J. (1978). Pupils and paradigms: A review of literature related to concept development in adoloscent science students. Studies in Science Education, 5, 61-84.

Feigenbaum, e. A. & McCorduck, P. (1983). The fifth generation artificial intelligence and Japan's computer challenge to the world. Reading, MA: Addison Wesley.

Feldman, D. H. (1980). Beyond Universals in Cognitive Development. Norwood, NJ: Ablex.

Fischer, G. (1988). Enhancing incremental learning processes with knowledge-based systems. In H. Mandl & A. Lesgold (Eds.), Learning Issues for Intelligent Tutoring Systems (pp.138-163). New York: Springer Verlag.

Gardner, H. (1987). The Mind's New Science. New York: Basic Books.

Gelman. R. & Gallistel, C. R. (1978). The Child's Understanding of Number. Cambridge, MA: Havard University Press.

Glaser, R. (1984). Education and thinking: The role of knowledge. American Psychologist, 39, 93-104.

Glaser, R. (1985). All's well that begins and ends with both knowledge and process: A reply to Sternberg. American Psychologist, 40, 573-574.

Glaser, R. (1988). Learning theory and theories of knowledge. In E. de Corte, H. Lodewijks, R. Parmentier, & P. Span (Eds.), Learning and Instruction (pp. 397-414). Oxford: Leuven University Press and Pergamon Press.

Glaser, R. & Bassok, M. (1989). Learning theory and the study of instruction. Annual Review of Psychology, 40, 631-666.

Godel, K. (1962). On Formally Undecidable Propositions. New York: Basic Books.

Goldstein. I. (1977). The Computer as a Coach: An Athletic Paradigm for Intellectual Education, MIT AI Lab. Memo.

Goldstein, I. & Papert, S. (1977). Artificial intelligence, language, and the study of knowledge. Cognitive Science, 1, 5-21.

Greeno, J. G. (1978). A study of problem solving. In R. Glaser (Ed.), Advances in Instructional Psychology (Vol. 1, pp.13-74). Hillsdale, NJ: Erlbaum.

Hayes, J. R. & Simon, H. A. (1977). Psychological differences among problem isomorphs. In N. J. Castellan, Jr., D. B. Pisoni, & G. R. Potts (eds.), Cognitive Theory (pp. 21041). Hillsdale, NJ: Erlbaum.

Hayes-Roth, F., Watrman, D. A., & Lenat D. B. (1983). Building Expert Systems. Reading, MA: Addison Wesley.

Holland, J., Holyoak, K. J. Nisbett, R. E., & Thagard, P. (1986). Induction: Processes of Inference, Learning, and Discovery. Cambridge, MA: MIT Press.

Johnson-Laird, P. N. (1983). Mental Models. Cambridge, MA: Havard University Press.

Just, M. A. & Carpenter, P. A. (1987). The Psychology of Reading and Language Comprehension. Boston, MA: Allyn & Bacon.

Kahneman, D., Slovic, P., & Tversky, A. (1982). Judgment Under Uncertainty: Heuristics and Biases. Cambridge, England: Cambridge University Press.

Kaplan, C. A. (1987). Computer Simulation; Separating Fact from Fiction. Pittsburgh: Carnegie-Mellon University.

Karmiloff-Smith, A. (1986). Stage / structure versus phase / process in modelling linguistic and cognitive development. In I. Levin (Ed.), Stage and Structure (pp. 164-190). Hillsdale, NJ: Erlbaum.

Kearsley, G. P. (1987). Artificial Intelligence and Instruction: Applications and Instruction. Reading, MA: Addison Wesley.

Kieras, D. E. (1987). What Mental Models should be Taught: Choosing Instructional Content for Complex Engineered Systems (Tech. Rep. No. 24 TR-87/ONR-24). University of Michigan.

Klahr, D., & Wallace, J. (1976). Cognitive Development: An Information Processing View. Hillsdale, NJ: Erlbaum.

Kuhn, D., Amsel, E., & O'Loughlin, M. (1988). The Development of Scientific Thinking Skills. Orlando, FL: Academic Press.

Kuhn, T. S. (1962). The Structure of Scientific Revolutions. Chicago: University of Chicago Press.

Kuhn, T. S. (1977). Objectivity, value judgment, and theory choice. In T. S. Kuhn (Ed.), The Essential Tension. Chicago: University of Chicago Press.

Kulkarni, D., & Simon, H. A. (1988). The process of scientific discovery: The strategy of experimentation. Cognitive Science, 12, 139-175.

Larkin, J. H., Reif, F., Carbonell, J., & Gugliotta, A. (1988). FERMI: A flexible expert reasoner with multi-domain inferencing. Cognitive Science, 12, 101-138.

Lesgold, A. m., Lajoie, S. P., Bunzo, M., & Eggan, G. M. (1988). Sherlock: A Coached Practice Environment for an Electronics Troubleshooting Job. Paper presented at the J. S> Macdonald Foundation Conference on CAI and ITS, CMU, Pittsburgh, PA.

Lewis, M. W., Milson, R., & Anderson, J. R. (1988). Designing an Intelligent tutoring system for high school mathematics ICAI: The teacher apprentice project. In G. Kearsley (Ed.), Artificial Intelligence and Instruction: Applications and Methods. New York: Addison Wesley.

Mandl, H. & Lesgold, A. (1988). Learning Issues for Intelligent Tutoring Systems. New York: Springer Verlag.

Mandl, H., Stein, N. L., & Trabasso, T. (1984). Learning and Comprehension of Text. Hillsdale, NJ: Erlbaum.

McCloskey, M. (1983). Naive theories of motion. In D. Gentner & a. Stevens (Eds.), Mental Models (pp.299-324). Hillsdale, NJ: Erlbaum.

McCloskey, M. & Kargon, R. (1988). The meaning and use of historical models in the study of intuitive physics. In S. Strauss (Ed.), Ontogeny, Phylogeny, and Historical Development (pp. 49-67). Norwood, NJ: Ablex.

Medin D. & Wattenmaker, W. D. (1987). Category cohesiveness, theories, and cognitive archeology. In U. Neisser (Ed.), Concepts and Conceptual Structure: Ecological and Intellectual Factors in Categorization (pp. 25-62). Cambridge, England: Cambridge University Press.

Merrill, M. D. (1988). The role of tutorial and experiential models in intelligent tutoring systems. Educational Technology, July.

Meyer, B. J. F. (1979). Research on Prose Comprehension: Applications for Composition Teachers (Res. Rep. No. 2). Department of Educational Psychology, College of Education, Arizona State University, Temple Arizona.

Miller, G. A. (1979). A Very Personal History. Paper presented at the Cognitive Science Workshop, MIT, Cambridge, MA.

Murphy, G. L. & Medin, D. (1985). The role of theories in conceptual coherence. Psychological Review, 92, 289-316.

Newell, A., Rosenbloom, P. S., & Laird, J. E. (1989). Symbolic architectures for cognition. In I. Posner, (Ed.), Foundations of Cognitive Science (pp. 93-132). Cambridge MA: MIT Press.

Newell, A. & Simon, H. A. (1972). Human Problem Solving. Engelwood Cliffs, NJ: Prentice Hall.

Nickerson, R., Perkins, D. N., & Smith, E. (1985). The Teaching of Thinking. Hillsdale, NJ: Erlbaum.

Norman, D. A. (1983). Some observations on mental models. In L. Stevens & D. Gentner (Eds.), Mental Models (pp. 7-14). Hillsdale, NJ: Erlbaum.

Osborne, R. J., & Wittrock, M. C. (1983). Learning science: A generative process. Science Education, 67, 489-508.

Palinscar, A. S. & Brown, A. L. (1984). Reciprocal teaching of comprehension fostering and monitoring activities. Cognition and Instruction, 1, 117-175.

Papert, S. (1980). Mindstorms: Children, Computers, and Powerful Ideas. New York: Basic Books.

Pask, G. (1976). Styles and Strategies of Learning. British Journal of Educational Psychology, 46, 128-148.

Piaget, J. (1930). The Child's Conception of Physical Causality. New York: Harcourt Brace.

Polya. G. (1954). Mathematics and Plausible reasoning. Princeton, NJ: Princeton University Press.

Polya G. (1957). How to Solve it. New York: Doubleday.

Rosch, E. H. (1975). Cognitive representation of semantic categories. Experimental Psychology: General, 104, 192-233.

Rumelhart, E. H. (1975). Schemata: The building blocks of cognition. In R. J. Spiro, B. C. Bruce, & W. F. Brewer (Eds.), Theoretical Issues in Reading Comprehension (pp. 99-135). Hillsdale, NJ: Erlbaum.

Rumelhart, D. E. & Norman, D. A. (1981). Accretion, tuning, and restructuring: Three modes of learning. In J. W. Cotton & R. Klatzky (Eds.), Semantic Factors in Cognition (pp. 37-60). Hillsdale, NJ: Er;baum.

Rumelhart, D. E. (1982). Comments on Cognitive Science. Unpublished Manuscript, University of California at San Diego.

Rumelhart, D. E. & Ortony, A. (1977). The representation of knowledge in memory. In R. C. Anderson, R. J. Spiro, & W. E. Montague (Eds.), Schooling and the Acquisition of Knowledge (pp. 99-136). Hillsdale, NJ: Erlbaum.

Samarapungavan, A. (1989, March). What Chldren From India Know about Observational Astronomy: A Cross-cultural Study. Paper presented at the annual meeting of the American Educational Research Association, San Fransisco.

Schank, R. C. & Abelson, R. P. (1977). Scripts, Plans, Goals, and Understanding. Hillsdale, NJ: Erlbaum.

Schneider, W. (1985). Development trends in the metamemory-memory behavior relationship: An integrative review. In D. L. Forrest-Pressley, G. E. MacKinnon, & T. G. Waller (Eds.), Metacognition, Cognition, and Human Performance. San Diego: Academic Press.

Schoenfeld, A. H. (1982). Measures of problem-solving performance and of problem-solving instruction. Research in Mathematics Education., 13, 31-49.

Shiffrin, R. M. & Schneider, W. (1977). Controlled and automatic human information processing: II Perceptual learning, automatic attending, and a general theory. Psychological Review, 84, 127-190.

Shortliffe, E. H. (1976). Computer-based Medical Consultations: Mycin. New York: American Elsevier.

Siegler, R. S. & Klahr, D. (1982). When do children learn? The relationship between existing knowledge and the acquisition of new knowledge. In R. Glaser (Ed.), Advances in Instructional Psychology. Hillsdale, NJ: Erlabum.

Siegler, R. S. & Richards, D. D. (1983). The development of two concepts. In J. Brainerd (Ed.), Recent Advances in Cognitive Developmental Theory (pp. 51-121). New York: Springer Verlag.

Simon, H. A. & Kaplan, C. A. (1989). Foundations of cognitive science. In I. Posner (Ed.), Foundations of Cognitive Science (pp.1-48). Cambridge, MA: MIT Press.

Sleeman, D. & Brown, J. S. (1982). Intelligent Tutoring Sytems. New York: Academic Press.

Smith, E. L.& Medin, D. (1981). Categories and Concepts. Cambridge, MA: Havard University Press.

Solomon, J. (1983). Thinking in two worlds of knowledge. In H. Helm & J. D. Novak (Eds.), Proceedings of the International Seminar: Misconceptions in Science and Mathematics (pp.127-132). Ithaca, NY: Cornell University.

Stead, B. F. & Osborne, R. J. (1980). Exploring students' conceptions of light. Australian Science Teachers Journal, 26, 57-84.

Stead, B. F. & Osborne, R. J. (1981). What is friction? Some children's ideas. Australian Science Teachers Journal, 27, 51-57.

Steinberg, E. r. (1989). Cognition and learner control: A literature review, 1977-1988. Journal of Computer-Based Instruction, 16, 117-121.

Sternberg, R. J. (1984). Mechanisms of cognitive Development. New York: W. H. Freeman.

Stone, P. J., Dunphy, D. C., Smith, M. S., & Ogilvie, D. M. (1966). The General Enquirer: A Computer Approach to Content Analysis. Cambridge, MA: MIT Press.

Tennyson, R. D. & Park, O. C. (1987). Artificial intelligence and computer-based learning. In R. M. Gagne (Ed.), Instructional Technology: Foundations (pp. 319-342). Hillsdale, NJ: Erlbaum.

Thorndike, E. L. & Woodworth, R. S. (1901). The influence of improvement in one mental function upon the efficiency of other functions. Psychological Review, 8, 247-261.

Tobias, S. (1987). Learner characteristics. In R. M. Gagne (Ed.), Instructional Technology: Foundations (pp. 319-342). Hillsdale, NJ: Erlbaum.

Turing, A. M. (1937). On computable numbers with an application to the entsheiddungs problem. Proceedings of the London Mathematical Society (Series 2, 42, pp. 230-265).

Van Dijk, T. A. (1980). Semnatic macro-structures and knowledge frames in discourse comprehension. In M. A. Just & P. A. Carpenter (Eds.), Cognitive Processes in Comprehension (pp. 3-32). Hillsdale, NJ: Erlabum.

Van Dijk, T. A. & Kintsch, W. (1983). Strategies in Discourse Comprehension. New York: Academic Press.

VanLehn, K. (1988). Toward a theory of impasse-driven learning. In H. Mandl & A. Lesgold (Eds.), Learning Issues for Intelligent Tutoring Systems (pp. 19-41). New York: Springer Verlag.

Van de Veer, G. c. & Beishuizen, J. J. (1986). Learning styles in conversation - a practical application of Pask's learning theory to human-computer interaction. In F. Klix & H. Wandke (Eds.), Man-Computer Interaction Research MACINTER-I (pp. 195-205). Amsterdam: Elsevier.

Vosniadou, S. (1989). Analogical reasoning and knowledge acquisition. In S. Vosniadou & A. Ortony (eds.), Similarity and Analogical Reasoning (pp. 413-437). Cambridge, England: Cambridge University Press.

Vosniadou S. & Brewer W. F. (in press). A cross-cultural investigation of children's acquisition of knowledge in observational astronomy: Greek and American data. In H. Mandl, E. DeCorte, N. Bennet, & H. F. Friedrich (Eds.), Learning and Instruction: European Research in an International Context. Oxford: Pergamon Press.

Wason, p. & Johnson-Laird, P. (1972). The Psychology of Reasoning: Structure and Content. Cambridge, England: Cambridge University Press.

Weber, R. P. (1985). Basic Content Analysis. Beverly Hills, CA: Sage Publications.

White, B. Y. (1983). Sources of difficulty in understanding Newtonian dynamics. Cognitive Science, 1, 41-65.

White, B. Y. & Frederiksen, J. R. (!987). Causal Model Progressions as a Foundation for Intelligent Learning Environments (Tech. Rep. No. 6686). Cambridge, MA: BBN Laboratories Inc.

White, B. Y. & Horowitz, P. (1987). Thinker Tools: Enabling Children to Understand Physical Laws (Tech. Rep. No. 6470). Cambridge, MA: BBN Laboratories Inc.

Winograd, T. (1975). Frame representation and the declarative-procedural controversy. In D. G. Bobrow & A. Collins (Eds.), Representation and Understanding: Studies in Cognitive Science. New York: Academic Press.

Wiser,M.(1988). The differentiation of heat and temperature: History of Science and novice-expert shift. In S.Strauss (Ed.),Ontogeny, phylogeny and historical development (pp. 28-48). Norwood, NJ:Ablex.

Wiser,M.,& Carey,S. (1983) When heat and temperature were one. In D.Gentner and A.L. Stevens (Eds.), Mental Models (pp.267-298). Hillsdale, NJ:Lawrence Erlbaum.

Wolf, F.M. (1987). Meta-analysis: Quantitative Methods for research synthesis. Beverly Hills, CA: Sage Publications.

MODELLING THE EXPERT TRAINING DEVELOPER

R.S. Perez amd E.C. Neiderman*

U.S. Army Research Institute
Alexandria, Virginia
USA

ABSTRACT

This paper presents the results of a study of problem solving in the domain of design and development of technical training. The study was conducted in two phases: phase 1, consisted of a structured interview of four "experts" in the field of training development, and phase 2 validated the Cognitive Model developed in the first phase. The results of Phase I will be presented in this paper.

Phase 1. Experts were selected using four criteria, including nomination by peers, years of experiences, awards from professional organizations, and hands-on experience. These experts were then interviewed using a structured interview form designed to elicit (expertise) how they designed and developed training materials.

A scheme for coding the protocols was developed (adapted, from Greeno, et.al. 1990) considering three aspects of the problem solving process involved in the development of training: subproblems, and types of knowledge used in the problem solving process. Data from the four interviews were used to develop a Cognitive Process Model that characterized expertise and the problem-solving processes used by our experts to design and develop training materials.

The data showed variation in the variables of interest among the "expert" developers. Although the amount of effort spent on the various subproblems, types of knowledge used, and use of operators differed during the development of training, a picture of common problem-solving strategies emerged. These common strategies described general features used by our "experts" during the training development process.

Phase 2. The second phase of the study will use eight participants, four expert training developers and four novices in a "think-aloud" instructional design task. The criteria for selection of the four experts will be the same as those used in phase 1 of the study. The criteria for selection of the four novice instructional designers is that they should have taken at least one course in instructional design and completed a master's degree in curriculum development, instructional design, or instructional technology.

---

* We would like to thank our "Experts" for their enthusiastic support of this project. We would also like to thank Professor John Guthrie and Ms. Shelley Weber for aiding in data collection; Ms. Julie Johnson and Rachel Brown for their comments on an earlier draft of this paper and for their help in coding and analyzing the protocols.

The analysis of these eight protocols are to be used to validate the Cognitive Model formulated in Phase 1 and to provide more detailed data for the design of automated training development job aids/tools.

1.0 INTRODUCTION

Beginning in the early 1960's, the U.S. Army, Navy, Air Force, and much of the industrial training community adopted a systems approach to curriculum development. This systematic approach, generally referred to in the Army as the Systems Approach to Training (SAT), came from the systems analysis or general systems theory of problem solving. This approach to instructional design and development is job performance oriented and is concerned with the identification of training requirements based on the analysis of a job/task data collected from the field.

The model that the SAT is based on is behaviorally oriented and stresses the identification of skills and knowledge's that students/trainees need to learn to perform satisfactorily a job/task. This approach was heavily influenced by the work of several learning psychologists: Robert Gagne, Leslie Briggs, Robert Mager, Robert Branson, Robert Glaser, and Lee Cronbach (Dick & Carey, 1978).

The systems approach to instruction as a model for the development of instruction had its initial impact on the design and development of programmed instruction. Since that time, hundreds of manuals and instructional materials have been developed using this approach.

1.1 Components of the SAT Model

The SAT approach consists of four major components: analysis, design, development, and evaluation (implementation and control).

The first component in the model is to conduct a job/task analysis, to determine what the training designer wants the trainees to be able to do when the trainees have completed the instruction. This first component analysis consists of multiple substeps: analyzing the job/task, selecting tasks/functions, constructing job performance measures, analyzing existing courses, and selecting the instructional setting.

The next component is design. After the designer has identified the instructional goals, the goal(s) must be analyzed in order to identify the subordinate skills that trainees must learn. During this component, the training designer must identify the concepts, rules, and any information which a trainee needs to perform the task. Steps in a procedural sequence, which must be followed to perform a particular process, also have to be identified. In addition to identifying the substance of the job/task, it is also necessary to identify the specific skills that trainees must have prior to training. The designer must also specify the general characteristics of the students/trainees which will influence the design of the instruction. These characteristics may include special interests, abilities (e.g., IQ), and aptitudes such as mechanical and perceptual aptitudes, motivational level, relevant experiences, and so on.

Based on the analysis of the job/task, the designer will then make statements of the entry behaviors and characteristics of the potential students, specifying the skills they will learn, the conditions under which these skills will be performed, the criteria for successful performance, and assessment instruments for determining whether students have attained the goals of the instruction. The substeps in this component are: develop objectives, develop tests, describe entry behaviors, and determine sequence and structure.

The next component in this model is to develop the training/instruction. Given the information from the preceding components, the designer/developer will be begin to formalize the design of the instruction. Formalization consists of the selection and refinement of an instructional strategy. The instructional strategy is then used as a "blueprint" to guide the production of the training. During this component, criteria for reviewing existing instructional materials are developed. For most developers and designers the process is iterative, where the instructional strategy is initially selected at the design phase and then changed, modified, or refined during the development phase. The substeps of this component are: specify learning events/activities, specify instructional management plan and delivery system, review/select existing materials, develop instruction, and validate instruction.

The last component is the evaluation of instruction where a series of formative evaluations are conducted to determine how effectively the instruction "works" and to collect data which may be used to identify how to improve the instruction. In some cases, the formative evaluation data will be used to not only revise the instruction but to completely redesign the instruction. The substeps of this component are: implement the instructional management plan, conduct instruction, conduct internal and external evaluation, and revise if necessary.

1.2 Problems with the SAT Model

Although the SAT approach to training development has produced hundreds of hours of training in the Army, its implementation has proven to be problematic for several reasons.

• First, the process itself is complex, labor intensive, time consuming, and costly.

• Second, and perhaps more importantly, for the SAT process to be performed efficiently, the developer must have a high level of expertise in the subject matter, knowledge of instructional design, and in experience in the application of learning principles.

• Third, several manuals and instructional materials have been developed to guide users through the SAT process, these documents are designed to be compilations of what Anderson (1982) calls declarative knowledge, which is, knowledge about facts, concepts, and principles of "the what" of the SAT process. For example, the Army's guidance in the SAT process is contained in a four volume series of pamphlets (TRADOC's 350-30) that depicts the four phases of the SAT process as consisting of at least eighteen substeps. All of the directions contained within this four-volume set describe the training development activities to be performed and in what sequence, and what forms need to be completed.

These guidance documents lack what Anderson (1982) has termed the "how-to knowledge" or procedural knowledge of the training development process. Thus, the quality of the training produced is dependent on the developer's knowledge base, skills, talents, and procedural knowledge. Moreover, since the SAT approach is predominantly based on a behaviorist model of learning, very little attention is paid to a description or identification of the cognitive process underlying the design of training.

• Fourth, the nature of the design and development of instruction as a domain/task is not well understood. It is not known, for instance, what knowledge', skills, talents or underlying cognitive processes expert training developers use in the design and development of training. Further, we do not understand the role that principles of learning, heuristics, planning procedures, decision-making rules, and problem solving skills play in that process. Moreover, many authors (Guidon, 1990; Greeno, Korpi, Jackson,& Michalchik,1990; Goel and Pirolli, 1989) have described and characterized the task/domain of design and development of training and instruction as ill-defined since there are few constraints, multiple correct solution paths, and few a priori design rules.

2.0 THEORY BASED DESIGN

Over the last twenty years several attempts have been made to develop a prescriptive comprehensive instructional design theory (Bloom, 1976, 1981; Rothkofp, 1981; Case,

1978a, 1978b; Lawson & Lawson, 1980; Markle, 1978; Ausubel, 1968; Bruner, 1960,1966; and Skinner, 1954, 1965). Prescriptive theories or models of instructional design prescribe when a model(s) of instruction should be used (Reigeluth, 1987). These theories were designed with the assumption that design principles and strategies are independent of the domain they are applied to, that is they are domain independent.

Most of these theories were either based on an analysis of widely used instructional methods to derive elementary components (e.g., use of feedback and reinforcement) or on a descriptive theory or set of learning principles. The successful application of these theories to the design of training /instruction has been met with varying levels of success.

The most recent theory of this kind has been the "Instructional Transaction Theory" advanced by M. D. Merrill and his associates at Utah State University (Merrill, 1991). This theory differs from earlier theories in that it is partly based on the extracted expertise of a instructional design theorist/practitioner using knowledge acquisition techniques and theory.

The principal reason that these theories have not fared as well as hoped is because, in our view, they are not specific enough in prescribing the instructional strategy and how to implement it to a given training situation. More importantly, they do not generally draw explicitly upon empirical principles of learning and when they do make use these of learning principles. The principles they select only apply to a narrow range of tasks. For example, instructional theories based on B.F. Skinner's principle of behaviorism(i.e., positive reinforcement) have a difficult time dealing with the teaching of complex problem solving.

Although many of these theories, are intended to be domain independent. They were, however, developed with academic subject areas in mind such as; social studies, mathematics, sciences, geography, history , and within an educational context. Where within this context the presentation of declarative and procedural knowledge of the subject content is presented in a general manner without specific context. Whereas, in the training context declarative and proceduralized knowledge must be and is presented within a specific context, the job/task to be performed.

The designer must be able to translate these general prescriptions of the theory to the specific set of circumstances of the training situation and then evaluate how appropriate the prescriptions are for that situation. This is not an easy task for designers, in the Army, who have limited experience, knowledge and familiarity with instructional design theories or principles of learning.

In addition, the developer must also make a multitude of judgments or design decisions about the importance of the task characteristics, student characteristics, and evaluate how these characteristics are matched to an appropriate design.

The process of training development does not end with the selection of an appropriate instructional strategy or design. Once the design strategy is selected, the developer then uses this "blueprint" to guide the development of all the components of the instruction (lesson materials, student and instructor guides, manuals etc.,). Moreover, the supposed utility of such theories, although limited by the factors indicated, is to provide a general framework to guide the design and development of instruction or training across many domains.

The type of specific guidance that is needed is for the translation of a general prescriptive theory to the specific instance. This capability requires according Baker(1989) and others the specification of a set of "micro-rules"(strategic skills). These rules would articulate the instructional strategy so as to provide specific guidance in the implementation of the instructional strategy and aid the designer in making design decisions at the lesson and unit level. Locatis and Park (in press) have argued that one of the difficulties that designers have in implementing an instructional theory is, although it is relatively easy to acquire the descriptive knowledge from text book descriptions of the design theory, it is very difficult to acquire the strategic skills needed to apply the theory to specific situations.

Furthermore, many of these theories are stated at such a general descriptive level that it is difficult for practitioners to use them. This is, in part, due to the fact that many of authors of these instructional design theories have not taken the time to field test their theory with the actual users for whom they are intended and therefore have not determined their usability. They are not designer referenced that is not based on a model of how expert designers actually design.

3.0 AN EXPERT BASED MODEL OF DESIGN

In our view, the systems and theory based approach to instructional design has been used with modest success in providing guidance and support to the design and development of effective instruction and training. At many levels of industry and the military, effective training and instruction is produced despite limitations in the development model. However, much of the materials used today are designed and developed by well-experienced individuals with expertise, or "expert" designers/ developers. How is this training produced; and what are the skills, knowledge's, and abilities of those designers who produced them? So what is the expertise required of an instructional designers?

One approach to answering these questions is to systematically study the design process used by experts in the field and to derive an expert based cognitive process model of instructional and training design.

3.1 Empirical Study of Expertise

The study of expertise has been an intriguing subject for investigation since the early 1960's. Early studies by de Groot (1966) and Chase and Simon (1973) on chess experts demonstrated difference between strong and weak chess players in their ability to correctly reproduce large patterns of chess positions after a few seconds in viewing the chess board. The primary research approach has been to compare the performance of experts with novices.

More recent studies of the performance of experts has been done by a number of researchers and in such varied areas as chess (Simon,1980); medical diagnosis (Clancey,1981,1988); medicine (Groen & Patel, 1988, 1991), geometry (Greeno, 1977); genetics (Smith & Good, 1984); physics (Larkin & Reif, 1976); economics (Voss, Blaise, Means, Greene, & Ahwesh, 1988); avionic troubleshooting (Gitomer, 1988); and electrical and mechanical troubleshooting (White & Frederiksen, 1985, 1986, 1987).

3.1.1 The Nature of Expertise Chi, Glaser, and Farr (1988) have characterized the findings of the research on experts as:

(1) Experts excel mainly in their own domain; their expertise does not transfer to other domains;

(2) Experts perceive large meaningful patterns in their domain. This perceptual ability is not reflective of superior perceptual abilities, but rather, it reflects an organization of the knowledge base;

(3) Experts are fast. They are faster than novices at performing the skills of their domain and they quickly solve problems with little errors. The explanation for this speed is that experts have acquired skills through many hours of practice making many of the skills automatic and thus freeing up memory capacity for the processing of other tasks;

(4) Experts have superior short-term and long-term memory. This is due to the automaticity of many of the skills which enable the expert to have greater resources for storage;

(5) Experts see and represent a problem in their domain at a deeper (more principled) level than novices. Novices tend to represent problems at a superficial level. Chi, Feltovich, and Glaser (1981) found that experts used principles of mechanics to organize categories, whereas novices organized their categories around the literal objects stated in the problem description;

(6) Experts spend a great deal of time analyzing a problem qualitatively. The research shows that at he beginning of a problem solving episode, experts typically attempt to "understand" a problem completely before initiating any action, whereas novices begin immediately to apply equations and solve the unknown. Experts analyze the problem qualitatively by building a mental representation of the problem from which they can infer relations that can be used to define the situation, and then they add constraints to further define the problem.

4.0 RESEARCH IN DESIGN

Although we found over 200 articles in our review of the literature that had to do with some aspect of design, a much smaller number of these studies focused on the generic nature of the design process as performed by experts (Guindon,1990; Guindon, & Curtis, 1990; Greeno, et.al,1990; Goel, & Pirolli,1989; Adelson & Soloway,1985; Carroll, Thomas, & Malhorta, 1979; Jefferies, Turner, Polson, & Atwood, 1981). A smaller subset of these studies (Guindon,1990; Guindon, & Curtis, 1990; Goel,& Pirolli,1989; Adelson & Soloway,1985; Carroll, Thomas, & Malhorta, 1979) have attempted to derive a generic model of the design process. The researchers in these studies explored the design process across several different domains as diverse as mechanical engineering, computer programming, architecture, and instructional design. We have included in our review only those studies that have applied knowledge acquisition techniques and include instructional design as a domain.

In the Goel and Pirolli (1989) study the objective was to develop a cognitive theory of design. They first made a distinction between design and nondesign problem solving by identifying major invariants in the design problem space. This was achieved by comparing the protocols of expert designers from three different domains (one each) mechanical engineering, architecture, and instructional design of which the designers were asked to provide a design solution to a design task. They also identified those design problem solving activities that are domain specific as opposed to those that are generic. The design tasks were different for each designer. The architecture design task was concerned with the design of an automated post office where human tellers were replaced with automated ones. The mechanical engineering design task was to design the automated teller machine (ATM) to be used in the automated post office. Unlike the first two tasks mentioned, the instructional design task set out to design a stand-alone text based instructional program to transition secretaries from typewriters to a word processing system. One major finding of this study is the definition of a generic problem space for designers.

The generic problem space is characterized by the designer engaging in several activities. These activities are: extensive problem structuring, extensive performance modeling, personalized or institutionalized evaluating functions and stopping rules, solution decomposing into leaky modules, using abstractions to transform goals to artifact specifications, making and propagating of commitments, and using symbol systems.

One of the most interesting findings of this study(Goel,& Pirolli, 1989) is that their designers did not engage in a linear sequence of short generate-evaluate cycles. Rather, all their subjects used what they refer to as a limited commitment mode control strategy (LCMCS). In the LCMS strategy the designer begins by focusing on one component of the problem at a time and generates an initial design for that component. They then focus on this proposed design solution, and evaluate it as an interim design solution, and either accept, reject or modify it. They may not **only** apply it to that component but **extend** it to other components of the design task, and then evaluate it as a design solution to the **whole design task,** and either accept, reject, or modify it. The authors found that the initial design solution is evaluated by the designer within three different contexts: (1) as an interim solution for a specific component of the design task; (2) as a working model of the final design; and, (3) as a final design solution for the design task.

The second study that is relevant to the present study was conducted by Greeno, Korpi, Jackson, and Michalchik (1990), as the first study reviewed they define the task of instructional design as problem solving within an information-processing-theory framework (Greeno & Simon,1988; Newell & Simon, 1972). In this study of the design process they observed eight novice (teacher- trainees) designers and recorded their think aloud verbalizations as they designed and developed instructional materials to teach how to operate a fictitious piece of equipment or general principles such as energy storage, extraction, and transportation. The eight subjects were trained in verbalizing their thoughts and instructed to "think aloud" (Korpi, 1988) while solving the design task. The analysis of the results of this study characterized the design process as consisting of three subproblems; (1) **Determine the materials** to be used in the design, the arrangement of these materials, and the details of their implementation; (2) **Types of knowledge** used in the design, knowing the domain of design and the objects of the thing to be taught; and, (3) Use of **interacting subprocesses**, as formulating the problem, including general goals and constraints, adding components to the design, and elaborating and modifying components that have been included in the design.

The limitations of these studies as with any study using knowledge engineering techniques of knowledge acquisition are; that the analyses are based on a small number (generally only three) of subjects' verbal protocols, there is a lack of a formal coding procedure, nor do they establish any inter-rater reliability among their coders. Thus, their conclusions are based on a small number of experts who may or may not be representative of design experts in general. The design process that these experts exhibit may be idiosyncratic. Because they did not use a formal coding scheme, the conclusions may only reflect the subjective judgments of the authors. The smallness of the sample raises questions about the generalizability of the results of these studies. Lastly, but perhaps more important is the limitation of these techniques themselves. Chi and Bjork(1991) have pointed out in their review of knowledge elicitation techniques "that different techniques produce different forms of knowledge, some declarative and some procedural, pg 70". As an illustration of their point, they contrast various techniques for eliciting declarative knowledge from experts. For example concept listing (Cook and McDonald,1987) can be used to assess the organization of the experts knowledge, however, this technique does not typically reveal procedural knowledge that is believed to be attached to conceptual knowledge. Another method they discuss is to simply ask experts to perform the task and observe their behavior in problem solving or making decisions.

The limitation of these technique is that they do not reveal what knowledge generated the sequence of observed behavior. Think-aloud protocols where the experts are asked to describe what they are thinking about while performing the task may reveal the knowledge underlying their behavior. We are, however, still limited to what knowledge is in the experts' working memory. Anderson(1982) has suggested that as we acquire a procedural skill declarative knowledge is compiled or "chunked" into procedures and is no longer available to the expert.

4.1 Research Objective

This is an exploratory study that has as its objective the formulation of a cognitive process model (expertise) of the design and development of instruction and training. This model could be used to guide the design of computer-based decision aids (e.g., expert systems)/tools, to aid the training developer.

We view the developer as a problem solver who has great discretion and creativity in arriving at a particular design. The design and development of training/instruction is an ill-defined problem-solving task. We used an information processing-theory framework to examine the training development process (Greeno, et.al, 1990; Greeno & Simon, 1988; and Newell & Simon, 1972).

4.2 Methods and Procedures

Expert knowledge or expertise has been defined as consisting of declarative knowledge (i.e., facts, concepts, and principles), formal rules (i.e. rules regarding how to use these facts), procedural knowledge (i.e.,skills), meta rules (i.e., rules for applying formal and procedural rules), heuristics (i.e., rules of thumb), and relations among all these (Wiggs & Perez,1988; Hart,1986; Hamill,1984). The method used in this study for deriving experts' knowledge is knowledge acquisition. Knowledge engineering is a set of methods used to extract knowledge in the form of rules and procedures from experts. These rules and procedures characterize how experts performs their job. The use of these techniques for eliciting knowledge, in our view, go beyond the standard task analysis and allow us to characterize the experience-based components of the their design expertise. The most common knowledge-acquisition technique used is the structured interview (Yaghmai & Maxin, 1984).

4.2.1 <u>Participant Selection Procedure</u> A primary methodological concern in this study was the selection of training development experts. Expertise is a "slippery" construct. We operationally defined training development expertise based on four criteria: 1) nomination by peers; 2) years of experience; 3) awards from professional organizations; and 4) hands-on experience.

The nomination by peers criterion included supervisory ratings, expert ratings, learner evaluation of materials, and teacher evaluation of materials. Years of experience was simply the amount of time spent as a training developer. The minimum amount of experience was preset at ten years. Awards from professional organizations included the number, type, and prestigiousness of the recognition. Hands-on experience included relevant education, teaching experience, and consulting.

These criteria were used to identify experts, in addition, we also required that the experts be articulate. Four experts agreed to participate in the study, and they were each interviewed for approximately eight, one and one half hour sessions. A two-hundred dollar honorarium was offered to each expert, only one of the participants elected to accept the honorarium.

4.2.2 <u>Procedures (Structured Interviews)</u> The four experts were interviewed using a structured interview designed to elicit how they developed training materials. Biographic information, including education, training, employment history, and types of materials developed was obtained. All interviews were tape recorded.

The structured interview focused on nine general aspects of training development: 1) description of materials; 2) content; 3) audience characteristics; 4) goals of the developer; 5) learner motivation; 6) features of the material; 7) training context; 8) development process; and 9) general/miscellaneous. There were approximately five questions for each aspect. See appendix one for the structured interview for training material experts.

The first session of the interview consisted of collecting biographical data, such as; formal education and training, teaching experience, consulting, work experience, types of materials developed, and subject matter expertise. Participants were also asked during the initial interviews to bring with them an example of their most successful design effort and were told that they would be asked questions about the development of these materials. During the interviews, the questions were used as a general guide and as probes for the interviewers. Questions were used as probes only when the expert did not address or mention an aspect of training development. We tried not to "lead", or elicit specific types of information from the experts. However, the interviewees were encouraged to elaborate when ever necessary. Two of the experts we interviewed commented that the design process was conducted in a team and they therefore had developed formal conventions and guidelines to insure the uniform execution of the instructional design. In these cases the interviewers would probe for more details as to how these conventions were developed and used by the design team.

Interviews concluded when the expert discussed all aspects of training development or had exhausted the topic. This usually required about four and one half hours.

The second series of interviews pertained to a difficult training design project. The experts had very little difficulty in identifying and recalling such examples in fact their recall of these events were much more vivid and detailed than their successful cases. In most cases the experts were able to provide copies of these projects, as in the successful examples, and the interviewers familiarized themselves with these materials prior to the interview. Reference to successful and unsuccessful examples by the interviewers were used during the interview process to cue the experts about the various aspects of the design process. For example, the question how did they determine the sequence of the instruction for a lesson was asked with reference to these specific examples.

Coding Scheme. A coding scheme, based on Greeno et. al. (1990) was developed. This scheme orginally considered three aspects of the problem solving process involved in training development: subproblems, types of knowledge, and problem solving operators. **Subproblems** are the component tasks on which a designer worked while constructing the design and is comprised of such things as determining the content, sequencing the instruction, and timing(pacing). **Knowledge types** are the types of information used in planning the training. These included content, pedagogy, task knowledge, and personal knowledge. Greeno proposed three types of knowledge used in the design task. These are: (a) knowledge about the design task itself; (b) knowledge about teaching; and, (c) knowledge about the content to be taught. **Problem solving operators** are the general problem solving processes used in the design solution. In coding the interview protocols a number of modifications were made to the coding scheme originally problem solving **operators** listed propose, modify, remove, elaborate/extend, recap/reflect/ evaluate/monitor, and justify we sampled it to include only propose and evaluate. We also reorganized knowledge types to correspond with the instructional design phases of plan, design, develop, evaluate, miscellaneous category, and added external knowledge sources such as references, experts, and novices. Lastly, we dropped the use of problem solving operators because it was difficult to obtain an acceptable level of consistency between raters.

Analysis of Verbal Protocols. Tapes from the interviews were transcribed verbatim. These verbatim transcripts were coded then independently by two coders who were trained in the use of the coding scheme. Following the coding, inter-rater reliability was established by using Kappa coefficient (Cohen, 1960). Kappa is a chance-corrected estimate of nominal scale agreement, originally developed to provide an estimate of the reliability of patient assignment to medical diagnostic categories. It used to provide a more conservative estimate of inter-coder agreement than a simple percent of agreement, which according to Fleiss (1981) yields an inflated estimate of the degrees of reliability. The Kappa coefficient was used to estimate the level of agreement of two coders. The Kappa coefficient for the combined categories was k= .74. Generally Kappa values that exceed .40 are considered an acceptable level of reliability, and those above .75 are considered excellent.

Thus, the estimate of agreement between our coders across all categories of the coding scheme as you can see is near the excellent range.

4.3 Results and Observations

Overall the four expert training designers are similar in characteristics to other experts that have been studied in other domains. Chi, Glaser, and Farr (1988) and others have characterized experts as having six distinct characteristics that set them apart from nonexperts(novices). In our interviews with our four experts, we too found that they shared many of the characteristics of the experts in other domains. They did not transfer their expertise to other areas, they tended to organize the design problem into meaningful patterns, they analyzed the design task in a qualitative manner, and they were perceived by their peers as having a "deeper understanding" of the design problem.

In this paper the four experts designers are designated as follows: GTD1, GTD 2, SMED1, and SMED. The designation of the letters GTD refers to the fact that these designers designed instruction and training irrespective of their domain. We analyzed the data produced by the structured interviews by coding the designers responses to our structured interview questions (See Appendix "A "for interview questions) using a modified coding scheme (See Appendix "B" for coding scheme, Greeno, et.al, 1990). These interviews generated over 18 hours of comments captured on audio tape. The first pass of the analysis of the experts response segments were then coded using a three part category; subproblem, knowledge type, and operator. A segments was defined as a verbal utterance in response to one of the questions. Thus, depending on the experts' elaboration a response to a question may have consisted of several segments with being coded using multiple codes. This preliminary analysis provided us with a rough estimate of the frequency of different coded response segments made by the experts to each of the questions asked by the interviewers. Table 1 presents a listing of the most frequent codes used. Thus, preliminary analysis was followed by a more extensive examination using the revised coding scheme. By and large the questions did elicit similar responses from each of our experts.

5.0 DESIGN PROCESS

Almost, all of our experts describe a process that they engaged in designing instruction. At the top level, this process reflected the Systems Approach to Training (analysis, design, development, and evaluation). However, at a more detailed level some of our experts did not really separate the design from the analysis steps.

All the experts agreed that the first step in the process was an attempt to analyze the job or content to be taught into goals and objectives of the task. In the next several examples the content areas varied from teaching methods for revising a course, writing instructions to teach trainees to use a software program, or how to use a sophisticated piece of equipment in chemical analysis to instructing teachers to use small group process to teach. However, the number of categories coded to a specific response varied as a function of the degree of elaboration by each expert.

For example, in asking" How did you choose the content for inclusion (in the instruction) from other books, current research, professional opinion, or curriculum guides?"

GTD 1: Indicated how she arrived at the content for instruction" *I built the workshop around a written case study of that course that is going badly. I found that what motivates these curriculum managers (trainees) is enrollment, that people will keep coming to a good course. I gave them ten years of evaluation data on this hypothetical course. I made the enrollment declining, so they would get disturbed about the design of the course and brain storm ways to change it. I asked them to analyze trends and causes for the failure of the course and recommend steps to take to improve it."*

GTD 2: *The content for the course is based on the types of jobs these people do. The content focus is driven by the client and the audience(trainees).*

SMED 1: *Content is usually available from recent research but it is usually not taught to the intended audience (trainees). The content is also not necessary for beginning and professional staffers in the occupational setting.*

SMED 2: *At the beginning of this study, there was a school identified (having) low achievers and being ethnically diverse and they (the principal) were looking for some new ideas, so the school asked me to come out there and make suggestions. This (these ideas) were (based on research on the use of) on metacognition theories in schools and(on the Research by) Paris and Brown's study as well as Duffy's notions of strategic reading and based on the review of this research. This seemed to be a real fertile area for dealing with 20% of the population(students) that doesn't do well in discovery and hands-on learning (instruction).*

Another example is the expert's response to the question" How did you sequence and structure the content of the materials?"

GTD 1: "*The hardest part of this (is) to determine the sequence I will use and thus my process (the developmental progression). Part of my teaching is to have them shadow through a design. I try to create a sequence in the training, even if we have to force in some things. I build the procedures and use concepts to support them. A new concept being introduced will keep their attention as we go along. I also try to allude to their own experiences because they are professionals and probably have about 50% of what I am trying to teach. So I tie the other 50% into their own experience. I have to tie (I also try to understand) what these things mean to them, not to me. I may (get to the point where) understand everything and each context I describe, but do they (trainees)?*

GDT 2: "*The sequence is determined by the process involved in teaching, the model which we are working (with). I am trying to get people(trainees) to deal with the process, which provides the sequence. I recommend that they review the style guide before beginning (to write) but it is more telling after they have already had trouble writing. I will find an (that after an) assignment (has been given) that each person in the class has done the assignment differently but the user of the program will be faced with these differing pieces from the writers. If the students are aware of this, then they will be more attentive to (how they write) the subject. I realize this by trying it out on students and watching their attention spans. So my sequence has also been determined empirically by experience."*

SMED 1: What sort of analysis would you be teaching the students to use?
" One of the more simple things would be determine the lengths of the x and y values and the absorbencies values of the peaks, some of these are obvious but some are missed by the naked eyes which the scale expansion helps make them big enough in proportion to random noise. We really aren't doing too many mathematical problems and are not doing any transformations or anything like that, we are simply using the ability of the computer to display an easy adjustable scale."

The rest of the transcript deals with the actual content of the specific lesson.

SMED 2 : *"Basic structure is the emphasis in the "whole." We do this with all of our materials. This shows a pattern which is the first thing that needs to be seen. You cannot present to kids a segment of something and then later elaborate. You capture the whole event first, and then using that as an outline, elaborate on the segments. The structure can be seen even while watching the training video. There the teacher says we've been working on these strategies all year, now we are going to use them while reading this story. She (the teacher) is promoting the idea of using the story as an arena for trying out the strategies.*

It's important for both the teachers and the children to see the "whole" in order to get a visual map and structure (it) in their minds.

We use games to help them learn this, expecting them to later be able to explain the strategies themselves.

This step was followed by an analysis of the characteristics of the potential students. The need as expressed by our experts to determine the characteristics of potential students and trainees was one the most frequent codes used by the coders. Examples, of their comments about the importance of student characteristics are:

All of our experts did express the need to motivate students and to keep their motivation high during instruction. The motivation that they were attempting to design into their instruction/training was intrinsic to the task. For example, SMED 1 and GTD 1 both spoke of the importance of getting the students/trainees interested in the instruction" getting them hooked in." This comment is particularly of interest, since most of the students were presumedly already externally motivated to learn. They were either taking the training to learn a job, keep a job, or maintain a high grade point average in their college major.

GTD 2: "*They will be motivated in the training if everything has some functional significance(to them), otherwise, they will perceive it(the task) as (consisting of)abstract concepts that do not relate to the tasks they will face."*

GTD 1: " *I found that what motivates these curriculum managers is enrollment, that people will keep coming to a good course.'* "

SMED 1: *Students are motivated to study this material because it is in a laboratory course in Chemistry: they want to know these principles because they(known) are useful in interpreting data from experiment and completing exercises in the chemistry practicum they are taking."*

The analysis of our experts responses revealed that the most frequently stated design activity they engaged in was the analysis of the functions and tasks of the job, motivation of the students in the content to be taught, and the characteristics of the students. The most frequently coded categories reflect this finding. The frequency of categories are listed and defined in Table 1. These activities more specifically are: (1) determining the design tasks' (e.g., job/content) goals and objectives;

GTD 1: "*I analyze the job description to get an idea of the students abilities."*(2) the target students characteristics, prior knowledge's, skills, and abilities;

GTD 1: *"I find that it is not the educational level that is crucial but experience (that is important for me in designing a course)"*

GTD 2:"*I also try to determine how long the (potential) students have been on the job. I learn about their backgrounds from phone interviews before the I design the workshops"* and, (3) determining the instructional resources, learning methods, and media to be used. The responses from the experts with regard to how they determine the instructional resources they used to design instruction and training varied from "the client provides them" to "a search of the literature." Learning methods used also varied significantly from the use of "a Logical Process Model" to the use of " developmental model sequential model. "When *you are moving someone through a developmental sequence, you have to determine which of your items are necessary to teach in order to move someone along from novice to expert, and which areas you can--catapult a student through-- to at least an intermediate level without much explanation."*

5.1 Use of Heuristics in the Design Process

They also made use of various heuristics in the design of their instruction. For example GTD 2, in reapplying to follow-up questions on sequence of instruction referred to the use of "size of step." When asked what they meant and how did they implement this concept in their design of instruction. They referred to that the application of "size of step" was based on George Millers famous principle of "seven plus or minus two" in reference to the capacity of working memory. Our expert's translation of this well known psychological principle is that "*each step of instruction should never be more than seven segments (segments being either sentences or phrases) of instruction long and that each step should result to an action that is observable.* " Of course there is more," to size of step" that our expert practitioner suggests. However, it the translation and application of this principle as a heuristic that is of interesting point

Another example of a heuristic, used by our experts was with regard to the inclusion of the teaching and presenting of concepts in their training. They prefaced their remarks by stating" *that very few, if any, of their trainees would sit still for training that presented a lot of formal concepts."* Therefore, when it was necessary to include concepts in the training the developed the rule of thumb of embedding these concepts within procedures of the job/task to insure that trainees saw these concepts as being job relevant." *concepts are to be hung on the procedures of the task for example, when teaching a particular word processing program they would include concepts like "delete" by embedding them within the set of actual commands that were used to execute the deletion of a word, then a sentence, a paragraph, and so on.*

CONCLUSIONS

A primary reason for conducting this exploratory study was to formulate a top level model of the design process of "Experts." Our conclusions are tentative; they await verification in the second phase of the study. However, overall the analysis of the content of the responses made by these four experts suggested that although the data showed variation in the variables of interest among the "Expert" developers in the amount of effort spent on the various subproblems, types of knowledge used. We did ,however, observe a commonality among the designers in the pattern of use of these features during the design process. At a top level the design process as depicted by our expert reflected the Systems Approach to Training (analysis, design, development, and evaluation). However, at a more detailed level our experts did vary with respect to their implementation of and how they described the design process. They vary with respect to the framework used in guiding design decisions. For example, GTD 1 almost exclusively used a human development framework in deciding the sequence of instruction; whereas GTD 2 relied exclusively on a task decomposition method. While others SMED 1 and SMED 2 are guided by the content structure or process they are teaching.

Half of the experts (2) we selected and interviewed were experts in a particular subject matter (chemistry and small group teaching) while the other half (2) were what we labeled, as generic training designers. In our analysis of the protocols we found differences between these two types of designers. The generic designers were more acutely aware of the design process and appeared to be very comfortable in describing their declarative knowledge (i.e., facts, concepts, and principles), formal rules (i.e. rules regarding how to use these facts), procedural knowledge (i.e., skills), meta rules (i.e., rules for applying formal and procedural rules), heuristics (i.e., rules of thumb), and relations among all these used to come up with a design solution. Whereas, our subject matter expert designers were not as explicit about their rationale for their design solutions. Our interpretation of this difference is that the subject matter experts rationale and procedures are embedded within the structure of the content. A chemist that has taught college chemistry for years has built up over the years of experience a very good understanding of what (topics) he should present first and how to teach them. The type of declarative knowledge, formal rules, procedural knowledge, meta rules, heuristics and relations among all these rules, may have become automatic to him and are no longer not accessible to his conscious memory. An example, of this is when we asked our expert subject matter experts to present and describe a successful and unsuccessful case of their design efforts they were far more consciously aware and thereby more elaborate of what they did wrong in their unsuccessful case than in their successful one.

In the present study we too found as Greeno, et.al (1990) that the design process consisted of at least three subproblems; (1) **Determine the materials** to be used in the design, the arrangement of these materials, and the details of their implementation; (2) **Types of knowledge** used in the design, knowing the domain of design and the objects of the thing to be taught; and, (3) Use of **interacting subprocesses**, as formulating the problem, including general goals and constraints, adding components to the design, and elaborating and modifying components that have been included in the design.

We also found that our expert designers did not engage in a linear sequence of short generate-evaluate cycles. Rather they used a strategy similar to that described by Goel and Pirolli (1989) where the designer begins by focusing on one component of the problem at a time and generates an initial design, for that component. They then focus on this proposed design solution, and evaluate it as an interim design solution, and either accept, reject or modify it. They may not **only** apply it to that component but **extend** it to other components of the design task, and then evaluate it as a design solution to the **whole design task,** and either accept, reject, or modify it. The authors found that the initial design solution is evaluated by the designer within three different contexts: (1) as an interim solution for a specific component of the design task; (2) as a working model of the final design; and, (3) as a final design solution for the design task. The results of our study did differ significantly from both the Greeno (1990) and Goel and Pirolli (1989) in that our designers reported that they spent a large portion of the design process concerned with motivation factors. The analysis of these protocols has been qualitative and descriptive. We are working towards building a model that is quantitative and predictive. The validation of that model and a more detailed account of that model is the objective of Phase 2.

REFERENCES

Adelson, B., & Soloway, E. (1985). The role of domain experience in software design. IEEE Transactions on Software Engineering, 11, 1351-1360.

Anderson, J.R. (1982) Acquisition of cognitive skill. Psychological Review, 89, 369-406.

Ausubel, D.P.(1968). Educational psychology: A cognitive view. New York: Holt, Rinehart& Winston

Baker, E.V. (1989) personal communication.

Bloom, B.S.(1976). Human characteristics and school learning. New York: Mc Graw-Hill.

Bloom, B.S.(1981). All our children. New York: Mc Graw-Hill

Bruner, J.S.(1960). The process of education. New York: Random House.

Bruner, J. S.(1966). Toward a theory of instruction. New York: W.W Norton.

Carroll, J. M., Thomas, J. C., & Malhorta, A. (1979) Clinical-experimental analysis of design problem-solving. Design Studies, 1, 84-92.

Case, R.(1978a). Piaget and beyond: Toward a developmental based theory and technology of instruction. In R. Glaser (Ed.) Advances in instructional psychology (Vol.1). Hillsdale, NJ. Lawrence Erlbaum Associates.

Case, R.(1978b). A developmentally based theory and technology of instruction. Review of educational research,48,439-463.

Chi,M.T.H, & Bjork,R.A., (1991) Modelling expertise. In Daniel Druckman and Robert A.Bjork (Eds,) In the mind's eye. (pp.57-79) National Academy Press Washington, D.C.

Chi, M. T. H., Feltovich, P. J., & Glaser, R. (1981) Categorization and representation of physics problems by experts and novices. Cognitive Science, 5, 121-125.

Chi, M. T. H., Glaser, R., & Farr, M. J. (1988) The nature of expertise. Hillsdale, New Jersey: Lawrence Erlbaum Associates.

Clancey,W.J.,& Letsigner,R.(1981). NEOMYCIN: Reconfiguring a rule based expert system for application to teaching. Proceedings of the Seventh International Joint Conference on Artificial Intelligence (pp.829-835). Los Altos, CA: Morgan Kaufman.

Cohen,J. (1960). A coeifficient of agreement for nominal scales. Educational and Psychological Measurement., 20,37-46.

Dick,W., & Carey, L.(1978) The systematic design of instruction. Scott, Foresman and Company Glenview ,Illinois

Druckman,D., & Bjork,R.A., (Eds,) (1991) In the mind's eye . National Academy Press Washington, D.C

Fleiss,J.L. (1981). Statistical methods for rates and proportions. (2nd.ed) NY: John Wiley.

Gagne, R. M. (1977). The conditions of learning and theory of instruction (3rd ed). New York : Holt, Rienhardt & Winston.

Gitomer, D. H. (1988). Individual differences in technical troubleshooting. Human Performance, 1(2), 111-131.

Goel,V., & Pirolli,P., (1989). Designing within information-processing theory: the design problems space. AI Magazine, Volume 10(1), 19 - 36.

Greeno, J. G. (1977). Process of understanding in problem solving. In R. Glaser (Ed.), Cognitive Theory (Vol. 2, pp. 43-83). Hillsdale, NJ: Lawrence Erlbaum Associates.

Greeno, J.G., Korpi,M.K., Jackson,D.N.,& Michalchik,V.S., (1990). Processes and knowledge in designing instruction. ONR Technical Report, Stanford University

Greeno, J. G., & Simon, H. A. (1988) Problem Solving and Reasoning. In R. C. Atkinson, R. J. Herrstein, G. Lindzey, & R. D. Luce (Eds.), Steven's handbook of experimental psychology: Vol. 2: Learning and cognition (2nd ed.) (pp. 589-672). New York: Wiley & Sons.

Groen, G. J., & Patel, V.L.(1988). The relationship between comprehension and reasoning in medical expertise. In M.Chi.R.Glaser& M. Farr (Eds.), The nature of expertise (pp.287-310). Hillsdale,NJ: Lawrence Erlbaum Associates.

Groen, G.J., & Patel, V.L.(1991). A view from medicine. In Mike U. Smith(Ed.), Towards a Unified Theory of Problem Solving.(pp.35-44). Hillsdale,NJ: Lawrence Erlbaum Associates.

Guidon, R. (1990) Designing the design process: exploiting opportunistic thoughts. Human-Computer Interaction, 5, 305-344.

Guindon, R., & Curtis, B. (1988). Control of cognitive process during software design: What tools would support software designers? Proceedings of the CHI '88 Conference on Human factors in Computing Systems, 263-286. New York: ACM.

Hamill, B.W. (1984). Psychological issues in the design of expert systems. (Report TR-ZEY-84-01) Laurel MD: Milton S. Eisenhower Research Center.

Hart, A.(1986). Knowledge acquisition for expert systems. New York: McGraw-Hill.

Jefferies, R., Turner A. A., Polson, P., & Atwood, M. E. (1981) The processes involved in designing software. In J. R. Anderson (Ed.), Cognitive Skills and Their Acquisition (pp. 255-283). Hillsdale, NJ: Lawrence Erlbaum Associates, Inc.

Korpi, M.(1988). Making conceptual connections: An investigation of cognitive strategies and heuristics for inductive categorization with natural concepts.(unpublished) PhD dissertation, Stanford University, Ca.

Larkin, J. H., & Reif, F. (1976). Analysis and teaching of a general skill for studying scientific text. Journal of Educational Psychology, 68(4), 431-440.

Lawson, A. E., & Lawson, C.A.,(1980) . A theory of teaching for conceptual understanding, rational; thought, and creativity. The Psychology of teaching for thinking and creativity (1980 AETS Yearbook,pp.104-149). Columbus OH: Ohio State University.

Locatis, C., & Park, O. K. (in press). Some uneasy inquiries into ID (Instructional Design) Expert systems.

Markle, S.M.(1978). Design for instructional designers. Champaign, Il: Stipes Publishing.

Merrill, M. D. (1983). Component display theory. In C.M. Reigeluth (Ed.), Instructional-design theories and models: An Overview of their current status. Hillsdale, NJ. Lawrence Erlbaum Associates.

Merrill, M. D. (1991) An introduction to instructional transaction theory. Utah State University.

Newell, A., & Simon, H. A. (1972) Human Problem Solving. Englewood Cliffs, NJ: Prentice Hall.

Reigeluth, C. M.,(Ed.) (1987) Instructional-design theories and models: An Overview of their current status. Hillsdale, NJ. Lawrence Erlbaum Associates.

Rothkofp, E. Z.,(1981). A macroscopic model of instruction and purposive learning: An overview. Instructional science,10,105-122.

Simon, H. A. (1980). Problem solving and education. In D. Tuma & F. Reif (Eds.), Problems solving and education: Issues in teaching and research (pp. 81-86). Hillsdale, NJ: Lawrence Erlbaum Associates.

Skinner, B. F.,(1954). The science of learning and the art of teaching, Harvard Educational Review,24 (2),86-97.

Skinner, B. F.(1965). Reflections on a decade of teaching machines. In R. Glaser (Ed.), Teaching machines and programmed learning., II. Washington,D.C. National Education Association.

Smith, E. E., & Good, R. (1984). Problems solving and classical genetics: Successful versus unsuccessful performance. Journal of Research in Science Teaching, 21, 895-912.

Voss, J. F., Blaise, M.L., Means, Greene, T.R., & Ahwesh, E. (1988) Informal reasoning and subject matter knowledge in the solving of economic problems by naive and novice individuals. In L.B. Resnick (Ed.) Knowing, Learning, and Instruction: Essay in honor of Robert Glaser(pp217-249). Hillsdale NJ: Lawrence Erlbaum Associates

White, B. Y., & Frederiksen, J. R. (1985, July). QUEST: Qualitative understanding of electrical system troubleshooting. ACM SIGART Newsletter, pp. 34-37.

White, B. Y., & Frederiksen, J. R. (1986). Progressions of qualitative models as foundations for intelligent learning environments (BBN Report No. 6277). Cambridge, MA: BBN Laboratories.

White, B. Y., & Frederiksen, J. R. (1987). Causal model progressions as a foundation for intelligent learning environments. (BBN Report No. 6277). Cambridge, MA: BBN Laboratories.

Wiggs, C. L., & Perez, R.S.,(1988). The use of knowledge acquisition in instructional design. Computers in human behavior, 4, 257-274.

Yaghmai, N. S. & Maxin, J. A.,(1984). Expert systems; A tutorial. Journal of the American Society for Information Science, 35,(5),306-316.

APPENDIX "A"

Structured Interview for Training Development Experts

Example Questions from Each Category of Structured Interview for Training Material Experts

Interviewer refers to a specific aspect during the demo:

1) **Describe or Demonstrate Materials (6 questions)**
How do you describe this feature of your material . . . etc.?

2) **Content Questions (6 questions)**
What factors determined which specific topics and subtopics you included etc.?

3) **Characteristics of the Audience (5 questions)**
How do you determine the background knowledge of the students that you prepare materials for etc.?

4) **Goals of the Developer (3 questions)**
What kind of learning are you trying to promote with these materials - recall, problem-solving, discovery, or other etc.?

5) **Motivation Questions (4 questions)**
How do you arouse curiosity regarding the content etc.?

6) **Specific Features of the Material (3 questions)**
When, why, and how do you use diagrams, pictures, flow charts, illustrations, drawings, graphics, metaphor, analogies, and examples in your materials etc.?

7) **Context Questions (3 questions)**
How do you make these materials meaningful to the student in the course or job situation etc.?

8) **Development Process (11 questions)**
Do you try-out the materials with a user? Who are they? What type of try-out do you use? How do you use the results of this try- out etc.?

9) **General Questions (4 questions)**
How does your material contrast with other materials in your field etc.?

APPENDIX B

Coding Scheme

I. INSTRUCTIONAL DESIGN PHASES

PLAN

A DETERMINE GOALS
B DETERMINE FUNCTIONS/TASKS TO BE TAUGHT
C IDENTIFY KNOWLEDGE, SKILLS, ABILITIES REQUIRED TO PERFORM TASK
D DETERMINE LEARNER CHARACTERISTICS
E DETERMINE CLIENT'S REQUIREMENTS AND CONSTRAINTS

DESIGN

F DETERMINE CONTENT/BREAK INTO OBJECTIVES
G DETERMINE FORMAT, SEQUENCING, TIMING
H DETERMINE INSTRUCTIONAL TRANSACTION, STRATEGIES FOR TRANSMISSION OF CONTENT (COLLABORATION, OPEN DISCUSSION, ANALOGY, LECTURE)
I DETERMINE MATERIALS OR FACILITIES (PERSON, COMPUTER, BOOK)
J CREATE STORYBOARD, TEMPLATE, MOCK-UP, MODEL, SIMULATION

DEVELOP

K DEVELOP FIRST DRAFT
L PILOT DRAFT
M REFINE DRAFT

EVALUATE

N FORMATIVE-EVALUATION DURING DEVELOPMENT
O SUMMATIVE-EVALUATION AFTER COMPLETION OR WHEN IMPLEMENTED WITH TARGET AUDIENCE

OTHER

P GENERALIZATION-COMMENT PERTAINS TO ANY OR NO PARTICULAR PHASE
Q MISCELLANEOUS

II. KNOWLEDGE SOURCES

PRIOR KNOWLEDGE

1 CONTENT OR SUBJECT MATTER
2 FORMAT, FEEDBACK, ORGANIZATION OF CONTENT
2A DESIGN PROCESS
3 PEDAGOGY, STRATEGIES, DELIVERY METHODS, LEARNING PRINCIPLES
4 REFERENCES TO PERSONAL EXPERIENCES OR MAKES PREDICTIONS BASED ON PAST EXPERIENCES

AWARENESS AND MONITORING

5 AWARENESS OR MONITORING OF GOALS, OBJECTIVES, CONTEXT OF INSTRUCTION
6 AWARENESS OR MON. OF DESIGNER'S ROLE, STRENGTHS/WEAKNESSES OF DESIGNER/THINGS I TYPICALLY DO/FAILURES/DIFFICULTIES/ THINGS I DO BEST
7 AWARENESS OR MON. OF DESIGN PRINCIPLES, DESIGN STRENGTHS/ WEAKNESSES/BENEFITS OF ONE FORMAT OVER ANOTHER
8 AWARENESS OR MON. OF AUDIENCE/STUDENT NEEDS/ CHARACTERISTICS/POINTS OF VIEW
9 AWARENESS OR MON. OF MOTIVATION/ATTENTION ISSUES
10 AWARENESS OR MON. OF CONSTRAINTS AND CLIENT REQUIREMENTS

EVALUATION

11 REFERENCE TO PERSONAL BELIEFS, THOUGHTS/VALUES/FEELINGS ABOUT DESIGN OR INSTRUCTION (GENERAL)
12 PERSONAL EVALUATION OF PRODUCT, DESIGN, METHODS, FORMAT, ETC. (PRODUCT-SPECIFIC)
13 NEWLY ACQUIRED KNOWLEDGE OR UNDERSTANDING BASED ON WHAT IS LEARNED FROM ONGOING DESIGN, RECENT PILOT OR END EVALUATION

EXTERNAL SOURCES

14 REFERENCES- BOOKS, OTHER TRAINING MATERIALS
15 EXPERTS, NOVICES

OTHER

16 MISCELLANEOUS-CANNOT BE CLASSIFIED IN OTHER CATEGORIES

ANALYTICAL METHODS FOR OPTIMIZING KNOWLEDGE-BASED INFORMATION STRUCTURES IN SIMULATION-BASED TRAINING

Bernhard Zimolong and Udo Konradt

Lehrstuhl für Arbeits- und Organisationspsychologie
University of Bochum
Bochum
Germany

1.0 INTRODUCTION

For obvious reasons, computer-aided instructions (CAI), training methods and intelligent tutoring systems, e.g. expert systems, are used to assist teachers in performing their tasks and functions. The extent to which it would be useful and desirable to replace a human teacher completely depends on the task to be accomplished, a problem which will be discussed at the end of this paper.

Intelligent tutoring systems play in important role in simulation-based training. The function of a tutoring system is to distribute tasks and study material among the students, present examples, observe and control learning progress and assist the students in their learning process by providing instructions, task definitions, feedback and supplementary material. It is also the task of the computer to assess student performance by establishing the number of problems solved and analyzing the quality of the solutions proposed. The PLATO computer system (Programmed Logic for Automated Teaching Operations), which was developed at the university of Illinois, is an example of a computer-aided instructional tool. This system is currently connected with 1,300 terminals operated at more than 200 places all over the United States. The original purpose of PLATO was to instruct students in individual courses, e.g. geography. Today this system is commercially available for instruction purposes in industry and the military. A description and evaluation of PLATO is given in Eberts & Brock (1987).

A later development of CAI is "intelligent computer-aided training (ICT)" which is frequently used together with a training simulator. According to Carbonell (1970) ICT is a system which not only provides knowledge of a special field of expertise but is also able to interact with the student user and assess the extent of his understanding on the grounds of the errors detected. Similar to the skills of a teacher, the system usually comprises three modules or elements:

- A knowledge module which stores domain-specific information relevant to the training course. Suitable procedures are applied for acquiring pertinent knowledge.

Advanced Technologies Applied to Training Design, Edited by
R.J. Seidel and P.R. Chatelier, Plenum Press, New York, 1993

• A learner module including the student's knowledge module which can establish the current level of knowledge on the basis of feedback information.

• A tutorial module which contains instructional strategies and selects them for each student.

The knowledge module contains domain-specific knowledge which is relevant to the training course, e.g. facts and rules concerning troubleshooting in electronic systems. Knowledge is represented by means of common standard procedures which, for example, are available in so-called expert system shells: production rules, semantic networks or procedural representations such as frames and scripts. In accordance with the instructional strategy selected, the knowledge module generates questions and problems for the student to deal with.

In the learner module, the individual knowledge structure of the student is established on the basis of the tasks accomplished and the answers given. The purpose of this process is to understand the personal knowledge structure of the learner with a view to analyzing and evaluating his errors, answers and solution-finding plans. The model should be able to tell why a particular error was made or a correct answer given.

The following methods are used to establish the knowledge structure of the individual learners:

• A list of all false and correct answers given by the student is prepared; its completeness is then checked and evaluated, e.g. whether 80% of all questions were answered correctly. This simple method, however, does not explain why a question was answered correctly or not.

• The correct and false answers are arranged in a network of rules and facts which is to represent the knowledge structure of the student.

• The student's correct/false answers and knowledge structures are compared with those of an expert, e.g. the teacher. The so-called overlay method allows common knowledge structures and knowledge gaps to be identified (Carr & Goldstein, 1977).

• Model generation: the strategies and plans used by the student to solve a particular problem are analyzed and combined to form a model. Although this method is by far the most sophisticated approach, it is also the one which has so far been tried and tested the least.

The tutorial module provides for communication with the student usually via text or graphical messages and in rare cases by means of a synthesized voice. It uses the learner model to decide upon the instructional strategy to be used, selects the tasks, provides aids when requested to do so and informs about the learner's behavior by way of feedback - including error messages. Eberts & Brock (1987) and Garg-Janardan, Eberts, Nof, Zimolong and Salvendy (1987) give a detailed description of the functions and capabilities of available intelligent instructional systems, some of which are used together with training simulators such as SCHOLAR, WHY, SOPHIE, WEST or STEAMER. Current criticism of available systems focuses on the following points:

• <u>System Completeness</u>
There are only a few ICT systems, e.g. SOPHIE and STEAMER, which provide explicit knowledge, learner and tutorial modules.

• Knowledge Module
It is a very cost-intensive and time-consuming task to enter knowledge into the database of the knowledge module. Although computer-aided knowledge acquisition procedures are available, they are only useful for retrieving specific knowledge, e.g. facts. Moreover, it is not clear which knowledge acquisition method is particularly well suited for what kind of knowledge (factual or procedural knowledge, object-specific or theory-related knowledge). Moreover, it has not yet been assessed which models are particularly useful for representing knowledge structures, e.g. rule-based, logic-based or structural representation models.

• Learner Model
As far as this model is concerned, it is particularly difficult to find useful ways of deriving a knowledge structure from the answers given by the student. As is the case with the knowledge module, suitable knowledge representation models must also be identified for the learner model.

• Tutorial Module
A tutorial module communicates with the student. At present, however, the natural language can only be used for small knowledge domains on computer systems. There are no precise decision rules as to whether dialog control should be exercised by the computer or by the student. Moreover, it has not yet been possible to design tutoring systems which can satisfactorily provide useful feedback by analyzing false answers and which enable the student to learn from his mistakes. The evaluation of false or unexpected answers still presents a major problem. SCHOLAR, for example, has a hierarchical knowledge base which allows the system to understand unexpected answers.

A discussion of the above shortcomings of intelligent tutoring systems suggests the following analytical approaches to optimum knowledge-based information structures for simulation-based training:

- Selection and improvement of knowledge acquisition procedures, e.g. for the knowledge and tutorial modules of an ICT.

- Selection and optimization of methods for modeling the user's ability to understand a particular subject matter.

- Selection of suitable forms of representation for modeling the information structures (knowledge structures) of the student and the expert (e.g. the teacher).

In the following, we will discuss the problem of knowledge acquisition methods which are suited for optimizing the knowledge module of an ICT system by means of simulation techniques. An evaluation of the available knowledge acquisition methods will be based on the following questions:

- How far is the method suited for analyzing problem-solving and automated knowledge?

- To what extent does this method allow declarative and procedural knowledge to be elicited?

- What are the instrumental and economic efforts involved?

2.0 APPROACHES TO A CLASSIFICATION OF EXPERT KNOWLEDGE

There are two basic approaches to a description of expert knowledge and activities and the underlying cognitive processes: whereas one approach uses a task-oriented description, the other tries to classify expert knowledge according to the knowledge structures involved in problem-solving processes.

Task-oriented approaches distinguish between various problem types which are derived from the individual problem situations. Breuker & Wielenga (1987) developed a detailed model showing the different types of tasks. Typical problems, for example, are interpretation of sensory data, prediction, diagnosis, design, planning, monitoring, debugging, repair, instruction and control.

By using this kind of approach it is possible to include cognitive requirements which are relevant to the execution of a task. Even after many years of training, for instance, planning and diagnosis require much more conscious control than process control. Zimolong et al (1987) prepared a taxonomic classification of process control tasks as they are performed in the control center of a chemical plant or in aircraft. This classification identifies the expertise required for the accomplishment of a particular task and gives examples of expert systems which are capable of performing this specific task to some extent.

Approaches which are based on knowledge structures distinguish between generic and domain-specific knowledge (Voß, 1990). Generic knowledge is generalized problem-solving knowledge of processing rules which control inference steps such as establishment of hypotheses, diagnosis and identification of symptoms. Domain-specific knowledge comprises the terminology used in a particular field of expertise, including pertinent facts and rules such as laws of nature, causalities, heuristics and associations.

Moreover, knowledge can be described as declarative, procedural or heuristic (Oswald & Gardenne, 1984). Declarative knowledge is the knowledge of facts. When, for example, a diagnosis of machine tool failures is to be made, such knowledge includes information about the technical configuration of the system, replaceable parts, possible malfunctions and other tools and materials used. Apart from this, diagnosis also requires knowledge of system functions, test procedures and repair and assembly work. Accordingly, knowledge - in contrast to a mere collection of data - also comprises procedural knowledge, i.e. rules and procedures required for knowledge processing. Strategic and heuristic knowledge is the knowledge of typical failures of particular types of machines, symptoms and conditional probabilities which determine the type and sequence of diagnostic steps to be taken. With the help of this knowledge, an expert can analyze a limited number of alternatives, systematically go through a check list or use a trial-and-error procedure. The fact that cost and benefit aspects - such as the difficulties involved in performing certain checks or the quality of the information obtained - can be taken into consideration is an important aspect of expertise.

Psychological learning models and problem-solving models assume that an expert draws on mentally represented knowledge which he temporarily holds active in his working memory in order to become cognitively able to cope with a particular situation. Mental representations refer to both permanently represented and briefly activated or generated information which is available in an information-processing system at a particular point in time (Tergan, 1989). Such representations include a description of the content and relational aspects of information which is or is to be represented. Information represented in terms of formal-propositional, verbal and analog-visual codes is organized like a mental model which describes to what degree this information has been activated for problem-solving purposes and which provides the basis for managing a particular situation (Tergan, 1989). In psychology of work, the term "operative representation system" (Operatives Abbildsystem) is used to describe this phenomenon (Hacker, 1986).

In practice, the classification of knowledge is usually performance-oriented and less based on psychological models. A list established by Welbank (1989) considers facts, causal relationships, rules, procedures, strategies and explanations and evaluates the suitability of manual acquisition procedures for establishing these knowledge structures. Bainbridge (1979) studied the knowledge structures required for executing process control and monitoring tasks. For this type of task, special emphasis was placed on networks of variables, strategical knowledge and decision sequences.

3.0 KNOWLEDGE ACQUISITION TECHNIQUES

Elicitation-oriented or model-oriented methods can be used for acquiring knowledge. Elicitation-oriented approaches attach particular importance to an optimization of knowledge elicitation with the underlying mental model usually being of minor importance. Nevertheless, the inference mechanisms of the knowledge-based system define such a mental model implicitly, e.g. when chaining rules. Model-oriented procedures focus on the development of expertise models with which expert performance can be described and - with the help of a knowledge-based system - mechanized. Accordingly, knowledge elicitation has so far played only a minor role in model-oriented knowledge acquisition techniques.

3.1 Elicitation-Oriented Methods

Knowledge can be extracted either directly from the human expert (primary analysis) or from data sources other than the human memory (secondary analysis). Secondary data material include dictionaries, books, reports and statistics, the content and data of which can be analyzed.

3.1.1 Surveys

In principle, either (oral) interviews or (written) questionnaires can be used for obtaining useful information from a single expert or from an expert group. In such a communication environment, a knowledge engineer must not only master a particular field of expertise but must also be able to consider social and psychological aspects of dialog management and motivate the respondent (Olson & Reuter, 1987). A successful survey requires that the expert generally be willing to cooperate and, what is more, that he has the linguistic competence of expression and differentiation.

Questionnaires: Questionnaires allow knowledge to be acquired in a standardized written form. The knowledge engineer can use expert answers given in a non-prescribed form as a reference material or use multiple-choice answers for validating existing knowledge bases. The advantage of questionnaires is that they can be easily and economically applied and analyzed. Since it is possible to question several experts at a time and to analyze their answers quickly, the time and cost inputs for both the knowledge engineer and the expert are relatively low. Developing and standardizing these procedures, on the other hand, is time-consuming.

Interviews: Interviews are the most common knowledge acquisition tool and can be used for all knowledge domains. In an interview, an expert is questioned about his knowledge of particular objects, relations between objects and inferences. The oral answers are then taken down in a protocol. David & Hakiel (1988) compiled a list of practical problems involved in the interview technique and on this bases established a catalog ofrecommendations on how to conduct interviews.

Interviews can be structured in different ways: in structured interviews, the same questions are asked in a prespecified order whereas in unstructured interviews the conversation itself determines the content and order of the questions. Semi-structured interviews are halfway between the two with the questions - although their content and order was specified beforehand - being asked in a way which is commensurate with the specific dialog situation. An example of this technique is the acquisition of strategic-diagnostic knowledge in the field of machine tools in Konradt & Zimolong (1990). For structured interviews to be conducted the knowledge engineer must be familiar with the knowledge domain in question.

This is not the case with an unstructured interview. The expert can largely talk about his particular field without having to answer a set of predefined questions. The knowledge engineer only asks certain general questions, e.g. Hoffman (1986) who tried to acquire knowledge from flight planners. Unstructured interviews allow aspects other than those planned to be considered and the importance of certain knowledge elements to be reflected. Similar to other methods which do not involve strict rules, an unstructured interview involves the risk of important aspects being ignored. Although interviews are a rather easy way of acquiring knowledge, substantial efforts are required to interpret and formalize the protocols. The special aspects to be considered in the evaluation and interpretation of verbal protocols are summarized in Ericsson & Simon (1985).

Sample surveys: Such surveys can make use of group discussions, brainstorming and the Delphi method. Brainstorming involves a group of experts being asked to generate useful information in a particular domain of knowledge. Due to the fact that this methods allows useful information to be collected spontaneously and associatively, it is particularly suited for investigating into new domains or for delimiting unstructured subject areas from each other. Sample surveys, however, produce disappointing results if the group members cannot be successfully encouraged to make spontaneous contributions to the discussion.

Although group discussions are related to the brainstorming technique, this method is more systematic in that a knowledge engineer performs the function of a moderator who presides and directs the expert discussion on a specific knowledge domain. Group discussions can serve as both an introductory approach to a specific domain and as a control procedure for validating knowledge-based systems (O'Keefe et al, 1987; O 'Leary et al, 1990).

Another sample survey technique is the Delphi method which involves a number of experts being questioned about a specific knowledge domain in a multistage process (Linstone & Turoff, 1975). The answer given by an expert is anonymously passed on to the other experts in the group who evaluate and supplement the information. The purpose of this procedure is to finally come to a generally accepted expert solution. Anonymity ensures that, on the one hand, ambiguous knowledge - i.e. conflicting knowledge - can be gathered and, on the other, that the tendency of individual group members to perform moderator functions or to dominate the discussion - which is natural in social situations - can be compensated to some extent.

3.1.2 Tests

Tests imply that an expert uses structured material to perform specific tasks in accordance with predefined instructions of how to proceed. Suitable test procedures for knowledge acquisition include structure formation techniques, similarity ratings and repertory grid techniques.

Structure formation techniques: Such techniques allow external knowledge to be visualized (Scheele & Groeben, 1984). First, relevant domain-specific concepts are either compiled beforehand or selected by the expert and written on cards. Then the expert takes these cards and combines them on the basis of the relations existing between the pertinent concepts, i.e. he establishes a network of concepts and relations (Bonato, 1988). Typical relations are "is part of", "and" "or". Since the concepts are increasingly resolved in the vertical direction, i.e. the expert begins with superordinate concepts and continues with the associated subordinate ones, a hierarchical structure is produced. This technique has the advantage that knowledge elements and their relations can be visualized and represented in a clear way. Accordingly, these structures are a kind of aid to the expert's working memory. What is more, this method allows the relations between concepts to be easily modified if necessary.

By using the structure formation technique, it may also be possible to represent the strategical knowledge of an expert. Konradt & Straeter (1990) have developed a computer-aided version of a structure formation technique which allows psychometric measures (in what order did the expert select particular cards, how long did it take him to select them, how did he arrange them and how long did it take him to insert them into the structure) to be established on-line during the structure formation process. The hypothesis underlying this method is that the structure formation process reflects how an expert uses his knowledge to solve a given problem. The main strategies used by the expert to perform a specific task can be derived from a reconstruction of the structure formation process.

Similarity ratings: This method involves a set of domain-specific concepts being rated according to similarities. Different concepts can be rated either qualitatively or quantitatively in pairs on the basis of rating scales (direct similarity rating) or by establishing groups of similar concepts (sorting technique). Cluster analysis and multidimensional scaling (MDS) allow the ratings obtained to be interpreted and graphically represented. Schvaneveldt et al (1985), for example, use a multidimensional scaling technique to elicit the knowledge structure of experienced and less experienced combat pilots. Similarity ratings are also used in computer-aided procedures, e.g. in the knowledge-supporting systems KSSO (Shaw, 1988).

Repertory grid techniques: The theory underlying the repertory grid technique (Kelly, 1955) assumes that a human being understands himself and his environment on the basis of a system of bipolar dimensions or constructs. Personal constructs can be established by means of a procedure which consists of data collection and analysis (for a detailed description see Fransella & Bannister, 1977).

First, an expert is asked to select between five and ten elements which are of major importance in his domain. Then these elements are used to successively form two- and three-construct pairs. The expert specifies descriptive details which allow him to describe in what these construct pairs mainly differ and in what they correspond. The resulting list of bipolar characteristics serves as the basis for a matrix in which the personal constructs and similarity ratings of the expert are reflected in rows and columns.

The repertory grid technique is a very time-consuming procedure. For this reason, it is essential that the participating experts are interested in this procedure and willing to cooperate. At the beginning, it is also a difficult and time-consuming task for the expert to learn and master a formalism with which his answers must comply. Due to standardized analysis rules it is, however, possible to extract knowledge rather quickly. Computer assistance (e.g. Shaw & Gaines, 1987) allows the efforts necessary to analyze the expert answers to be reduced significantly. An advantage of this technique is that the expert is not required to explain similarities and dissimilarities. He does not need to verbalize his knowledge but rather use his experience to establish relations in his domain from which a knowledge engineer - e.g. by using MDS - can infer the dimensions and networks underlying this relations. In this respect, such techniques are suited for constructing associative or semantic networks, the nodes of which consist of heterogeneous objects and rules of how to process them. One disadvantage inherent in personal construct techniques is that the objects must already be known beforehand.

3.1.3 Observation

Observation techniques as a knowledge acquisition tool register the reactions of an expert in given problem situations. Auditive (protocols) and visual (video tapes) techniques are used to record all important aspects of expert behavior systematically. The purpose of video tapes is to demonstrate in full detail and without interruption how an expert behaves and acts in a natural situation or in an experiment. A later analysis of the tape can help to clarify open questions. Observation techniques have the disadvantage that the experts to be observed tend to be skeptical about this method and may be influenced by the situation even if they do not object to being observed.

Thinking aloud: In this method an expert speaks out loud what he is thinking while solving a problem. These thoughts are recorded and analyzed. Frequently, this process is also videotaped in order to display what action accompanied a specific thought. The purpose of this technique is to identify cognition's and major problem-solving strategies which determine an expert's actions during the solution-finding process. This technique, which is traditionally applied in research into problem-solving, was used by Bereiter & Miller (1989) for eliciting knowledge in the field of automated manufacturing.

When an expert performs difficult or completely new tasks on which he has to concentrate all his attention, the requirement to say out loud what he is thinking can have an adverse effect on the problem-solving process. In this case, thinking aloud is tantamount to an additional task which can interfere with the actual problem-solving process. Studies have shown, though, that it can be much easier for a person to solve simple problems when he directly verbalizes possible hypotheses and procedural steps. For this reason, the use of this technique depends on the complexity of the task to be performed and on the knowledge and experience of the expert.

Scenarios: In a scenario, an expert is observed while performing a real or fictitious task. There are different kinds of scenarios (Cordingley, 1989): in a forward scenario simulation, an expert tries to find a solution on the basis of certain initial states and combinations of conditions. The procedural steps which are necessary to come to a solution are recorded or the solution-finding process is interrupted by the knowledge engineer whenever he needs precise information. Klein (1987), for example, used the scenario technique to explore how designers of particular parts are influenced by similar parts which are already available.

Whereas forward scenario simulation allows new problems to be constructed, backward scenarios use a solution to infer the underlying facts. For the critical incident technique an expert talks about an event or a case which has been of crucial importance in his knowledge domain.

A basic problem inherent in the use of scenarios for knowledge acquisition is the representativeness of the samples used for the behavioral study. It must be ensured that the selected cases are typical of the problem to be solved and that a sufficient number of problem configurations are included in the study. In the case of earlier scenarios it is possible that memory failures which result from the mechanisms used by the human information processing system to retrieve information can significantly reduce data validity.

3.1.4 Computer-Aided Methods

It was already at an early stage of development that due to the immense cost of knowledge engineering personnel substantial efforts were undertaken to enable experts to enter their knowledge into a system directly. An example of such a system is Advice Taker developed by John McCarthy (1968). The user enters predicate-logical knowledge into the system which interprets this information in order to make inferences. Generally, available computer-aided knowledge acquisition methods and tools are, however, still far from satisfying the widespread interest in this technology. Whereas operational tools are already available for comparatively simple knowledge acquisition tasks such as consistency checks in a knowledge base, they are yet to be developed for more complex tasks. Such tasks particularly include the elicitation of general knowledge from examples by means of machine learning methods. In the following, some selected examples of computer-aided methods will be described.

The great majority of computer-aided knowledge acquisition tools use repertory grid techniques. Examples thereof are ETS (Boose, 1985, 1986), PLANET (Gaines & Shaw, 1986), AQUINAS (Boose & Bradshaw, 1987; Kitto & Boose, 1987), FMS Aid (Garg-Janardan & Salvendy, 1987), KITTEN (Shaw & Gaines, 1987), KRITON (Diederich, Ruhmann & May, 1986) and KSSO (Gaines, 1987). In the following, two typical systems are selected from this group and described in detail.

The Expertise Transfer System (ETS) developed by Boose (1985, 1986) asks an expert - on the basis of diagnoses established beforehand - in which symptom two diagnoses differ from a third one. The results obtained are represented in a repertory grid which consists of a symptom/diagnosis matrix. The expert verifies and - if necessary - completes the repertory grid. The knowledge is then used to infer rules underlying the relations between symptoms and diagnoses which can be used by an expert system. Moreover, ETS is able to include the expert's opinion on the strength of the relation between a symptom and a diagnosis and his evaluation of the importance of individual symptoms for a diagnosis.

A later development of ETS is AQUINAS - a program which combines a number of different knowledge acquisition tools including knowledge elicitation methods based on the repertory grid technique, methods allowing the acquisition of different types of data, and methods allowing the acquisition of both correlative and hierarchical relations. A dialog manager (Kitto & Boose, 1987) assists the user in selecting those knowledge acquisition techniques which are particularly well suited for a specific domain. For this purpose, the dialog manager uses a taxonomy of generic application tasks and problem-solving methods according to Clancey (1985).

Computer-aided acquisition procedures can also contribute to the optimization of a knowledge base in another way. The use of knowledge editors is one method of verifying a knowledge base interactively (Musen, 1989). A knowledge editor is the interface between user and knowledge base. By using knowledge editors, a domain expert can access a knowledge base. Since knowledge editors can be used not only for debugging but also for entering structured knowledge, they can be regarded as simple knowledge acquisition methods. Today they are an integral part of the basic configuration of expert system development environments such as KEE, Knowledge Craft, ART, NEXPERT or Babylon.

The knowledge editor TEIRESIAS (Davis, 1983), for example, supports debugging and knowledge acquisition for the expert system MYCIN (Shortliffe, 1976) which was designed for use in the diagnosis and treatment of bacterial infection. TEIRESIAS allows the user to interact with the knowledge base in something approaching natural language, to identify and correct erroneous rules and to enter new rules. It checks whether the new rule is complete and whether it is consistent with available data and rules.

3.2 Model-Based Procedures

In the following, KADS (Knowledge Acquisition, Documentation and Structuring) is taken as an example of a model-based knowledge acquisition procedure which considers the mental model used by a problem solver (Breuker et al, 1988). The general purpose of KADS is to represent expertise at the knowledge level. This means that KADS models can be established independently of cognitive representation assumptions and implementation requirements. A description of problem-solver models comprises four levels.

a. Domain level:
Knowledge is the description of fundamental concepts and relations irrespective of their role in problem solving.

b. Inference level:
Knowledge is arranged according to the roles which it can play in an expertise model, i.e. it can either serve as a metaclass or a knowledge source. In the field of medical diagnosis metaclasses, for example, include symptoms, hypotheses and therapies. Accordingly, the domain-level concept of "appendix" must be assigned to the metaclass of "hypothesis." It is also possible for one concept to be assigned to more than one metaclass. Knowledge sources establish relations between metaclasses. Knowledge sources include the calculation of a higher metaclass, comparison of two metaclasses and goal-oriented manipulation of metaclasses.

c. Task level:
Whereas the purpose of the previous levels is to define the basic units or the vocabulary to be used in the problem-solving process, the task level focuses on

modeling the problem-solving process. For example, in diagnosis this includes the sequential analysis of individual error sources. Tasks and goals are the basic units used at this level.

d. Strategic level:
The purpose of this level is to model typical expert skills such as flexibility in the selection of problem-solving strategies, early detection of dead ends and errors. The latest KADS developments, however, do not yet provide for concrete ways of how to produce such a model.

3.3 Evaluation

The use of knowledge modules in ICT systems does not require that the knowledge underlying the performance of an expert be represented completely. Instead, those items of problem-solving expertise are chosen which are essential to an expert's perception, selection and decision-taking processes. However, psychological tools have yet to be developed which can specify for certain types of tasks and their cognitive requirements a technique which is able to cover explicit facts, rules, procedures, strategies and explanations. If knowledge acquisition is regarded as a process of constantly refining a knowledge model, it is evident that every acquisition method is suited only for particular steps in the knowledge acquisition process.

Timpe (1990) has developed a model which allows knowledge to be refined step by step. This model implies that global knowledge acquisition is followed by the establishment of a knowledge model and sophisticated knowledge management including knowledge evaluation. As far as ICT systems are concerned, this means that knowledge and learner modules can be refined gradually. Timpe's model is supported by empirical findings which prove that as yet there is no method available which alone can ensure acquisition of almost all knowledge structures (Timpe, 1990). A combination of different acquisition methods would also allow the respective disadvantages of various acquisition methods to be offset.

Global knowledge can be extracted from secondary material, i.e. books, dictionaries, reports and statistics. The knowledge contained in such external media gives the knowledge engineer an initial insight into the knowledge domain. Apart from the use of secondary material, various techniques such as observation, thinking aloud and simulation can be used for acquiring procedural knowledge.

The psychological characteristics distinguishing an expert from a beginner have already been discussed. Routine, however, results in a large degree of automation of knowledge and skills so that the expert is hardly able to articulate the planning and decision-taking processes underlying his work. Rasmussen (1983) and Hacker (1986) describe these control levels of human behavior as levels at which stimulus/response mechanisms function automatically - without the conscious attention of the individual. If at all, it is at the next higher level - the conceptual, perceptive or rule level - that information on strategies or errors can be provided (Zimolong, 1990). Simulation experiments provide an early insight into what information an expert selected and how he weighted relevant information and marginal conditions in the simulated problem-solving situation. Later, the expert is asked specific questions about his strategy so that it is possible to identify the underlying knowledge and processing structures, decision sequences and networks of variables and to infer structural changes in those cognitions which determine the expert's actions.

The first phase ends with the establishment of a preliminary knowledge model which in a second phase can be supplemented and modified on the basis of (oral) interviews, (written) questionnaires and association techniques. The problem with survey techniques is that they allow only that knowledge to be acquired which can be verbalized. These methods cannot be used when it is impossible or very difficult to express knowledge in words or when implicit knowledge is to be acquired (Berry, 1987). But it is precisely this knowledge which the knowledge engineer is primarily interested in when trying to elicit knowledge from experts, In brief, experts are distinguished for their perceptual capabilities, their ability to differentiate between important and unimportant information, to structure complex relations and to work by effective behavioral rules.

When an expert is asked to solve a complex problem involving extreme cognitive requirements or to perform a completely new task, it can be an aid to his working memory if he reasons in writing. Therefore, an expert who has to accomplish a very difficult task, e.g. to measure a wave, will verbalize much more knowledge in a written survey than in an interview (Rothe, Timpe & Warning, 1990). Another advantage of written over oral surveys is their being very economical tools in that they allow several experts to be questioned at a time. If difficult tasks are to be performed and if the experts are willing to cooperate, it is also possible to use a structure formation technique which involves the visualization of knowledge. The suitability of this technique depends on the expert's willingness to cooperate and the organizational availability of experts. It is particularly under real production conditions that loss of time due to lengthy surveys is costly and therefore economical techniques need to be applied for reasons of rationality.

In a third phase, sorting methods can be used to complete the sophisticated knowledge model composed of concepts and semantic relations. Individual ambiguous and conflicting knowledge structures can be clarified by the expert in group discussions or in multiple-choice questionnaires.

4.0 APPLICATIONS OF COMPUTER-AIDED SIMULATORS

The description and discussion of intelligent tutorial systems show that such systems are well-suited to assist teachers but that they can likewise be used to replace them. For a closer examination of this problem, it would be useful to have a look at the training requirements to be met. For what kind of training courses is computer-aided simulation-based training best suited? In the following, the training of skilled workers in the metal-working industry is taken as an example to show potential applications of tutorial systems.

In the metal-working industry, graphical-dynamical simulations are increasingly being used for computer-aided machining. According to Laur-Ernst (1986) simulators are particularly useful for introducing skilled metalworkers to information technology. The metalworker can be trained in programming and managing various dialog procedures, e.g. for loading and correcting programs. Suitable fields of training also include failure diagnosis and preventive maintenance. For this purpose, computer-based systems are being developed which allow the user to acquire knowledge of system configuration, to identify the causes of failures and to receive instruction on how to correct them. When using such a system, a person can improve his skills through on-the-job training. Zimolong & Konradt (1990) are developing a failure-diagnosing system which enables skilled workers to supplement their knowledge of the technical configuration of the machine. On the basis of simulated failures/causes, they learn to identify symptoms of failure and their causes both quickly and reliably. The knowledge base of this expert system was established on the basis of knowledge acquisition tools which can be applied under operating conditions. The purpose of this system is to enable the user to add his own data to the available knowledge base (Konradt & Zimolong, 1990; Zimolong & Konradt, 1990).

These are only two examples of how the introduction of new technologies has changed the qualification requirements to be met by skilled workers. Today's skilled worker must not only be familiar with the latest technological findings in his domain but also develop social and communicative abilities and skills, something which is becoming increasingly important in the wake of modernization. In the pertinent literature these extrafunctional abilities are discussed as “key qualifications” (Zimolong, Gottlieb & Rosemann, 1988). The qualifications needed in the field of machine tools have been established by Sonntag (1985, 1989) and others. As one might expect, skilled workers must now meet higher and changing requirements when it comes to knowledge of technical data, diagnosis and correction of failures and work organization. Moreover, their work has become more demanding as far as communication and cooperation requirements, independent planning and acting are concerned. According to Meyer (1987), the key qualification requirements to be met by skilled workers include:

- The ability to solve integrated tasks independently or as a member of an interdisciplinary expert group,
- Understanding of system operation and knowledge of basic facts such as production patterns and effects of intervention in the production process,
- Enthusiasm at work and the ability to develop and implement ideas and concepts of their own,
- Willingness and ability to learn autodidactically,
- Ability to communicate and cooperate with others.

Against this background, the requirements to be met by teachers have also changed and now include:

Knowledge of the domain and relevant procedures and strategies: domain-specific knowledge. The teacher has knowledge not only of how a specific machine or technical system functions but also of operational requirements. Procedural system management knowledge, for example, implies that the teacher is familiar with programming techniques, heuristics for optimal input sequences and debugging strategies.

Knowledge of how to convey knowledge: didactical competence. The teacher can draw on his pedagogical knowledge and experience of how to adequately convey domain-specific knowledge to selected personnel. This includes the ability to identify errors, ineffective approaches and suboptimum strategies, to make the learners aware of such deficiencies and to ensure more efficient learning by selecting suitable tasks and examples.

Knowledge of how to convey social competence. Useful methods and didactical approaches enable teachers to instruct learners how to plan and take decisions independently and to act responsibly in cooperation with others.

Knowledge of how to convey didactical competence. Teachers are able to select, motivate and didactically advise qualified personnel and to integrate their activities into the company's training program (management and transfer competence).

The requirements to be met by teachers show that if simulation-based training can replace teachers at all, then only to convey domain-specific knowledge. In any case, it would be far more desirable to use it as a means of supporting the teacher's didactical competence. Simulation-based training would then become a useful tool which allows a teacher to optimize technical-instrumental knowledge acquisition.

LITERATURE

Bainbridge, L., 1979, Verbal reports as evidence of the process operator's knowledge, International Journal of Man-Machine Studies, 11, 411-436.

Bereiter, S. R., and Miller, S. M., 1989, A field based study of troubleshooting in computercontrolled manufacturing systems, IEEE-SMC, 19, 205-219.

Berry, D. C., 1987, The problem of implicit knowledge, Expert Systems, 4, 144-151.

Bonato, M., 1988, Knowledge elicitation with structure formation techniques, In: J. Diederich, and T. Uthmann, (ed.), Knowledge Acquisition, Arbeitspapier No. 281. St. Augustin, GMD.
Boose, J., 1985, A knowledge acquisition program for expert systems based on personal construct theory, International Journal of Man-Machine Studies, 20, 21-43.
Boose, J., 1986, Rapid acquisition and combination of knowledge from multiple experts in the same domain, Future Computing Systems, 1, 191-216.
Boose, J., and Bradshaw, J., 1987, Expertise transfer and complex problems: Using AQUINAS as a knowledge-acquisition workbench for knowledge-based systems, International Journal of Man-Machine Studies, 26, 3-28.
Breuker, J. A., and Wielenga, B. J., 1987, Use of models in the interpretation of verbal data, In: A. Kidd, (ed.), Knowledge acquisition for expert systems, New York, Plenum.
Breuker, J. A., Wielinga, B. W., van Someren, M., de Hoog, R., Schreiber, G., De Greef, P., Bredeweg, B., Wielemaker, J., Billaut, J., Davoodo, M., Hayward, S., 1988, Modeldriven knowledge acquisition: interpretation models, Deliverable task A1, Esprit Project 1098, University of Amsterdam.
Carbonell, J. R., 1970, AI in CAI: An artificial intelligence approach to computer aided instruction, IEEE Transactions on Man-Machine Systems, 11, 190-202.
Carr, B., and Goldstein, I., 1977, Overlays: A theory of modeling for computer aided instruction, Cambridge, Mass., AI Laboratory, MIT.
Clancey, W. J., 1985, Heuristic classification, Artificial Intelligence, 27, 209-350.
Cordingley, E. S., 1989, Knowledge elicitation techniques for knowledge-based systems, In: D. Diaper, (ed.), Knowledge Elicitation, Ellis Horwood, Chicester, 89-175.
Davies, M., and Hakiel, S., 1988, Knowledge harvesting: A practical guide to interviewing, Expert Systems, 5, 42-50.
Davis, R., 1983, Interactive transfer of expertise, In: E. H. Shortliffe, and B. Buchanan, (ed.), Rule-based expert systems: the MYCIN experiments, Addison-Wesley.
Diederich, J., Ruhmann, I., and May, M., 1986, KRITON: A knowledge acquisition tool for expert systems, Proceedings of the Knowledge Acquisition for KBS Workshop, Banff, Canada.
Eberts, R. E., and Brock, J. F. 1987, Computer-Assisted and Computer-Managed Instruction, In: G. Salvendy, (ed.), Handbook of Human Factors, New York, Wiley, 976-1011.
Ericsson, K. A., and Simon, H. A., 1985, Protokoll analysis: Verbal report as data, London, MIT Press.
Fransella, F., and Bannister, D., 1977, A manual for repertory grid technique, London, Academic Press.
Gaines, B. R., 1987, Rapid prototyping for expert systems, In: M. Oliff, (ed.), Proceedings of the International Conference on Expert systems and the Leading Edge in Productions, Planning, and Control.
Gaines, B. R., and Shaw, M., 1986, Induction of inference rules for expert systems, Fuzzy Sets and Systems, 18, 315-328.
Garg-Janardan, C., and Salvendy, G., 1987, A conceptual framework for knowledge elicitation, International Journal of Man-Machine Studies, 26, 521-532.
Garg-Janardan, C., Eberts, R.E., Zimolong, B., Nof, S. Y., Salvendy, G., 1987, Expert Systems, In: G. Salvendy, (ed.), Handbook of human factors, New York, Wiley, 1130-1176.
Hacker, W., 1986, Arbeitspsychologie, Bern, Huber.

Hoffman, R. R., 1986, Procedures for efficently extracting the knowledge of experts, Technical Report, Office of Scientific Research, Bolling Air Force Base, Zit. In: Hoffman, R. R., 1987, The problem of extracting the knowledge of experts from the perspective of experimental psychology, AI Magazine, 8, 53-67.
Kelly, G. A., 1955, The psychology of personal constructs. Vols. 1 and 2, New York, Norton.
Kitto, C. M., and Boose, J. H., 1987, Heuristics for expertise transfer: An implementation of a dialog manager for knowledge acquisition, International Journal of Man-Machine Studies, 26, 183-202.
Klein, G. A., 1987, Applications of analogic reasoning, Metaphor and Symbolic Activity, 2, 201-218.
Konradt, U., Stræter, O.,1990, Ein computergestuetztes Strukturlegeverfahren zur Akquisition von Diagnosestrategien in komplexen Arbeitssituationen, Unpublished paper, Lehrstuhl fuer Arbeits- und Organisationspsychologie, University of Bochum.
Konradt, U., and Zimolong, B., 1990, Die Analyse von Diagnosestrategien bei Wartungs- und Instandsetzungsaufgaben in flexiblen Fertigungsstruktu-ren, In: D. Frey, (ed.), Bericht Ueber den 37. Kongress der DGfP in Kiel 1990, Hogrefe, Gœttingen, 322.
Laur-Ernst, U., 1986, Mit CNC-Simulation lernen - Mœglichkeiten und Grenzen, In: M. Hoppe, and H.-H. Erbe, (ed.), Rechnergest Uetzte Facharbeit, 106-118. Wetzlar, W. v. Siemens-Schule.
Linstone, H. A., and Turoff, M., 1975, (ed.), The dephi method, London, Addison-Wesley.
McCarthy, J., 1968, Programming with common sense, In: M. Minsky, (ed.), Semantic information processing, 403-418, Cambridge, Mass., MIT Press.
Meyer, N., 1987, Neue Anforderungen an die Ausbilder, In: 2. Fachtagungsbe-richt aus dem Forschungsprojekt "Mikroelektronik und berufliche Bildung - Gesamtauswertung", Fachtagung am 15. und 16. Oktober 1987 im Wissenschaftszentrum in Bonn "Innovation und Qualifikation", Fraunhofer-Institut fuer Arbeitswirtschaft und Organisation (IAO), 1987.
Musen, M. A., 1989, Automated generation of model-based knowledge acquisi-tion tools, London, Pitman.
O'Keefe, R. M., Balci, O., and Smith, E. P., 1987, Validating expert system performance, IEEE Expert, 2, 81-90.
O'Leary, T. J., Goul, M., Moffitt, K. E., and Radwan, A. E., 1990, Validating expert systems, IEEE Expert, 5, 51-58.
Olson, J., and Rueter, H. H., 1987, Extracting expertise from experts: Methods for knowledge acquisition, Expert Systems, 4, 152-168.
Oswald, M., and Gardenne, V., 1984, Wissen, Kœnnen und kuenstliche Intelli-genz, Sprache & Kognition, 3, 173-184.
Rasmussen, J., 1983, Skills, rules and knowledge, signals, signs and symbols and other distinctions in human performance models, IEEE-SMC, 13, 257-266.
Rothe, H.-J., Timpe, K.-P., and Warning, J., 1990, Methodenkritische Analyse der Erfassbarkeit von Fachwissen, Unpublished paper, Humboldt-University, Berlin.
Scheele, B., and Groeben, N., 1984, Die Heidelberger Struktur-Lege-Technik (SLT), Eine Dialog-Konsens-Methode zur Erhebung subjektiver Theorien mittlerer Reichweite, Weinheim, Beltz.
Schvaneveldt, R. W., Durso, F. T., Goldsmith, T. E., Breen, T. J., Cooke, N. M., Tucker, R. G., and DeMaio, J. C., 1985, Measuring the structure of expertise, International Journal of Man-Machine Studies, 23, 699-728.

Shaw, M. L., 1988, Problems of validation in a knowledge acquisition system using multiple experts, In: J. Boose, B. Gaines, and M. Linster, (ed.), Proceedings of the European Knowledge Akquisition Workshop, Sankt Augustin, GMD, 5-1 - 5-15.
Shaw, M. L., and Gaines, B. R., 1987, An interactive knowledge-elicitation technique using personal construct technology, In: A. Kidd, (ed.), Knowledge acquisition for experts systems. A practical handbook, Plenum Press.
Shortliffe, E. H., 1976, Computer-based medical consultations: MYCIN, New York, Elsevier.
Sonntag, K., 1985, Qualifikationsanforderungen im Werkzeugmaschinenbereich, In: K. Sonntag, (ed.), Neue Produktionstechniken und qualifizierte Arbeit, Kœln, Wirtschaftsverlag Bachem, 81-100.
Sonntag, K. H., 1989, Trainingsforschung in der Arbeitspsychologie, Bern, Huber.
Tergan, S.-O., 1989, Psychologische Grundlagen der Erfassung individueller Wissensrepræsentationen, Teil I: Grundlagen der Wissensmodellierung, Sprache und Kognition, 8, 152-165.
Timpe, K.-P., 1990, Experimentalpsychologie und technische Wissensverarbeitung - psychologische Aspekte bei der Entwicklung von Expertensystemen, Unpublished paper, Humboldt-University Berlin.
Welbank, M., 1987, Knowledge acquisition: a survey and British telecom experience, In: T. Addis, J. Boose, and B. Gaines, (ed.), Proceedings of the First European Workshop on Knowledge Acquisition for Knowledge-Based systems, Reading Unversity.
Zimolong, B., 1990, Fehler und Zuverlæssigkeit, In: C. Graf Hoyos, and B. Zimolong, (ed.), Enzyklopædie der Psychologie, vol. 2: Ingenieurpsychologie, Gœttingen, Hogrefe, 313-345.
Zimolong, B., Gottlieb, W., and Rosemann, B., 1988, Ziele und Aufgaben psychologischen Handelns in der Praxis. Tradition und neue Trends, In: E. Frieling, and F. Ruppert, (ed.), Psychologisches Handeln in Organisationen und Betrieben, Bern, Huber, 22-42.
Zimolong, B., and Konradt, U., 1990, Arbeitsbericht zum Forschungsvorhaben Entwicklung eines Ratgebersystems fuer die Stœrfalldiagnose in flexiblen Fertigungsstrukturen, SFB 187: Project A-2, University of Bochum.
Zimolong, B. and Konradt, U., 1991, Diagnosis strategies of maintenance personel in flexible advanced manufacturing structures, In: H.-J. Bullinger (ed.), Human Aspects in Computing, Amsterdam, Elesevier, 1051-1055.
Zimolong, B., Nof, S. Y., Eberts, R.E., and Salvendy, G., 1987, On the limits of expert systems and engineering models in process control, Behavior and Information Technology, 6, 15-36.

SECTION VI

This section discusses the definition and application of artificial intelligence/expert systems as a part of computer based instruction. It will place much emphasis on data from the scientific literature which can assist the alliance in understanding the value of applying this quickly emerging technology.

EVALUATION MODELS FOR EXPERT SYSTEMS IN MILITARY TRAINING*

Harold F. O'Neil, Jr.[1] and Eva L. Baker[2]

[1]School of Education, University of Southern California, Los Angeles, CA 90089 and Advance Design Information, Sherman Oaks, CA 91403
[2]National Center for Research on Evaluation, Standards, and Student Testing (CRESST), University of California, Los Angeles, CA 90024

ABSTRACT

This paper focuses on evaluation models for the dominant application of expert systems in training, that is, Intelligent Tutoring Systems (ITS) or Intelligent Computer-Assisted Instruction (ICAI). Here we provide a new evaluation model for such applications. Since intelligent CAI shares some common attributes with traditional CAI, evaluation technology approaches to the traditional will be integrated into our model. A case study of the model designed for an ICAI program in weather is provided. A wide range of evaluation techniques would be included, for instance, both quantitative and qualitative data collection and analyses. Instrumentation and data collection strategies are suggested along with student and instructor questionnaires. Finally, we make suggestions for measuring intelligence in ICAI.

INTRODUCTION

This paper focuses on evaluation models for the dominant application of expert systems in training, that is, Intelligent Tutoring Systems (ITS) or Intelligent Computer-Assisted Instruction (ICAI). We prefer the latter term as it denotes the application of Artificial Intelligence (AI) to computer-assisted instruction without specifying a particular instructional strategy, for instance, tutoring. Further, we view this intelligent application as not qualitatively different from computer-assisted instruction (CAI). Thus, many of the models/constructs/measures useful in CAI can also be applied to ICAI.

The basis for our approach to evaluation models is in part represented by our prior work in the area of evaluation and measurement issues in intelligent computer-assisted instruction (O'Neil & Baker, 1991) as well as ongoing work sponsored by the USAF Armstrong Laboratory (e.g., O'Neil & Jacoby, 1990). This ongoing work supplemented

* Presented at the NATO Workshop on Advanced Technologies Applied to Training Design, 22 October 1991, Venice, Italy. The research reported herein was supported by the U.S. Air Force's Armstrong Laboratory under subcontract to Expert-EASE by Advance Design Information. However, the views, opinions and/or findings contained in this report are the authors', and should not be construed as an official Department position, policy or decision, unless so designated by other official documentation.

our original work via a review of the extant literature on evaluation of computer-based instructional systems, courseware, intelligent CAI systems, and formative evaluation (O'Neil & Baker, 1990). Our evaluation models are instantiated in particular contexts, for example, the evaluation of an ICAI system in weather (see Figure 1).

Objective:

- To design and execute an evaluation plan for the Weather Trainer.

Input:

- Visits to Chanute AFB for interview with students and teachers, and administration of measures.
- Analysis of knowledge-based intelligent tutoring system for the weather training task.

Approach:

- Analyze interviews with students and teachers to determine elements for evaluation, particularly unanticipated outcomes.
- Prepare a qualitatively-oriented evaluation plan that addresses task objectives, fidelity of instructional process, student outcomes, and instructor satisfaction.
- Conduct an assessment on-site for the purpose of data collection, using think-aloud protocols for student instructional process and outcomes, and interviews with instructors.
- Analyze and prepare an evaluation report.

Results:

- An evaluation plan to guide the execution of the study.
- An evaluation report to assist in formative and summative judgments.

Milestone:

11.1 Completion of evaluation plan.

11.2 Completion of evaluation report.

Figure 1 Task 11: Evaluation

The ICAI system in weather was designed to teach the cognitive strategy of visualization, that is, the individual difference variable that "allows an individual to manipulate or transform the image of spatial patterns into other arrangements" (Ekstrom, French, Harman, & Derman, 1976). What makes our weather project interesting instructionally is the following: (a) This is the first ICAI system designed to teach a cognitive strategy (visualization) within a domain context (advection); (b) we measure both cognitive (visualization) and affective (state worry) outcomes; and (c) it is a potential test bed for instructional research in ICAI. An overall evaluation model must be responsive to these issues. This paper will not address the test bed issues.

The document is organized in four sections. In the above section, an introduction was included to provide background and context. The next section provides an overview of the formative evaluation model proposed for the study. The following section provides the evaluation technology and measuring instruments to implement the evaluation model. In the final section, we will discuss the measurement of intelligence in ICAI systems.

FORMATIVE EVALUATION MODEL

Multiple models and procedures are common within formative evaluation. Formative evaluation (Baker, 1974) seeks to provide information that focuses on the improvement of the innovation and is designed to assist the developer. Basic literature in formative evaluation was developed by Baker (1974), Baker and Alkin (1973), Baker and Saloutos (1974), and Scriven (1967). Formative evaluation also addresses the effectiveness of the development procedures used, in order to predict whether the application of similar approaches will likely have effective and efficient results. In that function, formative evaluation seeks to improve the technology at large, rather than the specific instances addressed one at a time. The formative evaluation approach is designed so that its principal outputs are identification of success and failure of segments, components, and details of programs, rather than a simple overall estimate of project success. The approach requires that data be developed to permit the isolation of elements for improvement and, ideally, the generation of remedial options to assure that subsequent revisions have a higher probability of success. While the evaluation team maintains objectivity, they typically interact with and understand program goals, processes, and constraints at a deeper level than evaluation teams focused exclusively on bottom line assessments of success or failure. Their intent is to assist their client (either funding agency or project staff) to use systematic data collection to promote the improvement of the effort.

An example of one approach to formative evaluation for ICAI is depicted in Figure 2. As is shown in Figure 2, formative evaluation begins with checking whether the design is congruent with specifications and ends with revision which includes new data collection on Steps 3-5. A particular challenge in ICAI is that in many cases there are no valid specifications as rapid prototyping outdistances the documentation.

1. Check ICAI design against its specifications.
2. Check validity of instructional strategies in tutor with research literature.
3. Conduct feasibility review with instructor.
4. Conduct feasibility test with student(s).
    - one-on-one testing
    - small group testing
5. Assess instructional effectiveness.
    - cognitive
    - affective
6. Assess unanticipated outcomes.
7. Conduct revision.

Figure 2. Formative evaluation activity.

EVALUATION TECHNOLOGY

Since ICAI shares some common attributes with CAI, evaluation technology appropriate to CAI could be used in ICAI (e.g., Alessi and Trollip, 1985, 1991; Gagné, Briggs, and Wager, 1988). The CAI lesson evaluation techniques in Table 1 present some formative techniques (quality review and pilot testing methods) and some techniques that could be formative or summative (i.e., validation). These activities were adapted from Alessi and Trollip (1991). Information of this sort is a necessary but not sufficient set for

ICAI evaluation. For the evaluation itself we would make a rating scale for each item in Table 1 consisting of a yes/no or go/nogo judgment for each item.

Table 1

CAI Lesson Evaluation Scale

	Rating		
QUALITY REVIEW	**Go**	**NoGo**	**NA**
Check language and grammar [e.g., appropriate reading level].			
Check surface features [e.g., uncluttered displays].			
Check questions and menus [e.g., making sure a choice is clear].			
Check all invisible functions [e.g., appropriate student records kept].			
Check the subject matter content [e.g., information is accurate].			
Check the off-line material [e.g., directions in operator manual are clear].			
Revise the lesson.			
Apply the same quality-review procedure to all revisions.			
PILOT TESTING			
Enlist about three helpers [i.e., representative of potential students].			
Explain pilot-testing procedures [e.g., encourage note-taking].			
Find out how much they know about the subject matter.			
Observe them go through the lesson.			
Interview them afterwards.			
Revise the lesson.			
Pilot-test all revised lessons.			
VALIDATION			
Use the lesson in the setting for which it was designed.			
Use the lesson with students for whom it was designed.			
Evaluate how the students perform in the setting for which you are preparing them.			
Obtain as much performance data as you can from different sources.			
Obtain data on student achievement attributable to the lesson.			
Obtain data on student attitudes toward the lesson.			

(Adapted from Alessi and Trollip, 1991, pp. 385-386)

However, computer-assisted instruction evaluation as it is currently practiced cannot be transferred directly to an application field such as ICAI. It is necessary but not sufficient. One approach to exploring the merging of existing technologies (ICAI applications with evaluation technology) is to shift points of view in order to determine where reasonable matches exist. A three-phase evaluation template is designed for use in ICAI project evaluation. The first phase of the evaluation includes an attempt to understand the "product" development cycle employed, the ideological orientations of the designers, and their stated intentions. A second phase of analysis involves reviewing the internal characteristics of the ICAI systems from two perspectives: first, the quality of the instructional strategies employed; and second, the quality of the content addressed. A third and major phase of the study is empirical testing of the program.

In empirical testing for our specific case (i.e., ICAI Weather Tutor), the intention was to document effects of the program with regard to individual difference variables among learners (i.e., spatial aptitude and state worry) and with regard to a broadly conceived set of outcome measures including achievement instruments. Based on these three major phases (theoretical, instructional, and empirical analyses), recommendations for the improvement of this particular project and for the ICAI design and development process in general would be developed. A wide range of evaluation techniques would be included, for instance, both quantitative and qualitative data collection and analyses. This process is a variant of Figure 2.

Evaluation Questions

The evaluation questions proposed for an evaluation model are presented below. These questions are a variant of Figure 2. In each of these, information related to the adequacy of the artificial intelligence (AI) components (i.e., knowledge representation, instructional strategy, and student model) is treated as appropriate.

1. What is the underlying theoretical orientation of the system under evaluation? To what extent does the program serve as a model for ICAI?

2. What instructional strategies and principles are incorporated into the program? To what extent does the project exhibit instructional content and features potentially useful to future military applications?

3. What are the learning outcomes for trainees? To what extent do trainees achieve project goals? Do trainees with different background characteristics profit differentially from exposure to the project? To what extent does the program create unanticipated outcomes, either positive or negative?

Each of these questions could be applied to the ICAI Weather project. Table 2 provides the evaluation design for the Weather project. The above evaluation questions with the dimensions of inquiry, measurement method, and data source are provided. The tasks (e.g., Task 1) that are mentioned in the data source column refer to tasks in the prime contractor's proposal. The state worry scale is a computer adaptation of the Morris, Davis, and Hutchings (1981) worry scale. Obviously, the evaluation *per se* cannot be complete until the Weather ITS is completed to allow for trainee testing. In Table 3, the student questionnaire is provided. This questionnaire would be administered following a trainee's experience with the ICAI Weather program. This questionnaire focuses on the following topics: Weather objectives, learner assessment of Weather Tutor effectiveness, assessment of Weather Tutor communication, and affective factors. The instructor questionnaire is a modified form of the student questionnaire and would be administered following trainee and instructor experience with the Weather Tutor (see Table 4). Following data collection, both the processes and outcomes of the formative evaluation would be documented in the evaluation report.

MEASUREMENT OF INTELLIGENCE IN ICAI

A conceptual issue remaining in the formative evaluation of ICAI systems is how does one measure the level of intelligence. If one conceptualizes the solution in terms of the major software subsystems in ICAI, then we believe progress can be made on this issue. The major subsystems are the expert model for domain knowledge, the tutor for instructional knowledge, and the student model for knowledge concerning the student's background and progress through the program.

In many cases the expert model in ICAI is an expert system for domain knowledge. The software simulates varying levels of human expertise (both process and outcome), that is, novice, apprentice, journeyman, and expert. Its major feature is its knowledge representation approach versus various program control strategies (i.e., use of heuristics and an inference engine). The knowledge representation techniques can vary, but rule-based, semantic nets, and object-oriented systems are frequently used. Further, there are specialized man-machine interfaces (for example, explanation facilities), and many such expert systems are created using expert system shells. Evaluation approaches for such knowledge-based systems are basically variations of a Turing test. Human benchmarking has also been applied to such evaluations (O'Neil, Baker, Jacoby, Ni, & Wittrock, 1990).

Table 2 **Instrumentation and Data Collection Strategy**

Evaluation Question	Dimensions of Inquiry	Measurement Method	Data Source
1. What is the underlying theoretical orientation of WEATHER? To what extent does the project serve as a model of development for ICAI?	• cognitive and content underpinnings of WEATHER lesson	content review	initial and final design specifications (Task 1)
	• theoretical view of learning and instruction	content review	developed Tools (Task 2)
	• ICAI development process	interviews	project developers
2. What instructional strategies and principles are incorporated into the program? To what extent does the project exhibit instructional content and features potentially useful to AF application?	• instructional strategies	program review	instructional strategy report (Task 3)
	• subject matter content	program review	Knowledge base (Task 4); WEATHER Data (Task 6); Courseware (Task 7); WEATHER SMEs
	• AF needs	interviews	WEATHER SMEs, Armstrong staff
3. What are the learning outcomes for students? To what extent do learners achieve project goals? Do students with different background characteristics profit differentially from exposure to the project? To what extent does the program create unanticipated outcomes, either positive or negative?	• visualization ability; content knowledge in advection	multiple-choice test	French Kit, visualization test; posttest (WEATHER students)
	• background characteristics (WEATHER prerequisites, visualization ability)	questionnaires, French Kit	WEATHER students
	• level of worry	state worry scale	WEATHER students
	• reactions to WEATHER	questionnaires in Tables 3, 4	WEATHER students and instructors
	• transportability of technology	content review	software integration and testing (Task 8)

Table 3

STUDENT QUESTIONNAIRE[a]

Student Name ______________________________

Please circle your answer on the following questions.

A. WEATHER OBJECTIVES

1. Were the purposes of WEATHER presented to you?	YES	NO	DON'T REMEMBER
2. Were you told the purposes of WEATHER by your instructor?	YES	NO	DON'T REMEMBER

B. LEARNER ASSESSMENT OF WEATHER EFFECTIVENESS

3. How effective do you feel that WEATHER was in helping you:	Not Effective		Somewhat Effective		Very Effective
A. To improve your visualization strategies?	1	2	3	4	5
B. To improve your knowledge of advection?	1	2	3	4	5
C. Other (explain)	1	2	3	4	5

A. Why? ______________________________

B. Why? ______________________________

C. ______________________________

4. Would you recommend WEATHER to other students?	YES	NO

Why? ______________________________

5. Would you change WEATHER to make it more effective?	YES	NO

If yes, how? ______________________________

[a] Modified from Student Questionnaire in Baker, Bradley, Aschbacher, & Feifer (1985).

C. LEARNER ASSESSMENT OF WEATHER COMMUNICATION

	Rarely or Never		Sometimes		Almost Always
6. In your opinion, how often did WEATHER not correctly identify your errors?	1	2	3	4	5
7. Was the feedback from WEATHER					
A. Easy to understand?	1	2	3	4	5
B. Appropriate?	1	2	3	4	5
C. Sufficient?	1	2	3	4	5
D. Related to class lectures?	1	2	3	4	5
8. Did any feature of WEATHER frustrate you?	1	2	3	4	5

If yes, how? ______________________________

____

____

D. AFFECTIVE FACTORS

	Not at All		Somewhat		Very Much So
9. How much did you enjoy using WEATHER?	1	2	3	4	5
10. To what extent did WEATHER motivate you to spend more time learning WEATHER on your own?	1	2	3	4	5
11. To what extent did WEATHER help you feel more comfortable with:					
Using the computer?	1	2	3	4	5
Using visualization?	1	2	3	4	5
12. Did the graphics enhance WEATHER's effectiveness?	1	2	3	4	5

If yes, how? ______________________________

____

____

13. How easy was it for you:	Not Easy		Moderately Easy		Very Easy
To use the computer?	1	2	3	4	5
To find the information you needed on the screen?	1	2	3	4	5

14. Other Comments: We would welcome any other comments or reactions you may have had to WEATHER.

____

____

____

Table 4

INSTRUCTOR QUESTIONNAIRE[a]

Instructor Name ____________________________

Please circle your answer on the following questions.

A. WEATHER OBJECTIVES

1.	Were the purposes of WEATHER presented to you?	YES	NO	DON'T REMEMBER
2.	Were you told the purposes of WEATHER by your supervisor?	YES	NO	DON'T REMEMBER

B. INSTRUCTOR ASSESSMENT OF WEATHER EFFECTIVENESS

3. How effective do you feel that WEATHER was in helping students:

	Not Effective		Somewhat Effective		Very Effective
A. To improve their visualization strategies?	1	2	3	4	5
B. To improve their knowledge of advection?	1	2	3	4	5
C. Other (explain)	1	2	3	4	5

A. To improve their visualization strategies?

Why? ____________________________

____

____

B. To improve their knowledge of advection?

Why? ____________________________

____

____

C. Other (explain)

____

____

____

4. Would you recommend WEATHER to other instructors? YES NO

Why? ____________________________

____

____

5. Would you change WEATHER to make it more effective? YES NO

If yes, how? ____________________________

____

____

[a] Modified from Student Questionnaire in Baker, Bradley, Aschbacher, & Feifer (1985).

C. INSTRUCTOR ASSESSMENT OF WEATHER COMMUNICATION

	Rarely or Never		Sometimes		Almost Always
6. In your opinion, how often did WEATHER not correctly identify student errors?	1	2	3	4	5
7. Was the feedback from WEATHER					
A. Easy to understand?	1	2	3	4	5
B. Appropriate?	1	2	3	4	5
C. Sufficient?	1	2	3	4	5
D. Related to class lectures?	1	2	3	4	5
8. Did any feature of WEATHER frustrate you?	1	2	3	4	5

If yes, how? ______________________________

____

____

D. AFFECTIVE FACTORS

	Not at All		Somewhat		Very Much So
9. How much did your students enjoy using WEATHER?	1	2	3	4	5
10. To what extent did WEATHER motivate your students to spend more time learning WEATHER on their own?	1	2	3	4	5
11. To what extent did WEATHER help your students feel more comfortable with:					
Using the computer?	1	2	3	4	5
Using visualization?	1	2	3	4	5
12. Did the graphics enhance WEATHER's effectiveness?	1	2	3	4	5

If yes, how? ______________________________

____

____

13. How easy was it for your students:	Not Easy		Moderately Easy		Very Easy
To use the computer?	1	2	3	4	5
To find the information they needed on the screen?	1	2	3	4	5

14. Other Comments: We would welcome any other comments or reactions you may have had to WEATHER.

____

____

____

With respect to tutors, the software subsystem should implement instructional knowledge. Instructional knowledge consists of relating task strategies (the *what* of learning) to instructional strategies (the *how* of teaching). Task strategies are strategies that an expert uses to solve a problem in a domain, where a domain equals task or subject matter. For example, in the domain of troubleshooting an electrical system, the task strategies would be change cost, split-half, and malfunction probability. In contrast, domain-independent instructional strategies (how to teach task strategies) are independent of task but appropriate to a class of learning outcomes (for instance, cognitive strategy). The instructional design implications that follow are twofold: (a) it is mandatory to specify what is the outcome of the learning (for example, problem solving), and (b) various outcomes (for example, procedural learning or problem solving) have specific learning conditions, that is, instructional strategies.

Thus, instructional knowledge should be explicit in some form of knowledge representation in which rules reflect, for example, the state of the art in instructional science as well as the desirable properties of a human tutor (see Table 5). In summary, the student model should reflect the state of the art in cognitive science (for example, individual differences in metacognitive skills) and should have some sort of explicit knowledge representation (for example, rule-based).

Table 5

Desirable Properties of a Human Tutor

1. The tutor causes the problem-solving heuristics of the student to converge to those of the tutor.
3. The tutor chooses appropriate examples and problems for the student.
4. The tutor can work arbitrary examples chosen by the student.
6. The tutor is able to adjust to different student backgrounds.
7. The tutor is able to measure the student's progress.
8. The tutor can review previously-learned material with the student as the need arises.

(Adapted from Gamble & Page, 1980)

Our view from an AI perspective is that the major attributes of all these subsystems should have both necessary and sufficient conditions in order to be "intelligent." The common necessary condition is an explicit knowledge representation. The sufficient condition varies by subsystem: (a) For the expert model, it should simulate various levels of human expertise and could be tested by a Turing test. (b) For the tutor, the sufficient condition should reflect the state of the art in instructional knowledge. (c) For the student model, the sufficient condition should reflect the state of the art in cognitive science (for example, individual differences). A complementary view of this issue has been provided by Schank (1991).

WHERE ARE WE NOW?

A formative evaluation model has been suggested to evaluate an intelligent computer-assisted instruction system in weather. This ICAI system has been designed to implement a visualization strategy. However, a computer-based training system has been implemented, and the formative evaluation of the CBT system remains to be completed. Our next step is to implement and test our formative evaluation design in an ICAI system.

REFERENCES

Alessi, S.M., & Trollip, S.R. (1985). *Computer-based instruction: Methods and development.* Englewood Cliffs, NJ: Prentice-Hall.

Alessi, S.M., & Trollip, S.R. (1991). *Computer-based instruction: Methods and development* (2nd ed.). Englewood Cliffs, NJ: Prentice-Hall.

Baker, E.L. (1974). Formative evaluation in instruction. In J. Popham (Ed.), *Evaluation in education.* Berkeley, CA: McCutchan.

Baker, E.L., & Alkin, M.C. (1973). Formative evaluation in instructional development. *AV Communication Review, 21*(4), 398-418.

Baker, E.L., Bradley, C., Aschbacher, P., & Feifer, R. (1985). *Intelligent computer-assisted instruction (ICAI) study* (Final Report to Jet Propulsion Laboratory). Los Angeles: University of California, Center for the Study of Evaluation.

Baker, E.L., & Saloutos, W.A. (1974). *Formative evaluation of instruction.* Los Angeles: University of California, Center for the Study of Evaluation.

Ekstrom, R.B., French, J.W., Harman, H.H., & Derman, D. (1976). *Manual for kit of factor-references cognitive tests.* Princeton, NJ: Educational Testing Service.

Gagné, R., Briggs, L.J., & Wager, W.M. (Eds.). (1988). *Principles of instructional design.* New York: Holt, Rinehart & Winston.

Gamble & Page. (1980). *International Journal of Man-Machine Studies, 12,* 259-282.

Morris, L.W., Davis, M.A., & Hutchings, C.H. (1981). Cognitive and emotional components of anxiety: Literature review and a revised worry-emotionality scale. *Journal of Educational Psychology, 73*(4), 541-555.

O'Neil, H.F., Jr., & Baker, E.L. (1990). *An evaluation plan for intelligent computer-assisted instruction in weather* (Technical Deliverable 4 to Expert-EASE Systems, Inc.). Sherman Oaks, CA: Advance Design Information, Inc.

O'Neil, H.F., Jr., & Baker, E.L. (1991). Issues in intelligent computer-assisted instruction: Evaluation and measurement. In T.B. Gutkin & S.L. Wise (Eds.), *The computer and the decision-making process* (pp. 199-224). Hillsdale, NJ: Lawrence Erlbaum.

O'Neil, H.F., Jr., Baker, E.L., Jacoby, A., Ni, Y., & Wittrock, M. (1990). *Human benchmarking studies of expert systems.* Los Angeles: University of California, Center for Technology Assessment.

O'Neil, H.F., Jr., & Jacoby, A. (1990). *Design of an instructional strategy to teach visualization in an advection context in intelligent computer-assisted instruction.* Sherman Oaks, CA: Advance Design Information, Inc.

Schank, R.C. (1991). Where's the AI? *AI Magazine, 12*(4), 38-49.

Scriven, M. (1967). The methodology of evaluation. In R.W. Tyler, R.M. Gagné, & M. Scriven (Eds.), *Perspectives of curriculum evaluation. AERA Monograph Series on Curriculum Evaluation, No. 1.* Chicago: Rand McNally.

SEMIFORMAL REPRESENTATIONS IN HYPERMEDIA FOR ANALYZING RICH AND COMPLEX DOMAINS

Daniel M. Russell

Xerox Palo Alto Research Center
Palo Alto, California
USA

Abstract

Semiformal knowledge representations, such as those often implemented in hypermedia, can support analyses of rich and complex data. As opposed to more traditional analysis representations that are fairly rigid and difficult to change, semiformal representations support searching for an analytic framework. In particular, semiformal representations (a) are re-interpretable as new perspectives on the database are created, (b) allow for incremental discovery, formalization and representation of knowledge about the domain, (c) support references from the analysis back to original source materials for verification and confidence-building, and (d) allow representations to be annotated with multimodal commentary appropriate to the analysis task, simplifying their use and comprehensibility. Experiences with IDE (a semiformal analysis environment) for analytical tasks in education and training illustrate these points.

1.0 The Analysis Problem

Confronting a large amount of qualitative or textual data and trying to make sense of it is a common problem. Analysis, in its most general sense, is manipulating sets of data and creating a representation of the data in a more concise and meaningful fashion. While the specifics of analysis vary widely from domain to domain, some common elements remain: understanding what is in the data set, searching for individual items, organizing the data, and creating symbolic representations of the data set as a whole. There are also common analysis problems across domains. Analysts invariably have too much data, in multiple formats, coming from multiple sources.

Although there are many kinds of analysis, in this paper I restrict *analysis* to mean "searching for a representation that captures salient aspects of a domain." That is, the outcome of an analysis is a representational structure describing a theory of the data under inspection. The problem of searching for a representation to act as an analytical theory is common whenever an analyst has to make sense of a large body of observations. Data sets such as video records, verbal protocols, notes from group meetings, email discussions, or field observations all have to be analyzed by constructing some representation structure to organize them into a coherent pattern.

Advanced Technologies Applied to Training Design, Edited by
R.J. Seidel and P.R. Chatelier, Plenum Press, New York, 1993

Building a representation is a difficult task. Finding an appropriate representation (that is, one that facilitates the problem solving tasks it's intended to) is largely a search process. But premature commitment to a particular representation often leads to blockage in the search for an effective answer. [Anderson, 1980] Most representations tend to be formalisms that heavily encode the knowledge being represented. That is, formalisms abstract behavior and observations into symbols that decouple the representation from its observational grounding. Creating a representation is, in large part, moving back-and-forth from the representational abstraction to ground observations, verifying that the symbols capture the phenomena within the paradigm.

After a formal representation is constructed, it is still often hard to use. This isn't surprising, since cognitive loading and memory limitations restrict human use of formal symbol systems. [Anderson, 1980] [1] Because of this, specifications for complex artifacts are notoriously difficult to write precisely and well. These problems stem from the way humans communicate and think about procedures: we are inherently context-encoding and context-using in our problem-solving [Hall, et al, 1989], a property that is not normally supported by our representations. When analyzing a domain and trying to create a representation, it is easy to underspecify the meaning of a formal construct. Creating a formal symbol system is difficult, especially if contextualizing information is lost during the analysis and if the formalism choice determines a particular interpretation. Fortunately, this need not happen.

2.0 Semiformal Knowledge Representation

In this paper I argue that hypermedia is peculiarly useful for analysis since it encourages and supports a particular style of representation. A representation is *semiformal* when it combines different types of representational structures at differing levels of interpretability into a single representation. Semiformality is a property of the composite representation, measuring how precise and semantically determined a representation is. Intuitively, a representation grows increasingly formal as its range of possible semantics grows smaller.

Formal representations are often machine computable, while informal representations tend to be human-useful.[2]

A semiformal representation is particularly useful in analysis tasks since it naturally encodes multiple views of a domain, and can be re-interpreted as global understanding of a domain or problem shifts over time.

Hypermedia & representation: Hypermedia is by definition a rich environment: a typical hyperspace may include natural language text, graphics, video, sound, or other essentially (currently) non-interpretable entities. When building representations in a hypermedia, we tend to include "too much" or extraneous material (from the perspective of a program that will use the resulting hyperspace). Over time, however, the desire grew to build tools that could assist the user with navigating and manipulating the hyperspace. Instead of simply having a complex representation for human use, we now find ourselves writing interpreters

---

[1] Which is why people tend to use external memory aids, such as paper and pencil, when doing any significant problem-solving with formal symbol systems. Without the use of memory chunking aids (paper & pencil, mnemonics or special algorithms) most people find formal symbol manipulation quickly stretches their cognitive capabilities.

[2] A similar notion, "semistructuredness," in [Malone, 1987], and [Harp & Neches, 1988] refers just to the format of a representation, and not the associated interpretation that's implied. Something is semistructured when it has encapsulated bits of informality; a representation is semiformal when an interpretation (or interpretor) is associated with sections of a representation.

that can use the hypermedia representations despite the extra materials in the symbol structures. Thus, we create programs that use the hypermedia representations we construct despite their richness.

Formality, Semiformality, Informality: John Searle writes [Searle, 1990] that a representation is "formal" if it is syntactically formed and can be parsed into constituents that may be assigned a semantics. An informal representation, by contrast, implies that it cannot be broken up into meaningful fragments and assigned a semantics. By this definition, hypermedia representations are *semiformal,* since portions of the hyperspace may be uninterpretable for a given interpreter. However, note that this is true only for a given, specific interpreter. A program may still use portions of a representation that are strictly uninterpretable as the basis of an inference (e.g., to deduce that an author wrote something, you need only know that there is a book, you don't need to read it). More to the point, a single representation might have zero or more interpreters, more than a single interpretation, and differing qualities of interpretation (from simple-minded to full and richly competent).

The *interpretation* of the symbol structure is, in the last analysis, operationally defined by the program that operates on it. Interpretation is computation that is contingent on the content and form of a symbol structure. The portions of a symbol structure that may affect a computation determines its interpretability. For example, a video snippet in a node might affect a computation by its presence or absence, but the snippet itself is non-interpretable unless some computation can examine its contents and answer some question about its internal structure. Thus, there are relative degrees of interpretability: it may be simple to determine the length of a video snippet (the number of frames / the frame rate), but more difficult to identify the actors on the video by name.

By contrast, formal representations assume a particular interpreter and interpretation. Such representation / interpretation schemes are powerful and efficient. They are often complete, and have great inferential power. But for representation analysis purposes, they embed assumptions about what is being represented by abstracting away unnecessary details.

Example: MYCIN's representation of disease abstracts away aspects of contextual use information about diagnostic knowledge, but simply relates symptoms and findings to predictions about possible infections. For MYCIN's particular task, context knowledge about diagnosis was not immediately applicable, and was not included in the MYCIN's knowledge representation system. In their knowledge engineering process, special attention was paid to which features of the problem description were salient, and only those were chosen to be represented in the MYCIN formalism.

Note, though, that not all of the MYCIN knowledge base is strictly formal. Portions of each rule in the rule base have author, time-date stamps, and some commentary included. Although never used as a semiformal component of the system (i.e., never used by anything other than as comments to future rules-writers), the MYCIN community perserved much contextual information in the knowledge base for future reference.[3] In essence, a function of the knowledge base was kept in a semistructured, informal form for use by human interpreters. Similarly, LASSIE -- a program analysis tool for helping programmers understand a large system -- uses the informal parts of a program text to determine what a function does, and to build meaningful pointers into the program for the programmer. [Devenbu, et al., 1991]

Like MYCIN and LASSIE, most representations in common use are semiformal. Programs have comments, and good knowledge engineers leave broad paths of commentary to guide subsequent knowledge workers in their use of the knowledge base. A semiformal

---

[3] [Clancey, W., personal communication, 1991]

representation recognizes that interpretations are associated with representations. Even when the interpreters are incomplete or missing, there is still an advantage to keeping information semiformally.

Analysis: Given this background, we begin to understand why a semiformal representation is especially well suited for analysis.

Many analysis tasks can be viewed as a search for an analytical framework, or in our terms, a representation. It is well known that an appropriate representation choice simplifies problem solving, or in this case, an analysis. [Polya, 1945], [Anderson, 1980] Possibly the major problem in finding an appropriate representation is overcoming a premature commitment to an analysis framework (aka "functional fixedness,"e.g., [Glucksberg & Danks, 1968]). Semiformal representations, by keeping the binding between representation and interpretation easily alterable, removes a barrier in the process of discovering a working analysis representation.

Very often, analytic techniques with predefined categories are used to study and organize information in a domain. In writing tasks, outliner programs are sometimes used to organize topics without a pre-determined framework. However, in both cases, the question of interpretation is removed from the work to be done. Formal, predefined categories provide a set of abstractions to the analyst. If the categories describe the domain well, then the analysis problem is fairly straightforward, and can be captured within that predefined framework. In such cases, the problem was not to discover a suitable analysis representation, but to correctly use an existing one. Creating an outline, by contrast, is a true search for new analysis of domain information. But outlines (as commonly formed) are representationally inexpressive, and capture only hierarchy information. The interpretation of outline elements is hidden, held implicitly in the writer's understanding of the outline.

3.0 Examples

The use of semiformal representations in hypermedia for analysis is best illustrated with a few examples. Each example shows one of the major strengths of a semiformal representation as a medium to support representation creation and evolution.

3.1 - Re-interpretation The central feature of a semiformal representation is that the representation structure is a mix of wholly interpretable (formal), partially interpretable and completely un-interpretable sections. Re-interpreting a representation can be useful when:

- new, more powerful interpretive tools come available (and change the level of interpretability of a portion of the representation);
- new information comes to light causing a re-evalution of data according to a new basis;
- criteria shift, and what was important now isn't (or vice-versa), or when combinations of information assume a new relevance. (e.g., you can view a database under different perspectives that are a function different interpretations of the data.

In each case, a fundamentally different view of the database can be expressed. In standard formal models, a record has an interpretation associated with it (e.g., the "height" slot is a real number indicating a person's height in inches), and re-interpretation according to a new basis requires shifting the underlying semantic model of the database. A "view" in relational database terms is a recombination (a "virtual table") of existing data. [Date, 1991] A re-interpretation allows for a completely new analysis of the data in the database beyond relational operations.

Example: In June, 1988, IDE 1.6 [4] was used by a group of instructional analysts to create an analysis of commercially available laser printers. This IDE analysis was built to aid instructional designers in understanding the essential characteristics of the different laser printers as they created a course on laser printing fundamentals. The original design of the resulting hyperspace relied heavily on describing printing systems in terms of functions, subsystems, nominal failure modes, and so forth. IDE supported the analysts in creating new cards, prompting for describing information in each card slot, and then linking cards together to form a large (roughly 3,000 cards) hyperspace network. (An example of the type of analysis cards used is shown in Figure 1.)

However, once the hyperspace was created, it became clear that it would be possible to re-interpret the network in terms of conceptual clusters, rather than as simple laser printer functional system descriptions. To that end, we wrote a clustering tool that translated each card in the analysis hyperspace into a word-stemmed descriptor by passing the textual description of each card through a morphological analyzer. (See Figure 2.) Concept clusters were then formed by grouping cards with similar word-stem profiles, and with similar positions within their hyperspace context (i.e., card type, incoming and outgoing links) into similar buckets. After a bit of tweaking of bucket size parameters, and similarity thresholds, we created a set of over fifty buckets that held subgraphs of the incoming hyperspace that had been identified as similar by this algorithm.

In showing the results of this clustering to subject matter experts, we found that several of the clusters identified higher-level abstractions than the experts had considered in their analyses. This occurred because the word-stemmer understood synonyms for this domain, and was able to find cards (by examining the text descriptions) that matched with synonym group profiles of other cards occurring in similar contexts (i.e., the surrounding cards and links).

Essentially, the first representation was made with a particular task in mind. Later, using the previously uninterpretable English language text in card descriptions, and partially interpretable relations between cards, clusters could be found.

3.2 - Evolving a representation Semiformal representations are particularly easy to modify since they do not carry the overhead of associated formal and fragile structure interpreters. An interpreter in a semiformal system must work with what is available within the symbol structure and be able to ignore encapsulated (and currently uninterpretable) symbol structure fragments. This robustness makes hypermedia representation systems particularly useful for representational analysis. Hypermedia representations support representation engineering by *not providing* a predetermined semantics, but rather, through simplicity and richness, give a representation analyst freedom to work. Within a hypermedia, it is easy to prototype symbol structures. This is partly due to the simplicity of the hypermedia interface, the ability to easy extend the set of "datatypes" (node formats), and perhaps most importantly -- to the ability to create multiple, overlapping representations in a single hyperspace (reification of non-independent objects) without conflicts.

Of course, the downside of these properties is that it is also easy to create representations that are intractable to work with formally. In ours and others experiences with creating representations, however, the largest hurdle is not specifying the formal properties of a representation, but rather, creating a representation system that captures the knowledge about a domain in a form that is understandable by the authors. [O'Neill & Morris, 1989]

---

[4] In this paper, we refer to two different versions of IDE -- 1.6, which ran in Interlisp-D on Xerox processors, and the now current edition of IDE, version 2.0, implemented in FoxBase+ on the Macintosh. Both versions are comparable.

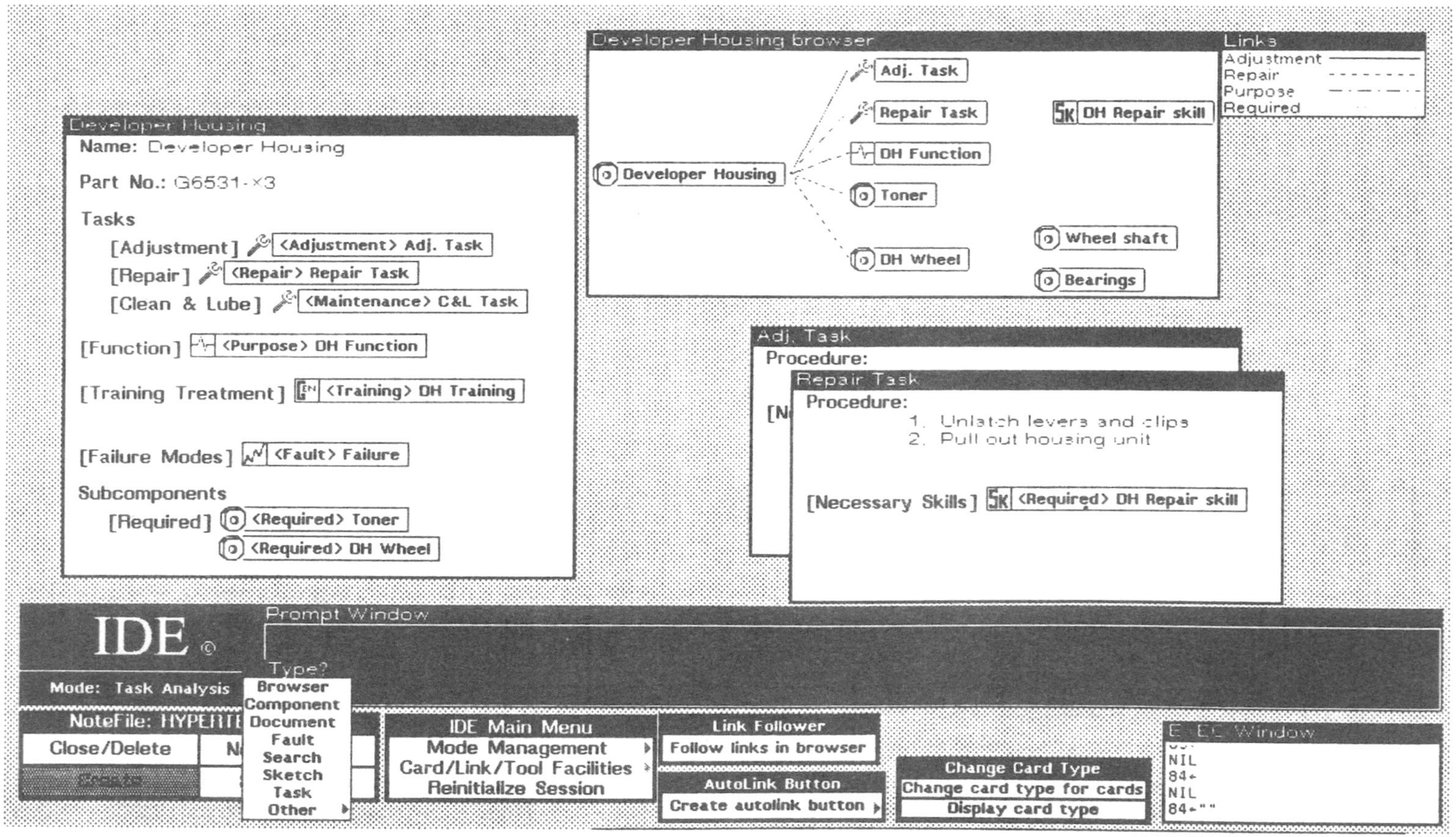

Figure 1. Sample cards from a laser printer analysis. Cards captured information about functions, subsystems, components, failure modes and training treatments.

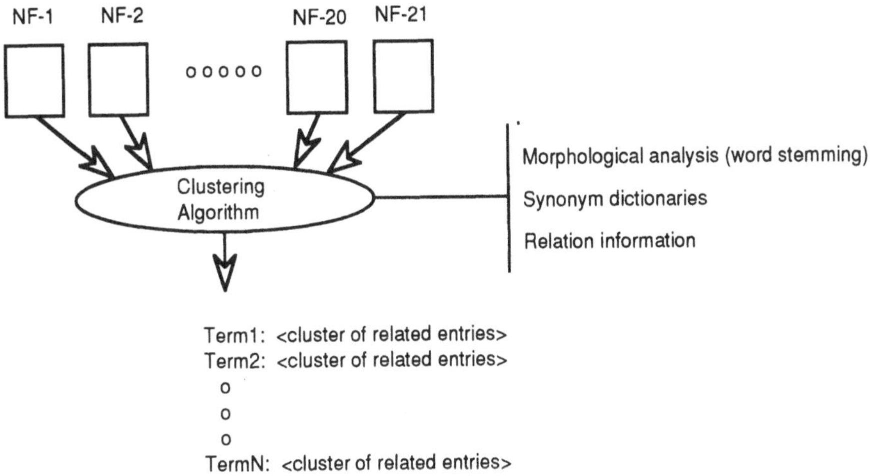

Figure 2. Re-analyzing a large hyperspace from a new perspective required a new interpretation of card fields. Each of the 21 notefiles (NF) represented a semi-formal analysis of a laser printer. Previously uninterpretable information from cards became usable by a using a word-stemming morphological analysis with a clustering algorithm, producing newly identified clusters of similar nodes in the hyperspace.

Example: During March, 1991, DARPA sponsored a weekend workshop on Sharable Knowledge Bases. IDE was used to capture and organize the arguments in the email discussion leading up to the workshop. In addition, the workshop was on-the-spot real-time transcribed into a semiformal database. Since this was the initial attempt at using IDE for real-time meeting facilitation, we quickly discovered the shortcomings of trying to categorize and structure a fast-paced discussion. At the outset, the representation had 8 cardtypes and 12 linktypes in an extended gIBIS style structure. [Conklin & Begeman, 1988] While this structure was fine for an *ex post facto* analysis of mail traffic, it was inadequate for capturing a lively discussion, especially when a speaker would begin to make a support statement that evolved into a counterargument. Consequently, many of the notes were mis-categorized, or were linked into the note structure incorrectly in light of subsequent discussions.[5]

Figures 3-7 tell a small story of how this representation was modified to accommodate a new understanding of the domain. In brief, as the analyst was examining a set of cards that described different possible experiments, it became clear that there was a distinction in the cards that was not clearly expressed in the current representation. Using the browsing and collection tools, the analyst defines a new category, and moves some of the experiment cards into the new classification.

The representation changed radically over time. As shown in the next several figures, examining the data collected at the workshop led to a change in the way the analysts wanted to group the domain data. The representation shifted to handle the new Outcome category and data items were regrouped accordingly.[6]

5 In this setting, IDE was not the main focus or organizer of the discussion. Our goal was to see how much of the meeting could be meaningfully captured in IDE structures in real-time without interference. As a consequence, the participants did not significantly alter their normal style of argument and orient to the IDE structures that were being constructed. This is very different than Conklin's experience, where gIBIS was the main organizer for a design problem.

6 A new version was declared every few days, whenever the representation changed significantly from its predecessor as the result of adding new card or link types.

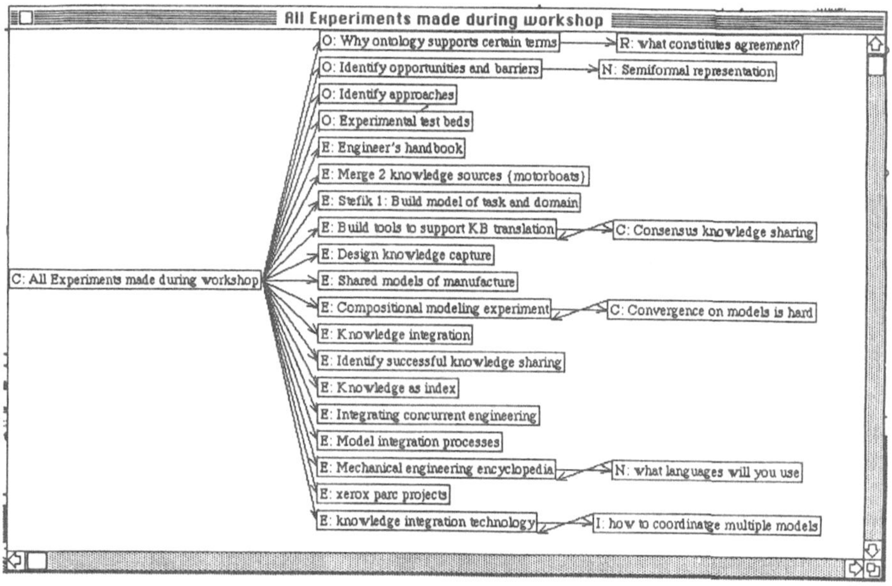

Figure 3. This graph browser shows a portion of the set of proposed Experiments from the SharableKnowledgeBases project. There are many proposed experiments, a very flat organization of the proposals, and little discrimination between subtypes.

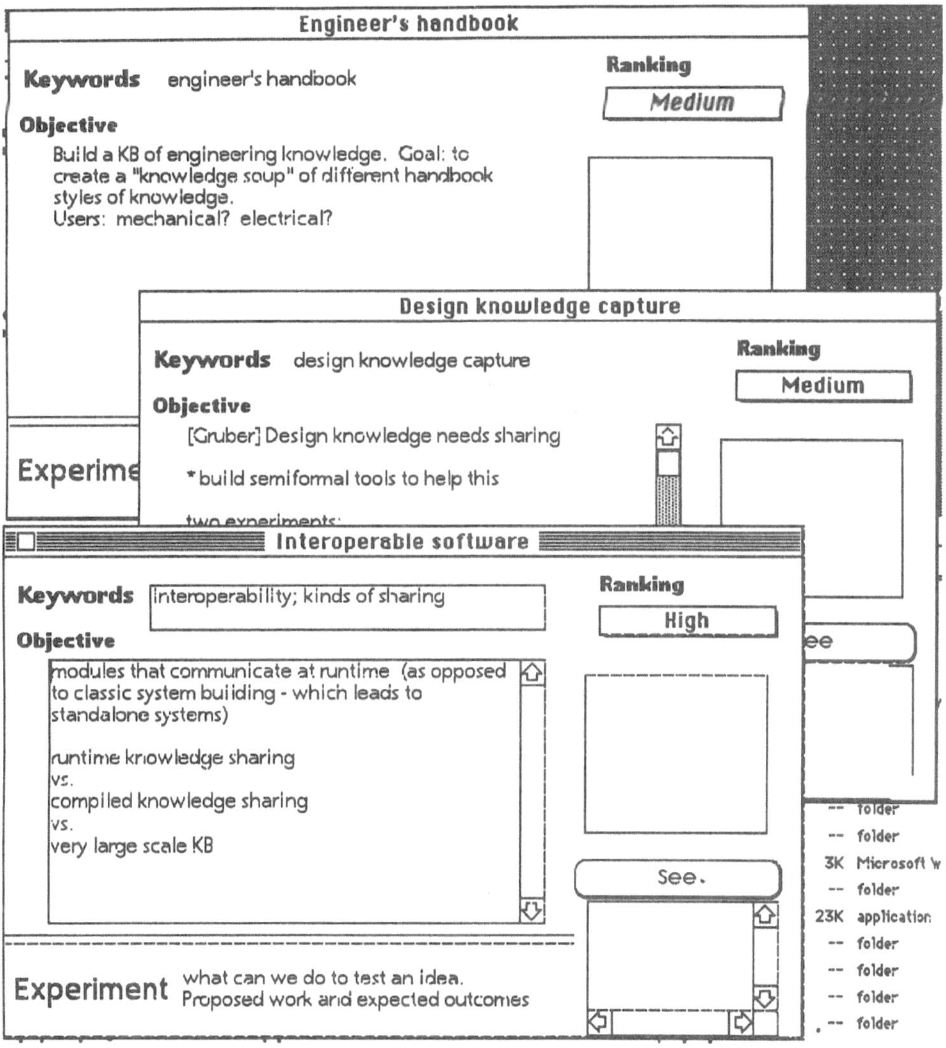

Figure 4. When examining the Experiment cards, it becomes clear that there too many cards, some of which might seem to properly belong in a new category. These cards that are different seem to be about something other than just experiments, but predicted outcomes of experiments.

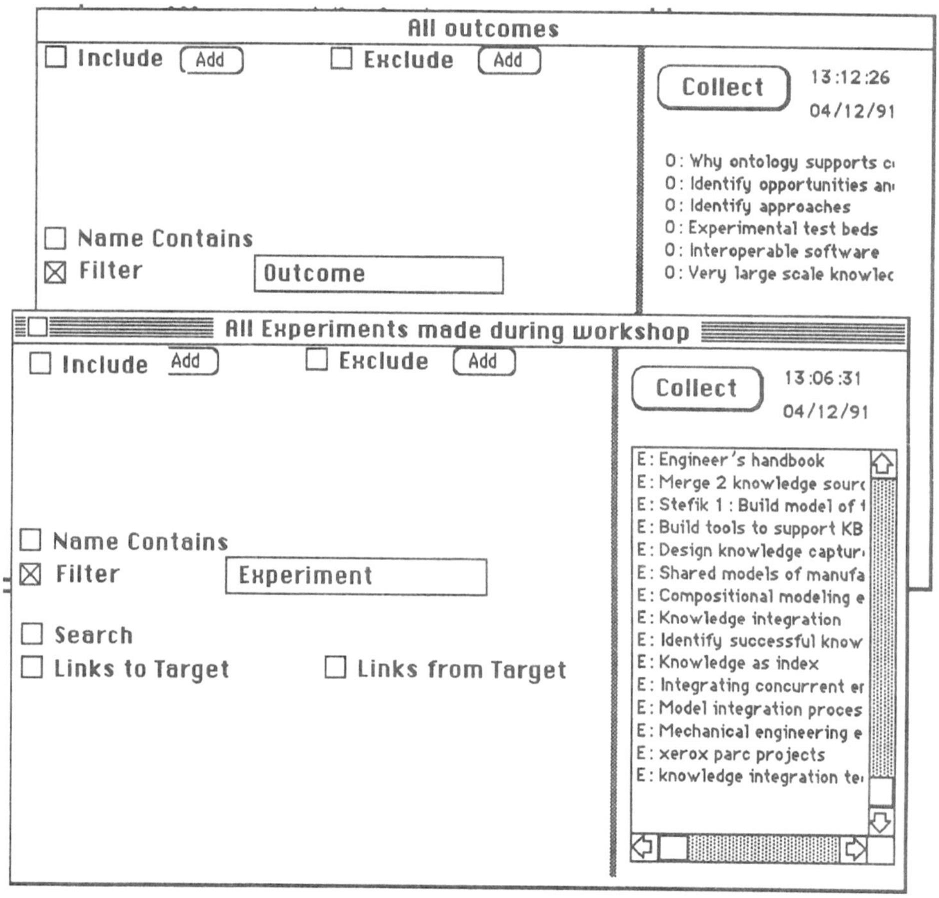

Figure 5. In this Collection card, the analyst has separated out the Experiment cards that don't quite fit into the new category, Outcome.

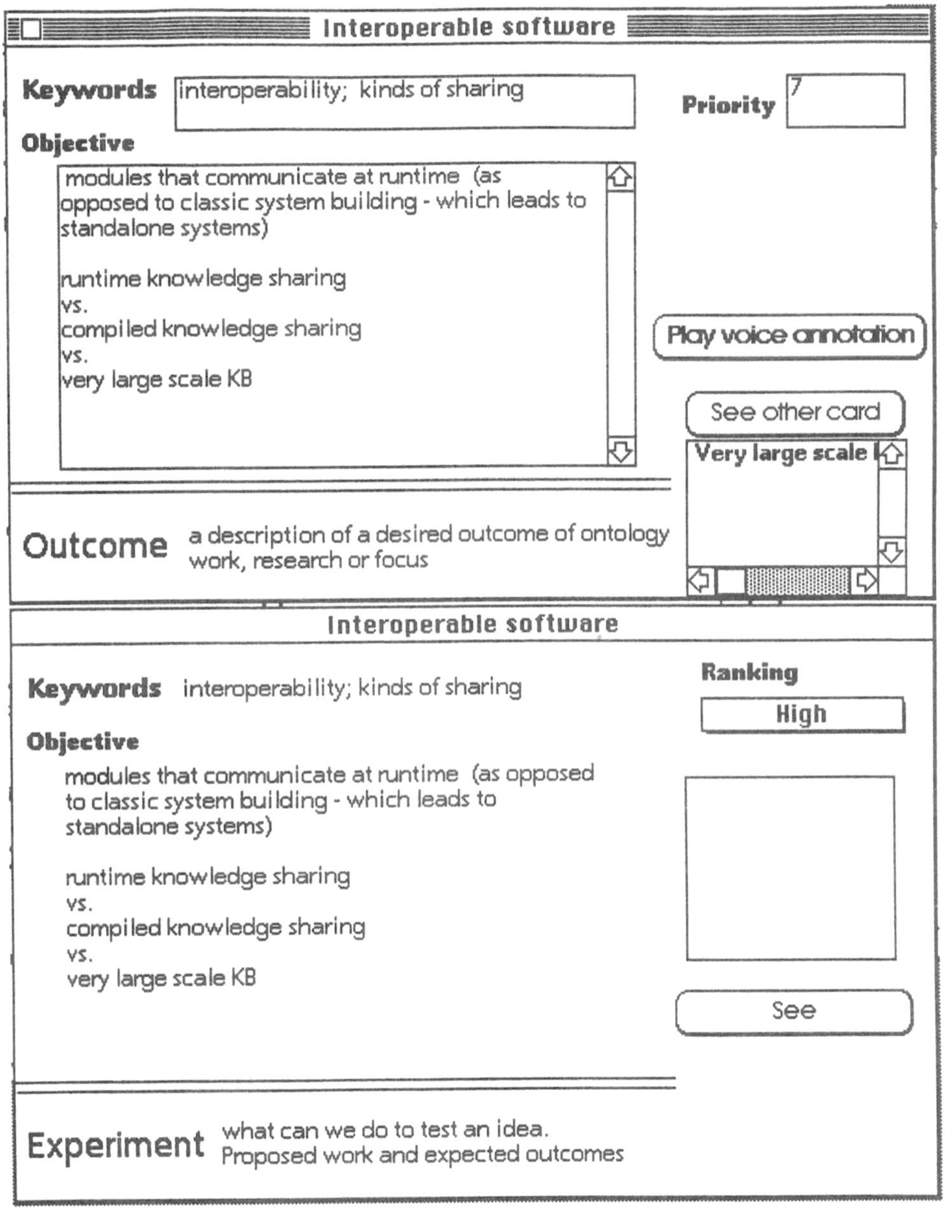

Figure 6. The user has created a new card type "Outcome" and coerced all of the cards in the Collection to be of this new type. Here is the old cardtype and the new cardtype side-by-side. Although the appearances are similar, there are a few differences between "Experiments" and "Outcomes," and a new type of card that carries a different interpretation of the text. For this analyst, there is value in discriminating between these two different readings of the Objective text field.

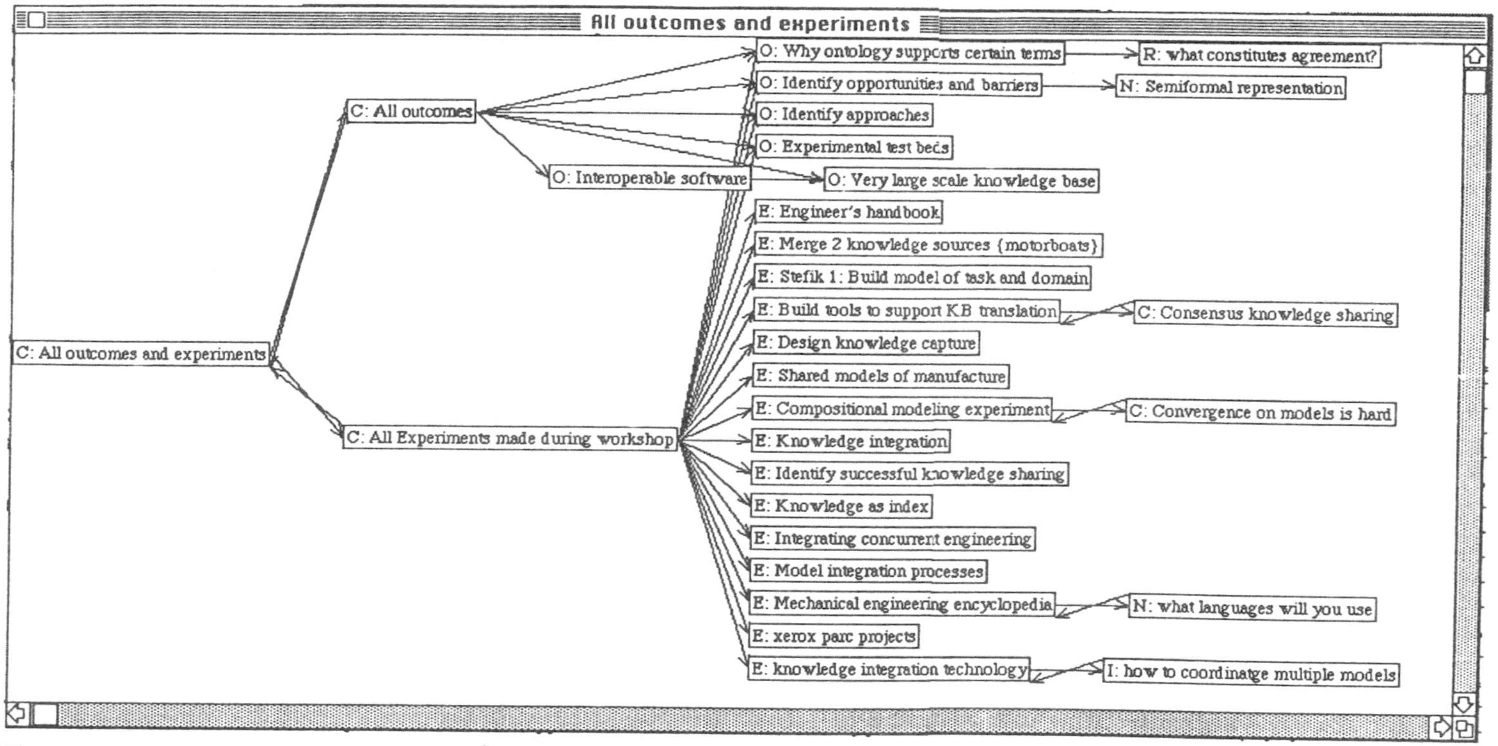

Figure 7 After re-structuring the hyperspace, the Experiments have been divided into two categories: Experiments and Outcomes.

As Figure 8 shows, in the early stages of analysis, a representation shifts to accommodate new understanding about how it should be organized. In general, the number of card and link types increase to a plateau[7] as the analyst searches for a framework that captures a set of understandings about the domain. At the same time, the average number of slots with formal interpretations also rises as the analyst begins to create abstractions that capture precise features of the domain.

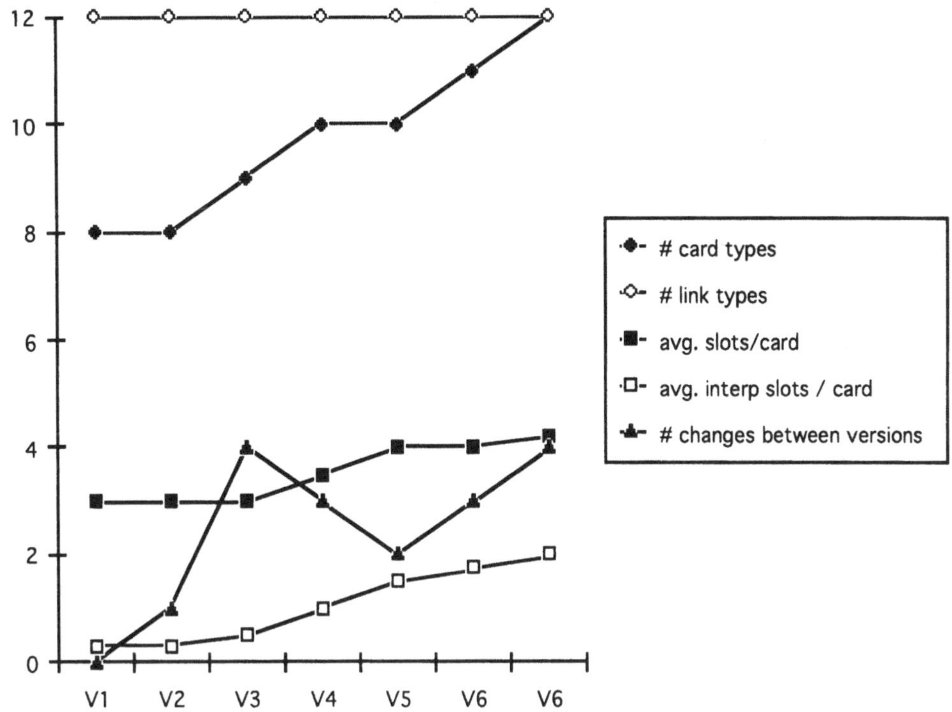

Figure 8. A representation changes with time. This table shows how node and link types were added and deleted as the analyst came to understand more about the domain. Here, the first 6 versions of the representation changes are shown. As versions change, the average number of slots with defined interpretations increases.

In searching for an analysis framework, using a semiformal representation keeps the set of possible interpretations open, making shifts in understanding easy to build into the representation.

3.3 - Richness of semiformal representations: As defined above, a semiformal representation may contain information that is not directly interpretable by any formal

[7] Although in this graph the number of linktypes stays constant, this is an artifact of this section of the data. Later (starting at version 10) the number of linktypes also begins to increase.

semantics, but is useful to a human user. For instance, representations might include relations that have no formal interpretation, but inform a human reader about the correctness of the data.

Example: Validating a representation means working with the representation to check ordinary and extreme cases, returning to source materials, and constantly evaluating how well the representation captures the phenomena and data of interest. In addition, many knowledge engineers have found that convincing the customers of a representation's validity to be a major hurdle.

When we used IDE to identify the major concept clusters in the Laser Printer domain (the example from section 3.1), we had many occasions when it was necessary to backtrace a conclusion (a cluster) to its original source materials. The source materials were semi-structured documents created by a variety of users distributed across several locations. And (as we discovered) some source data was deemed not as reliable as others by the subject matter experts evaluating the results of the cluster analysis. It was critical that our analysts be able to revert back to the original source documents, and not only read them, but to discover who wrote them, when, and where. Thus, the richness of the sources and their links with the derived conclusions were important in validation.

4.0 Summary

Analysis of complex and rich data is rarely simple or straightforward. Often, the major problem is to "understand" the essential structure and features of the data. In research analyses, this is equivalent to searching for a representation structure to organize the data, trying to interpret the data according to some theory, or simply trying to find patterns in the data.

We argue that semiformal representations have several properties that support these kinds of analyses.

A semiformal representation fundamentally allows re-interpretation of the database as new views and information becomes available. By contrast, more formal representations have eliminated most ancillary information by abstraction and have shifted the interpretation of data into the interpreter, and away from the data source materials. While formality leads to precision and efficiency, formal representations are also frequently difficult for humans to use (due to cognitive load and memory limits), are unconvincing to experts (who no longer see any contextual information surrounding the expressed data), and are usually difficult to re-interpret under new perspectives on the data.

Semiformal representations are rich in redundant or apparently insignificant information. Yet we have found that it is this richness that allows re-interpretation, validation, and ease-of-use on the part of an analyst creating a representation that captures and organizes a large and complex body of data.

Acknowledgements

Many thanks to Tom Gruber, Sanjay Mittal, Peter Pirolli, and Mark Stefik for helping me to understand these issues.

References

Anderson, J. R. *Cognitive Psychology and its Implications*, W. H. Freeman, San Francisco (1980).

Barwise, J., Etchemendy, J. "Visual information and valid reasoning," CSLI Technical Report, Stanford University, (1991).

Carlson, D. A., Ram, S. "Hyperintelligence: the next frontier", CACM v 33, n 3, (March, 1990).

Conklin, Jeff, and M. Begeman, "gIBIS: A hypertext tool for team design deliberation", MCC Technical Report STP-016-88, (1988).

Date, C. J. *An introduction to database systems*. Addison-Wesley, Reading MA, (1981).

Devenbu, P., Brachman, R. J., Selfridge, P., G., Ballard, B. W. "LASSIE: A knowledge-based software information system" CACM v 34:35-49, (May, 1991).

Glucksberg, S., Danks, J. H. "Effects of discriminative labels and of nonsense labels upon availabilty of novel function" Journal of Verbal Learning and Verbal Behavior, 7, 72-76, (1968).

Hall, R., Kibler, D., Wenger, E., Truxaw, C. "Exploring the episodic structure of algebra story problem solving" Cognition and Instruction, 6 (3), 223-283, (1989).

Harp, B., Neches, R. "NOTECARDS: An everyday tool for aiding in complex tasks" Information Sciences Institute Reprint Series, ISI/RS-88-204 (March, 1988).

Irish, P. M., Trigg, R. H. "Supporting collaboration in hypermedia: issues and experiences" in *The Society of Text: Hypertext, Hypermedia and the Social Construction of Information*, MIT Press, (1989).

Jordan, D., Russell, D. M., Jensen, A. M., Rogers, R. A "Facilitating the development of representations in IDE" Proceedings of Hypertext '89, Pittsburgh, PA, (1989).

Lynch, K. J. "An overview of the Arizona Analyst Information System," University of Arizona MIS Department Technical Report (1990).

Malone, T. W., Grant, K. R., Kum-Yew Lai, Rao, R., Rosenblitt, D. "Semistructured messages are surprisingly useful for computer-supported coordination," ACM Transactions on Office Information Systems v. 5, n. 2, 115-31 (April, 1987).

O'Neill, M., Morris, A. "Expert systems in the United Kingdom: An evaluation of development methodologies" Expert Systems 6(2), 90-98, (1989).

Polya, G. *How to solve it: A new aspect of mathematical method.* Princeton University Press, Princeton, NJ. (1945).

Searle, John R. "Is the Brain's Mind a Computer Program?" Scientific American, 26-31 (January, 1990).

Smith, B. C., "Formality & Separation" Xerox PARC Technical Report, (1991).

MODELING EXPERTISE IN TRAINING SYSTEMS

Bruce Roberts

Bolt Beranek and Newman Inc.
Cambridge, MA USA

Abstract: The effectiveness and generality of a training system can be increased by incorporating rich models of the devices and processes that it is intended to teach. The models provide the structure for several kinds of instruction: textual and graphical explanations, demonstrations, exploring or browsing, and responding helpfully to user actions. The models integrate all parts of the training system: knowledge representation, behavior simulation, and graphical presentation. The models are enriched further when important functional abstractions, in addition to the concrete concepts describing a device or process, are explicitly included. These abstractions introduce general concepts that are transferable to other situations and other domains.

Instruction designed around models can package the training system's capabilities into several different activities to provide a variety of learning experiences. Instructional activities must motivate, inform, and guide the user toward greater competence, greater confidence, and better performance on the job. Keeping the system's capabilities distinct from how they are combined into instructional activities ensures a modular and adaptable training system.

These principles are illustrated by examples drawn from an intelligent tutoring system currently under development.

I. Introduction

This discussion concerns the class of training systems designed to teach students about large, complex, human engineered artifacts. Such artifacts typically incorporate one or more processes; i.e., they *do* something. Competent personnel are required to design, operate, and repair them. From among these different tasks we have chosen to focus on repair and on the nature of the expertise required to find faults in a complex system, although the models that support this task would also find use in training systems for the designer or operator.

We advocate a commitment to a style of training system that itself exhibits the competence one wants students to construct for themselves. The internal representation of the artifact and its behaviors, the cornerstone of the tutor's own understanding of "what" and "how", are made available to the student. Directly through inspection and presentation, and indirectly through carefully crafted activities that he or she initiates, the student experiences the domain through the eyes of the tutor. A simulated world is only the starting place. The focus of the tutor should be understanding, not just task rehearsal in a safer, cheaper, more portable learning environment.

Advanced Technologies Applied to Training Design, Edited by
R.J. Seidel and P.R. Chatelier, Plenum Press, New York, 1993

With this perspective in mind, the title phrase "modeling expertise" is seen to have two connotations, both of which are intentional.

1) The first connotation is meant to suggest that our goal as tutoring system designers should be to develop good models of expertise, to incorporate those models explicitly in the tutoring system, to articulate these models to students both graphically and textually, and to design learning activities by which students are exposed to these models.

2) The second connotation is meant to emphasize modeling as a form of expertise, and therefore one of the skills we would like a tutoring system to impart to its students. The ability to develop, recognize, and apply simplifying mental models to a complex artifact is an important asset to a troubleshooter. These models organize the knowledge one has about normal system operation, failures, and troubleshooting techniques.

/ discussion that follows.

Widespread adoption of object-oriented languages has brought to tutoring system designers the ability to construct more easily introspective and inspectable systems, whose own organization is reflected in the learning environment and therefore is accessible to a student trying to understand a complex artifact. In addition to the benefits of an object-oriented language for the software development process, which in itself can be substantial, the adoption of an object-oriented paradigm makes it easier to build useful models of devices and processes comprising a complex artifact.[1]

Expertise, in the form of models designed to mimic the mental models used by human troubleshooters, finds expression in several parts of an intelligent tutor. A simulation is an obvious beneficiary of an object-oriented design in which the models used to simulate the system's behavior are also used to describe the system to the user.

Another class of model appearing in tutoring systems is the explanation model. These are closely related to the simulation models in that they describe the structure of devices, but do not necessarily participate in the simulation per se; i.e., they do not always describe behavior.[2] Explanation models extend the descriptive power of the tutor by providing alternate perspectives on the simulated system. The perspective can be either a fragmentary (and often simplified) view onto a more complex component which serves as an abstraction, or an aggregate view which summarizes several components by treating them as a composite.

From the standpoint of the software development process, graphical models also benefit from being expressed in an object-oriented language. Whereas simulation models describe the structure and behavior of a device, graphical models describe the appearance of a device and its interaction with the user. Some of the earliest examples of the use of object-oriented

---

[1]This development of object-oriented languages owes a debt to the frame and semantic network knowledge representation paradigms that have been explored so fruitfully in the AI community. The application of the object-oriented paradigm to modeling expertise can benefit from the insight provided by that research. Although the current collection of commercial object-oriented languages do not support the full range of features popularized by some of the experimental knowledge representation languages, they do embody many of their essential features. They simultaneously address pragmatic goals of being computational efficient and providing programming support to aid system development. Some object-oriented languages allow extensions that permit one to incorporate additional features reminiscent of knowledge representation languages. The notion of "facets", properties on slots, is one example of an extension. We reveal a bias by referring the interested reader to CLOS (Common Lisp Object System) as an excellent example of a modern object–oriented language. (Keene, 1989; Kiczales, des Rivieres, and Bobrow, 1991)

[2]The issue of multilevel simulation, in which one simulates a device at more than one level of detail, is bypassed in the current discussion. The appeal of such an arrangement is that only portions of the complete device would need to be simulated at any one time, depending on where the student's attention is focused and on what is visible via the interface. The remaining behavior is summarized by simulating it in aggregate whereever possible. The current discussion treats the so–called explanation models solely as pedagogical aids introduced into the tutor to provide better descriptions and feedback for the student.

programming to build graphical interfaces ably demonstrated the suitability of this approach (Stevens, Roberts, & Stead, 1983).

The following section discusses several ways in which explicit models contribute to the design of a tutor. The final section illustrates how these models support the learning activities of a tutor and thereby directly contribute to meeting its pedagogical goals.

To illustrate the discussion, examples will be drawn from a tutor, called the FAPA Tutor, which is currently under development.[3] The tutor's users are Airmen in the US Air Force whose job is to operate and repair the automatic test stations used in maintenance shops for analyzing and isolating faults in the avionics modules of F-15 and F-16 aircraft. Although the test stations are effective at diagnosing faults in the modules, the test stations themselves are prone to failure. When this happens Air Force personnel are faced with the problem of troubleshooting these complex systems. An automatic test station occupies several equipment racks filled with drawers devoted to generating stimulus signals (power supplies, signal generators), routing and conditioning these signals (relay networks, amplifiers, filters), and measuring the resultant signals produced by the avionics unit being tested (voltage meters, frequency counters). The FAPA Tutor is named after the programming language in which the tests that control the test station are written. The individual statements and organization of a test are a critical source of guidance during the troubleshooting process.

II. Models

This section addresses the central role that models—explicit representations of the structure, behavior, and appearance of the domain—play in the operation of a tutoring system.

II.a. Simulation Models

The purpose of a simulation in a tutoring system is obvious: to provide a realistic counterpart to the actual device within a learning environment. If the aggregate behavior of the device were all that mattered, many simulation techniques would suffice. However, tutoring systems place an additional burden on the simulation. It must be inspectable. Since the purpose of the tutor is to teach troubleshooting, the simulation must respect the component structure of the device. In addition to the aggregate behavior, the detailed behavior the constituent parts must be observable. This need to retain the component structure in simulations embedded in tutoring systems has led to the widespread adoptance of object oriented languages for this purpose.

The FAPA Tutor simulation reuses some of the technology built for one of its predecessors, a simulator written as part of a intelligent tutor for the US Army HAWK radar system (Massey, de Bruin, and Roberts, 1988). The approach has proven itself in both applications, and is summarized here to illustrate an effective use modeling in tutor simulations. In this type of simulation all behavior is localized inside devices; the behavior of the whole is determined by propagating changes through fixed interconnections between the devices.

Figure 1 shows a fragment of a simulation model consisting of two *devices*, A and B. Devices have an internal state (with associated defaults) and a repair state (with associated faulty states). Devices have input and output *ports*, named points of connection with other devices. The connection between an output port of one device and an input port of another device is explicitly represented by a *terminal*. Terminals transmit *signals*, which represent symbolically whatever is being propagated from device to device.

---

[3]Development of the FAPA Tutor at BBN was supported by the US Air Force Armstrong Laboratory Human Resources Directorate. My colleagues in this work are Allan Collins at BBN, John Frederiksen at the Educational Testing Service, and Barbara White at the University of California, Berkeley.

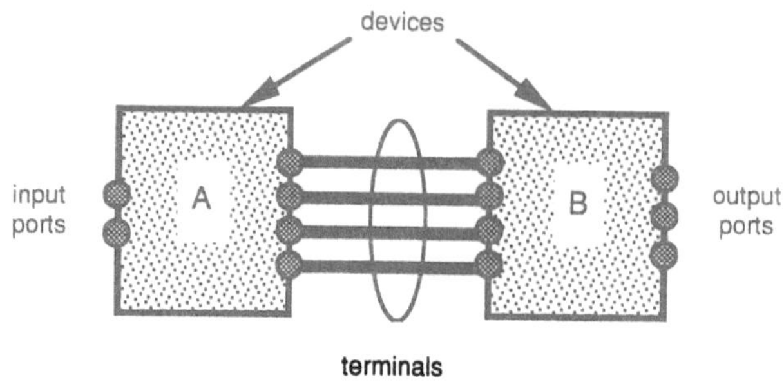

Figure 1. Devices

Signals are represented as a set of arbitrary properties. In the FAPA Tutor, typical properties consist of measurable physical properties (e.g., amplitude, frequency, waveform, phase, noise, modulation), and status ("present" or "absent"). In the radar model, other properties identified the functional properties encoded by a particular signal (e.g., target range).

Device state is represented similarly. It can be as complex as necessary but is typically a symbolic indication of the device's operating mode and parameter settings. In the FAPA Tutor, typical device states are the set relays on a relay card, the programmed output of a stimulus generator, or the settings for frequency counter.

Operations on devices are defined as methods (or messages) obeying a well-defined protocol. The intrinsic behavior of a device is defined by a SIMULATE method, which is responsible for computing a new set of output signals and a new internal state from the current values of the input signals and state.[4] A simulation engine is responsible for noting any changes in signals and simulating any other devices whose inputs have been affected. In this way, effects are propagated throughout the simulation.[5]

Other operations define actions that can be performed on a device; these are changes effected from outside the simulation, and usually correspond to points of human intervention (e.g., apply power, close a relay, set a switch, adjust a knob). Finally, some device operations mediate the interaction between the simulation, the tutor, and the user interface (e.g, install a fault, highlight the device on a particuliar diagram, report the effects of simulating the device, describe the current state, produce a textual explanation).

When a simulation is designed around recognizable constituents, it provides not only a working model of the whole, but can be decomposed into comprehensible parts, each of which can articulate its role in the overall simulation.

Simulations must also be written with pedagogical goals in mind, not simply the desire to mimic the observed behavior of an actual device. In the FAPA Tutor the principal learning activity focuses on finding and tracing signal paths through the test station. The precise electrical properties of the signal were deemed less important than the ability to create, propagate, and reason about signal paths. Examples of the tutor's reasoning about signal path are its ability of locate all the components that lie along a path, to identify the origins of

---

[4]In fact, it is sometimes necessary to propagate effects upstream, from the input of one device to the output of another. Electrical short circuits are the prime example of a situation where a change downstream affects devices upstream. The simulation engine is capable of handlings this situation as well.

[5]This characterization of a device-oriented, discrete simulation is clearly not limited to simulations of electrical devices like the automatic test station. It is equally applicable to any situation in which there is some identifiable "stuff" that can be defined by a set of properties, recognizable places where its properties can change, and pathways through which it moves. Chemical processes, steam plants, information data-flow models, and transportation systems all are candidates for this kind of simulation.

the signals affecting a component, and to locate the nearest measurement that has been taken upstream or downstream from a particular point. The introduction of signal paths is a dynamic process requiring simulation; they can not be predefined. In particular, a test station contains many switching components that route signals; therefore, many signals paths are possible, but only a few are implicated in any one test program. Moreover, some component faults (e.g., a stuck relay) indirectly alter signal paths. For these reasons the FAPA Tutor simulation includes a richly elaborated model of signal paths.

Signal paths in the FAPA Tutor are an example of objects introduced into the domain model because of their cognitive significance. They represent an abstraction that appears repeatedly in the language of experts, and that intrinsically provides an important perspective on the operation (both correct and faulty) of the test station. Common troubleshooting strategies depend on being able to reason about signals. It was necessary to introduce them as explicit objects, maintained by via the simulation, in order to create for students the illusion that they are real and important entities to identify and understand.

The explicit representation of signal paths supports several features of the tutor's operation. The tutor uses its knowledge of signal paths to depict the operation of the test station; paths can be highlighted on any diagrams through which they pass. As a test program is run, the path being created can be highlighted step by step to indicate the immediate and derived consequences of each step. Measurements taken by a student can be critiqued by comparing the locus of the measurement to the active signal paths and the position along it relative to any previous measurements. These mechanisms are supplemented through the use of signal

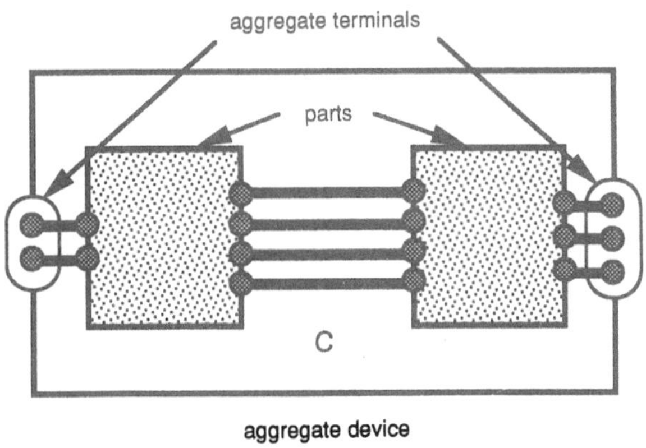

Figure 2. Aggregates

path terminology in the tutor's direct explanations, feedback, and wrapup. In this way, the tutor introduces students to the meaning and utility of this concept.

The notion of a path is an example of a pervasive abstraction, one with broad applicability in thinking about repair of many kinds of devices. Other abstractions, described in the next section, do not participate directly in the simulation but provide important perspectives onto the devices which comprise it.

II.b. Explanation Models

We have suggested how a rich model of the domain can contribute to the competence and performance of a tutor by capturing, organizing, and revealing expert thought. But the most important models in a tutor are not necessarily the ones that correspond literally to real-world entities. Although these literal models are adequate to describe a simulation and to serve as the referents for the control panels, schematics, and images that are typically part of the interface to the tutor, they fail to capture a critical type of understanding that permeates expert discussion; namely, a functional decomposition of the domain.

Hierarchical functional models allow one to picture a complex device in varying levels of detail and to adopt different perspectives on the behavior of a single component or module. Reasoning about a complicated device is simplified by recognizing of the aggregate behavior of its subsystems.

The FAPA Tutor simulation device model includes the notion of *aggregate devices* and *aggregate terminals*. These constructs are used to define natural groupings of devices into functional units, which have well-defined composite behavior. Aggregate devices have named *parts*. In all other respects they are like the basic devices described above. They have input and output ports and can be connected to other devices. Figure 2 shows an aggregate device C consisting of the two basic devices pictured previously. The parts of an aggregate device can be either a basic devices or other aggregate devices. Aggregate terminals also have parts, which are either basic terminals or other aggregate terminals. Any number of levels of decomposition can be represented using this structure.

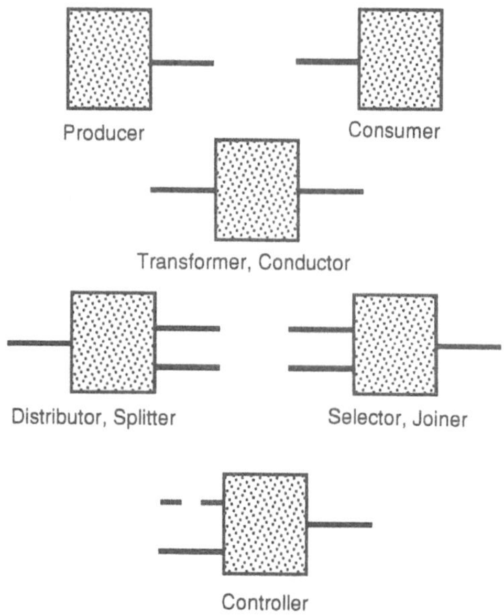

Figure 3. Abstract Devices

Using this mechanism for defined aggregates, a rich set of hierarchical mental models can be incorporated into the tutor. Some aggregates correspond to the physical boundaries in the test station; e.g., all the components on a single circuit card, or in a single drawer. Other aggregates correspond to functional groupings, components that work together to perform a functional role in the internal behavior of the test station; e.g., a collection of components, cards, and cabling that implement a signal conditioning module. To distinguish these perspectives, a tutor must contain functional models which augment the device definition.[6] Both basic and aggregate devices can have associated functional models.

[6]Object-oriented language aficionados will recognize the usefulness of multiple inheritance for associating functional models with definitions of devices and aggregates. Viewing X as a Y, or a Z, ... is simplified by having X "inherit" the characteristics of Y, and Z, ...

Figure 3 shows some of the abstract functional models that can be used to describe devices found in an avionics test station. Some of them obviously belong to the tradition of engineering systems analysis; others are more specialized to the domain of FAPA Tutor. The most general device has any number of inputs and outputs. This corresponds to the basic simulation building block described above. Common generalizations are the PRODUCER and CONSUMER, which have only no inputs, or no outputs, respectively. Important specializations invoke restrictions on the numbers of inputs and outputs (ONE-TO-ONE, ONE-TO-MANY, MANY-TO-ONE, MANY-TO-MANY). In the test station, a major distinction between modules captures whether it is a ROUTER, which alters the topology of the interconnections, or a TRANSFORMER, whose outputs are a modification of its inputs. Two important specializations of ROUTER are the SELECTOR (also MANY-TO-ONE) and the DISTRIBUTOR (also ONE-TO-MANY). The switching versions of these are called the JOINER and the SPLITTER, respectively. An important specialization of TRANSFORMER is the class CONDUCTOR, which passes signals without modification. Another common pair of specializations of the TRANSFORMER are the AMPLIFIER and the ATTENUATOR, which affect only the amplitude of a signal. A FILTER is a TRANSFORMER that affects only the frequency. A CONTROLLER has a control input whose signal determines the form of the transformation applied to other inputs. And so on.

The same device may satisfy several functional classifications, corresponding to the different but valid perspectives from which it can be viewed; for example, Power vs. Signal, Control vs. Process, or High Frequency vs. Low Frequency. In the FAPA Tutor simulation, a waveform signal generator is a PRODUCER from the Signal perspective, but a CONSUMER from a Power perspective.

Functional models themselves can be aggregates. For example, feedback loops are a prevalent composite object in signal processing systems. A feedback model has four parts: SPLIT, FORWARD PATH, JOIN, and FEEDBACK PATH. Each part of this model maps onto some portion of the simulation. In doing so, functional models introduce new nomenclature not just for the components of a system, but for the connections (terminals and signals) as well.[7]

Why is this ontology useful? What is the purpose of introducing all these abstractions? A useful knowledge representation heuristic is to avoid representing distinctions that don't matter; a representation should be designed with its use clearly in mind. (Kieras, 1988) These mental models serve several important needs for expert troubleshooters. They aid the envisioning process by summarizing and simplifying the structure and the behavior of devices. They suggest the kinds of faults that could occur, and the impact of these faults on other devices. In the FAPA Tutor, they help associate particular statements in the test program with their affected stimulus, routing, conditioning, or measurement devices. They simplify the application of troubleshooting strategies (e.g., by defining signal paths and places along the path to search for faults).

The most potent characteristic possessed by these abstractions is their generality; once learned, they can be used in many different contexts. They can provide the bridge between skills learned in one context and a problem presented in another context. They provide the framework on which otherwise disparate experiences and facts can be organized, remembered, and adapted to new circumstances.

II.c. Graphical Models

Extreme attention is being lavished on human-computer interfaces. Some powerful metaphors for conveying computer operations and expressing user actions have been promulgated and become widely accepted and imitated: "desktop", "dialog box", "direct

[7] Consequently, implementations must provide some means for handling coreference, which guarantees that references to the same object grounded in different models do in fact access the same object.

manipulation". These metaphors embody a recognition that visual techniques promote highly useful and perspicuous forms of interacting with a computer (and having it interact with you). Intelligent tutors naturally benefit from the user's ability to directly interact with simulated objects depicted on the screen. To directly interact is to create the illusion that portions of the screen (as defined by what has been drawn there) correspond to recognizable entities of the domain of the application, and that pointing to one of them is tantamount to identifying and operating on that object in the domain. This technique presumes a rich model of the domain is available to support the many references one would expect a student to make by pointing to parts of the pictures, diagrams, text, graphics, and other presentations that appear on the screen. This form of highly reactive interface requires a comprehensive model of the domain in order to support the illusion.

Despite the need for closely coupling graphics and simulation models, it is a mistake to collapse the definitions of a simulation object and the details of its appearance into a single model. The same simulation object can be depicted in more than one way. Moreover, it is common for the same object to appear (in various guises) on multiple diagrams or images. Even non-graphical presentations of an object can be treated uniformly if the textual presentation of the object (in the form of a name or other linguistic reference) is considered to be just another kind of presentation.[8] The linkage between the viewable presentation and its internal reference mimics the relation between the syntactic and the semantic aspects of a linguistic reference. The simulation (or explanation) model serves as the semantic referent for the visible (syntactic) depictions thoughout the interface.

As is the case for explanation models, the needs of cognitive fidelity must be balanced against the desire for physical fidelity. An accurate picture or image of a complex system may may obscure identities and relationships that a more conceptual diagram readily reveals. The goal of physical (or perceptual) accuracy should not dominate the design of tutoring systems. Unlike the actual equipment, tutoring systems have the burden of teaching (or making learnable) its operation, not just exposing students to a virtual image of it.

The FAPA Tutor uses functional schematics to depict the functional decomposition of the test station as described by the simulation and explanation models. These are idealized diagrams of the actual components and their interconnections. The FAPA Tutor lets a student perform troubleshooting operations on these diagrams as well as on more detailed diagrams. The diagrams provide an abstract visual environment that mirrors the abstract conceptual description of the test station. Of course, in an actual troubleshooting situation, a student will be working in more detailed context, in which being able to read formal schematics is a required skill; but exposure to other cognitive troubleshooting skills can be hindered by burying them beneath the complexity of literal reconstruction of the actual context.

For any presentation, the connection to the underlying semantic model it depicts must be consistent and pervasive. This leads to several problems that need to be addressed when using graphical presentation models. The aggregates and abstractions in the semantic model tend to manifest themselves graphically as overlapping and embedded regions on the diagram. Interconnections produce potentially confusing visual paths corresponding to "wires" and "cables" between the devices. Labels add another visual artifact. The behavior of the user interface itself depends on identifying and disambiguating these graphic objects.

A primary function of a reactive graphical interface is answering the implied question: "At what am I pointing?". The visual features of the presentations are explicitly stored in the graphical object, where they can be used to map any point on the screen to the object

[8]The style of designing interfaces around presentations of application data structures is best supported, among all the currently available interface development software tools, by a package called CLIM (Common Lisp Interface Manager) (McKay, York, and McMahon, 1989).

containing it. This mapping cannot be static, since the context provided by what is being asked for by the tutor can limit the meaningful responses to a subset of the objects drawn on the screen. The ability to highlight or otherwise emphasize a graphical object is needed to provide feedback to the user about the tutor's interpretation of his or her pointing gesture, and to provide a means by which the tutor can refer to models during an explanation; e.g., to indicate a particular signal path by shading all of its participating devices and connections on a diagram.

Understandable explanations must pay attention to what is visible on the screen. Explicit graphical objects make this possible. In the FAPA Tutor, explanations that contain references to models adapt their wording to take into account the context of explanation by adopting these principles: mention everything that is visible, don't mention what is not visible, and use terminology consistent with the labeling of the diagram.

III. Learning activities

Models of expertise, no matter how well conceived and explicitly rendered, do not alone make a tutor. Learning activities must be designed to create situations for the student to see demonstrated and to practice recognizing and applying these models. Moreover, a principled plan of instruction is needed to allow students to guide themselves through the conceptual space of models and their application to troubleshooting.

Two learning activities are implemented in the current version of the FAPA Tutor: a Main Task, which emphasizes troubleshooting the test station, and an Auxiliary Task, which emphasizes envisioning the signal paths established by a given FAPA test program. Within each activity students are presented with a well-defined problem, instructions for working on the problem, and a constrained set of possible interactions with the tutor. Several graded problems are available for each task. Help is available should students lose track of where they are in the session, what actions are possible, or how they should think about approaching the task. A tutoring session consists of working on a series of problems within either the Main or Auxiliary Task, or both. Feedback is provided after each student action to critique the action and suggest subsequent actions. At the end of each problem, a general summary is provided to allow the student to assess their own performance and thereby choose a new problem (or an alternative task) appropriate to their skill level.

The Main Task involves troubleshooting the faulty test station. By virtue of replicating the salient features the real troubleshooting situation, we find that students are well motivated to use the tutor, and see the relevance of this experience to their own job speciality.

An important troubleshooting paradigm—supported in the FAPA Tutor by the hierarchical description of the domain and conveyed through the structure of the learning activity—consists of the following guidelines:

- Troubleshoot from the top down; identify the functional block responsible for the observed behavior before delving into the details of the fault at the drawer and component level.

- Identify the signal paths implicated by the FAPA test program and limit the search to components along the path.

- Narrow the search by measuring signals; resist swapping components or board until the measurements have isolated the fault.

These basic strategic troubleshooting principles are embodied in the Main Task in the form of the constraints on the student's actions, the feedback generated in response to these actions, and in the modal style of the interface.

The student is given this strategy to follow: First determine the signal paths established by running the FAPA program, and then PROBE (measure) along those paths, comparing the measured signal properties with the expected ones. Proceed in this way to localize the fault within a particular section of the test station, and only then HYPOTHESIZE that that device or connection is the culprit. PROBE and HYPOTHESIZE are independent modes controlled by the student.

Only a single diagram is visible at one time. Each problem always begins with only the highest level diagram visible. This is replaced by another diagram, which shows more detail about the suspect section of the test station, when the student successfully localizes the fault on the more abstract functional diagram. Figure 4 shows a possible succession of diagrams for a fault in the "Buffer Amplifier". It is first located in the "Measurement Switching Matrix" on the top diagram, then to the "Routing and Conditioning" module in the middle diagram, and finally to the "Buffer Amplifier" in the bottom diagram.

In PROBE mode, clicking on any connection shown on the diagram measures its signal. Clicking on an aggregate connection measures all of its active signals.[9] A probe reports the properties of a signal at any point in the test station without requiring knowledge of the measurement instrument or its settings, and without having to determine the precise location of the test point. In addition to reporting the measurement, each probe evokes feedback which assesses whether the location of the probe was well chosen; namely, was it on one of the active signal paths established by the test program, or could it have been inferred from the outcome of previous measurements. Feedback when the probe is on a signal path simply encourages the student to continue probing along the path.

Typical "off path" probe feedback:

> You probed ..., which is not on any signal path. Examine the FAPA program to determined which signal paths are in use, and probe along them to localize the fault.

Typical "redundant" probe feedback:

> You already know that there is no signal coming out of the ....

In HYPOTHESIZE mode, clicking on any part in the diagram asserts that you believe it is broken. The FAPA Tutor responds by saying whether you are right or wrong. If you incorrectly hypothesize the fault, you are allowed to choose another part. If you correctly hypothesize the fault, but are still working at a functional block diagram level, the current diagram is replaced by one showing more detail, and the tutor reverts to PROBE mode. You must then continue troubleshooting at the more detailed functional or physical component level. When you correctly hypothesize the fault at the physical component level (i.e., no further decomposition is possible), the problem is solved and a summary of your performance is given. In addition to testing the hypothesis, each hypothesis evokes feedback which assesses whether the location of the hypothesis was well chosen; namely, was it on one of the active signal paths established by the test program and, if so, was it warranted by having adequately narrowed down the search by measuring the inputs and outputs of the part. Unwarranted probes reflect an element of guesswork on the part of the student.

Typical "on path (unwarranted)" hypothesize feedback:

> To hypothesize that ... is faulty, you should know the status of both its input and output signals. You have information only about its inputs.

This learning activity takes advantage of explicit functional models to provide experiences impossible to have using real equipment. Notice that in the preceding description of the Main Task not only the test station but the student's troubleshooting actions also have been abstracted from their real-life counterparts. Measuring a signal is as simple as pointing to the connection which carries it; and only the relevant signals are reported. Testing an hypothesis is as simple as asking the tutor. In actual troubleshooting, verifying a hypothesis

[9] Active signals are those established by the current FAPA test program, not all the signals potentially carried by a connection.

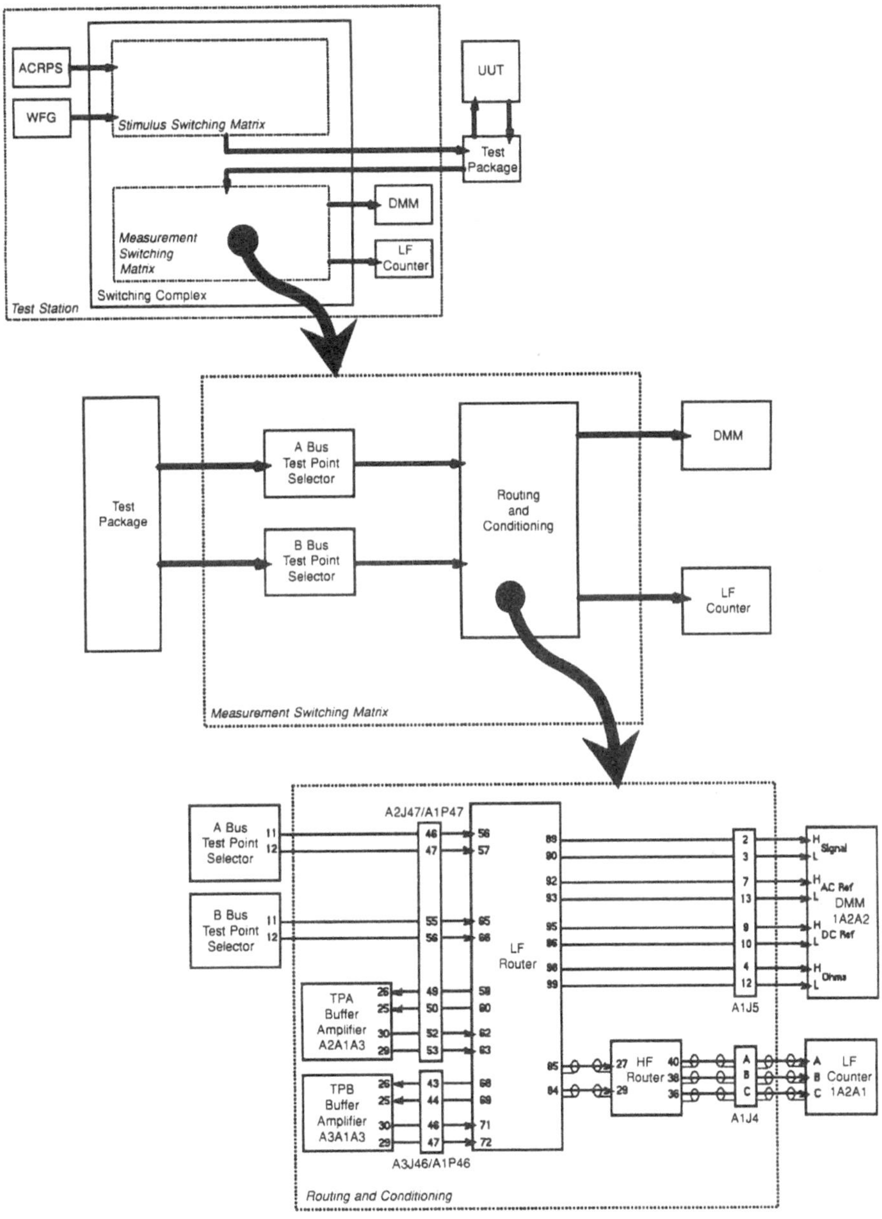

Figure 4. A Succession of Hypotheses Isolates the Fault.

can be a complex and time consuming process (shut down the test station, replace a component, and re-run the test program). We have reduced these steps to their essence in order to emphasize the strategic aspects of the troubleshooting process and the use of the FAPA program as a source of guidance. In a more realistic setting, the pragmatic aspects of the troubleshooting task can overwhelm the cognitive aspects. We have minimized the pragmatic aspects of measuring signals and swapping components in order to emphasize the more cognitive aspects of the troubleshooting process. One can think of this approach as providing "conceptual part–task" training. (Frederiksen and White, 1989)

A troubleshooting tutor designed around embedded expertise differs significantly from an intelligent system whose only requirement is to perform troubleshooting tasks itself. The tutor must present the expertise in form that is learnable and that contributes to its users becoming better troubleshooters themselves. The suggested approach is to design the tutor so that its own operation reflects the style of troubleshooting and the form of the expertise one wants to impart to the students. This rules out techniques that, however effective computationally, fail to correspond to actual human troubleshooting practice. Regardless of the learning activities designed around them, troubleshooting expert systems may not be appropriate starting points for a tutor.

A computationally opaque troubleshooting expert system could be used as a baseline against which to compare the performance of a human troubleshooter. Tutors are frequently designed which juxtapose a student's performance with its own and use the differential as a source of feedback and evaluation of the student's performance. But such a design begs the question of whether the student is indeed engaged in the same task as the tutor throughout a session. It presumes to identify differences in troubleshooting behavior as being associated with the student's own goals, despite the limitations of the typical human-machine dialogue to convey nuances of intentionality.

A more transparent, and therefore inspectable, tutor allows a different source of performance feedback and evaluation for its student. The tutor's actions are available for inspection and can be understood by the student on his or her own terms. Any coaching is accompanied by the assumptions implicit in their formulation; i.e., its content reflects the criteria associated with the overall performance of the task. This approach shifts the burden of evaluation onto the student by being explicit about the task, the metrics associated with it, and the implications of each action with respect to these metrics. It fosters the skill of self-evaluation—an important meta-cognitive skill—by making the criteria and the implication of the action available to the student, rather than the predigested results of a tutor.

A further advantage of this style of tutor feedback is that it can accommodate multiple sources of commentary on a student's actions. Freed from having to commit to knowing what a student is thinking, the tutor can offer several, possibly contending, points of view. Complex tasks like troubleshooting often produce disagreement about what the next best action should be, even among experts. This disagreement arises from valid but conflicting approaches to troubleshooting. Alternate points of view that affect the troubleshooting process include the repair history or statistical reliability of components and the costs of an action in time, personnel, or danger. When multiple models of the troubleshooting process are able to contribute feedback simultaneously, the reality of the troubleshooting process is more accurately presented to the student.

REFERENCES

Frederiksen, J. and White, B. (1989) An approach to training based upon principled task decomposition. *Acta Psychologica, 71*, 89-146.

Gentner, D. and Stevens, A. (Eds.) (1983) *Mental Models.* Hillsdale, NJ: Lawrence Erlbaum Associates.

Keene, S. (1989) *Object-oriented Programming in Common Lisp.* Reading, MA: Addison-Wesley.

Kiczales, G., des Rivieres, J., and Bobrow, D. (1991) *The Art of the Metaobject Protocol.* Cambridge, MA: MIT Press.

Kieras, D. (1988) What mental model should be taught: Choosing instructional content for complex engineered systems. In J. Psotka, D. Massey, & S.A. Mutter (Eds.) *Intelligent Tutoring Systems: Lessons Learned.* Hillsdale, NJ: Lawrence Erlbaum Associates.

Massey, D., de Bruin, J., and Roberts, B. (1988) A training system for system maintenance. In J. Psotka, D. Massey, & S.A. Mutter (Eds.) *Intelligent Tutoring Systems: Lessons Learned.* Hillsdale, NJ: Lawrence Erlbaum Associates.

McKay, S., York, W., and McMahon, M. (1989) "A Presentation Manager Based on Application Semantics". In UIST '89 (Proceedings of ACM SIGGRAPH Symposium on User Interface Software and Technology).

Means, B. and Gott, S. (1988) Cognitive task analysis as a basis for tutor development: Articulating abstract knowledge representations. In J. Psotka, D. Massey, & S.A. Mutter (Eds.) *Intelligent Tutoring Systems: Lessons Learned.* Hillsdale, NJ: Lawrence Erlbaum Associates.

Stevens, A., Roberts, B. and Stead, L. (1983). The use of a sophisticated graphics interface in computer-aided instruction. IEEE Computer Graphics, 3(2).

White, B. and Frederiksen, J. (1989) Causal model progressions as a foundation for intelligent learning environments. *Artificial Intelligence.*

SECTION VII
DATA BASES FOR ADVANCED TRAINING DESIGN

One of the most serious deficiencies in training that continually drives its cost upwards is the tendency to develop new rather than rely on effective "old" training approaches. This chapter will discuss the various data bases that are being used in the United States and elsewhere and how they might be used internationally. It is scheduled to be the last paper in order to allow the participants to understand what is available in terms of training technology and therefore appreciate what a fully integrated data base structure can provide for them.

A NATO DATABASE ON ADVANCED TECHNOLOGY TO AID TRAINING DESIGN

J.D. Fletcher, E.A. Alluisi, and Paul R. Chatelier

Institute for Defense Analysis
Alexandria, VA 22311 and
Paul R. Chatelier
DMSO
Alexandria, VA 22311

I. BACKGROUND

One way for the NATO community to exchange information and stimulate research on advanced technology to aid training design is to establish a database on this topic. Such a database will give researchers, training developers, system developers, manpower and personnel planners, policy makers, and others concerned with military manpower, personnel, and training issues a shared forum for exchanging information and advancing the state of the art.

Much of the groundwork needed to create a NATO database on advanced technology to aid training design has been completed. Databases on training technology already exist in some NATO countries, and some of these databases may be accessible to all NATO participants. Because of the savings in costs and time to be gained by building on these existing databases, an assessment of what they are and how they might be used is a reasonable first step in establishing a database for NATO. This assessment was the topic of the final session of the RSG-16 Workshop, "Advanced Technologies Applied to Training Design." This chapter briefly summarizes the workshop and pursues its assessment of what can and should be done to implement a NATO database on advanced technology to aid training design.

From this perspective, five of the questions that should be addressed are the following:

- What are the objectives of the NATO database? In other words, what sorts of users would access the NATO database to do what?
- What kind of database is needed to meet NATO requirements and to satisfy the needs of its users?

Advanced Technologies Applied to Training Design, Edited by
R.J. Seidel and P.R. Chatelier, Plenum Press, New York, 1993

- What databases on training technology are already available to the NATO training community?

- Which (if any) of the data elements in these databases should be included in the NATO database?

- How should the NATO database be accessed by its users?

Section II of this chapter discusses what sort of NATO database is needed to provide a source and a forum for information on advanced technology to aid training design. Some databases that are now available to the United States Department of Defense (DoD) are discussed in Section III. Some alternatives for implementing the NATO database are discussed in Section IV. Conclusions are presented in Section V.

II. THE NATO DATABASE

A database is made up of data elements, but it is generally considered to be more than just a receptacle for data. The key notion behind a database is that its data elements are linked to each other and/or to abstractions that build up structures of data elements in a hierarchical fashion. These associative links are usually called relations, and today's databases are generally expected to be relational. A user who accesses any one data element in a database is immediately linked to other, related data elements. The fact that these data elements are related and the ways in which they are related is additional "knowledge" that is carried in the database along with whatever knowledge is borne by the data elements themselves.

Another expectation of today's databases is that these relations must be explicitly defined. Data elements that are located in the same places in records of the same type are linked implicitly by virtue of where they are stored. Databases, on the other hand, provide more flexibility and variety for creating relations than does a collection of homogeneous records. Data elements need not be stored in the same places, and the records in which they are stored may be widely diverse.

Finally, databases reduce duplication. Since their data elements are linked, they do not have to duplicate them to make them available for all the purposes and in all the records to which they apply. Even though information may be added to the collection of records incorporated in a database, the amount of space occupied by the database may be substantially less than that required for the records, since duplicates can be eliminated. More information may be made available in less space, for less cost.

The price paid for the increased information, increased flexibility, and reduced cost of a properly designed database, is that the data elements can and must be linked explicitly rather than implicitly. Database software is intended to aid in this process, and is designed to make the creation and use of database links (or relations) both convenient and efficient. Such software would naturally be applied to establish a NATO database that uses data elements drawn from existing, diverse databases dispersed across various organizations in participating NATO countries.

The architecture that evolves most naturally from this perspective is a NATO directory of databases such as that discussed by Seidel and Chatelier in the Introduction to this book. Such a database could be accessed on-line, directly, or indirectly. It would benefit directly from the maintenance and updating provided to the parent databases and would require few resources of its own for these purposes. Although such a database would be an offspring dependent on the quantity and quality of its parents, the countries that support the parent databases have a major stake of their own in the quality of these databases. It does not seem unreasonable to rely on their sustained self-interest to maintain the quality of the parent databases.

What are the objectives of the NATO database? In other words, what sorts of users would access it to do what?

As suggested earlier, the NATO database could be used by researchers, training developers, military system developers, manpower planners, and policy makers.

Researchers would usually be concerned with research, development, and evaluation of innovative approaches to all aspects of training. They could use the database to find attempts to pursue technical opportunities that are similar to their own. They could also find potential customers and/or sponsors for the capabilities and benefits they expect to produce and generally learn better how to respond to the needs of those who use the products of their research.

Training developers could use the database to find the costs, effectiveness, and applicability of approaches that may be available for meeting their requirements for analysis, design, development, implementation, and evaluation of the training programs that prepare military personnel to perform their jobs and missions. As new approaches enter the state of the art, developers will be able to learn about and apply them sooner. They will also be able to locate already developed training technologies and packages -- especially authoring and interactive courseware packages -- that they can import more or less directly to meet their training needs. They will be able to communicate lessons they have learned to other users. They will be able to articulate their requirements more directly to researchers and their capabilities more clearly to system developers.

System developers and system program managers could use the database to understand better (ideally in a quantifiable and predictable manner) the personnel, training, and human factors implications of their system designs. Especially, they will be able to use the database to determine explicit tradeoffs among the selection, classification, training, human factors, supply, and safety implications of the system design alternatives they must assess and among which they must choose. They will also be better able to coordinate both their designs and the personnel system procedures they require with those of their NATO allies.

Manpower planners will be able to use the database to better determine the force size and mix needed to meet future requirements. Training requirements have a major impact on the numbers of people needed on hand to perform specific military missions. The NATO database will help planners to better trade-off the resources they must invest in force size with the resources they must invest in training and to coordinate these trade-offs with their allies.

Policy makers are concerned with the proper conduct of training in their Services, the cost-effective use of scarce training funds, the ability of their training to produce military effectiveness, and the coordination of the activities of their services with those of their NATO allies. They will be able to use the NATO database to better assess the strengths and weaknesses of their manpower, personnel, and training systems and to better determine directions for managing these systems both within their own countries and in concert with their NATO allies.

Transcending all these uses and users is the fact that this is a NATO database. Its emphases differ from those of a country that is solely concerned with its own internal military training problems. The NATO database should emphasize relations among data elements that support the coordination of training activities in allied countries seeking to improve their mutual defense. Explicit identification of these relations and their inclusion in the NATO database will be information that it uniquely adds to that already provided by the parent databases. The authoring system interface capabilities discussed by Park and Seidel in the first chapter of this book are examples of the sort of relations that could and should be included in a NATO database.

What kind of database is needed to meet NATO requirements and to satisfy the needs of its users?

The best way to answer this question may be to start with the elements that are as standard in training systems as they are in any other:

- Analysis The decisions that more than anything else determine the effectiveness of a training system are those concerning what and what not to teach. A NATO database should contain a family of data elements covering such matters as front end analysis, job skill requirements analysis, job knowledge requirements analysis, and career design. These data elements should supply information on technologies, techniques, and approaches for establishing training objectives, standards, and conditions. Information that affects training analysis such as personnel characteristics and flow, projected and present mission requirements, and quality of the civilian workforce that serves as the source for military manpower could also be included among these data elements.

- Design Given a set of training objectives, standards, and conditions, it is the business of training design to determine how best -- most efficiently, or most cost-effectively -- to meet them. Training approaches, training media and their mixes, training delivery systems, and training scheduling, along with their productivities measured in terms of progress toward specific training outcomes (such as speed of response, accuracy of response, retention, transfer ability, insight, and motivation to continue) and the resources they require should be covered by this family of data elements.

- Development. Once training design is set, the problem of how to implement the design and produce the necessary training materials presented by the recommended training media is next. This family of data elements should cover matters such as production schedules for training materials, authoring languages for interactive courseware, aids for producing and editing graphics, desktop publishing for training materials, quality-cost tradeoffs in audiovisual production, and the like.

- Implementation. No matter how responsive training materials are to the training requirements, how well they are produced, and how nearly optimal the mix of presentation media and delivery methods, any training system can be scuttled by poor implementation. Data elements in this family should cover instructor preparation, instructor aids, physical environments for training, training schedules, software packages for administrators, and other matters of this sort.

- Evaluation. Data on the costs and effectiveness of existing training systems and training approaches as well as data on technologies, techniques, and approaches used to assess training systems and training approaches should be covered by data elements in this family. Evaluation concerns the costs and effects of various approaches in achieving the intended training outcomes. Evaluation and design are tightly linked with design providing direction for evaluations and evaluations providing feedback for design.

Although this early stage of development can not generate a comprehensive list of the data elements needed in a NATO database on advanced technology to aid training design, the coverage and utility of data elements that may be proposed for inclusion can be reflected against the above systems framework. Among the data elements that may reasonably be proposed for a NATO database are those in databases that already exist and that could be made available to the NATO community. The best place to begin may be with a brief review of the databases that are already available to the United States Department of Defense (DoD).

III. DOD TRAINING DATABASES

Some, perhaps most, of the data elements in the databases discussed in this section could be used to create a NATO database on advanced technology to aid training design. It should be emphasized, however, that no formal agreements have been established to permit access to these databases and their data elements by anyone outside the DoD. Such agreements will have as much impact on which DoD data elements are used in a NATO database as their technical quality and utility. This section assays these DoD databases under the assumption that many of them could be made available to the NATO community if there were good reasons to do so. It addresses the following question:

What databases on training technology may already be available to the NATO training community?

Presently there are at least 8 data bases available to Department of Defense (DoD) training planners and managers in the United States. One is maintained by the Defense Technical Information Center, one by the National Library of Medicine, one by the Defense Audiovisual Policy Office, and the remaining 5 are maintained by the Defense Training and Performance Data Center (TPDC). These data bases are the following:

- Manpower and Training Research Information System (MATRIS)
- National Library of Medicine (NLM) AuthorBase
- Defense Automated Visual Information System (DAVIS)
- TPDC Defense Instructional Technology Information System (DITIS)
- TPDC Fielded Training Equipment Data Base
- TPDC Course Training File
- TPDC Correspondence Course File
- TPDC Teletraining Data Base

Each of these databases, among others, was discussed during the RSG-16 workshop. The following comments describe these data bases and their data elements:

• The Manpower and Training Research Information System (MATRIS) is maintained by the Defense Technical Information Center (DTIC) through a facility located at the U.S. Navy Personnel Research and Development Center. It collects and reports data on all "people related" research and development (R&D) projects supported by the U.S. DoD, and its primary users are R&D personnel. Its data elements are collected from summaries of work units that are reported to the U.S. Congress under the four categories of Human Factors and Safety, Manpower and Personnel, Education and Training, and Simulation and Training Devices. Most of these work unit summaries are further classified into the R&D

budget categories of Basic Research, Exploratory Development, Advanced Development, Engineering Development, and Studies and Analyses.

MATRIS operations are more proactive than those of many data collection and reporting activities. Its staff members may initiate database reports, help requesters better define their needs, and maintain close and frequent contact with researchers and developers in the manpower, personnel, training, and human factors communities. The MATRIS organization has in place an online data retrieval and reporting capability available to remote users via standard telephone dial-ups using modems.

• The U.S. National Library of Medicine (NLM) AuthorBase is a database of authoring systems. These are systems used to enter interactive courseware onto computers and other delivery media for interactive instruction and are designed to meet the requirements of these specialized purposes. These systems range from simple to complex and from those that use menu driven selection routines for instructional presentations to sophisticated programming languages. Their impact on the productivity of training materials developers, on the quality of the materials they produce, and on the portability of these materials across different training delivery platforms is of natural interest to those concerned with training technology, and the primary users of AuthorBase are training system developers.

• The Defense Automated Visual Information System (DAVIS) was established by DoD Directive 5040.2, "Visual Information." It holds information concerning materials that use one or more visual media, such as films, videotapes, and videodiscs to display training presentations. DAVIS is intended to facilitate life cycle management and accountability for these materials, and DAVIS data elements provide information on their format, use, and distribution. It is primarily used by the producers of audiovisual materials, and its primary users in the training world are training system developers.

• The TPDC Defense Instructional Technology Information System (DITIS) was established by DoD Instruction 1322.20, "Development and Management of Interactive Courseware (ICW) for Military Training." DoDI 1322.20 states that courseware consists of all training materials including the curriculum database and all discs, tapes, books, charts, and computer programs necessary to deliver a complete training program. Interactive courseware is computer controlled courseware that relies on student input to determine the pace, sequence, and content of the instruction being delivered.

DITIS was created and is maintained to support two DoD-wide requirements, the first of which requires maintenance of an inventory of DoD ICW programs and the second of which requires that this inventory shall be queried before any new ICW program is developed so that duplication of effort will be avoided and the DoD can take advantage of DoD ICW investments already made. The primary users of DITIS are training system developers and training policy makers.

DITIS records are created or updated when ICW is proposed, under development, completed, revised, or terminated. DITIS information includes program information, program description, software requirements, hardware requirements, and management information.

• The TPDC Fielded Training Equipment Data Base provides information on separate units of training equipment, any combination of training equipment that provides a capability that is not available when the equipment is used separately, or any combination of training equipment, like a mobile training set, that is grouped for training administration purposes. The training equipment may be a simulator, training device, training aid, or a capability embedded in operational equipment. Its data elements include the costs of the training equipment, the functional category of personnel it trains (e.g., maintainers, operators, tacticians), its primary operating environment (e.g., air, ground, sea, space), and the weapon system or general training area it supports. Its primary users are training system developers.

- The TPDC Course Training File includes records for every formally convened course provided by the U.S. Military Services. Although it does not include most crew, group, team, and unit training, most training provided by operational organizations with primary missions other than training, and most factory training for new systems or newly modified systems, this database contains information on more than 24,000 courses taught at one time or another by the U.S. Military Services. It data elements include a student type code (e.g., enlisted, office, civilian), DoD occupation code, number of training days required, and location. Its primary users are system developers and manpower planners.

- The TPDC Correspondence Course File contains data elements on all the correspondence courses offered by the U.S. Military Services. Its data elements include required materials, eligibility requirements, occupational area, target audience, and course length. Its primary users are system developers and manpower planners.

- TPDC Teletraining Data Base contains data elements on all teletraining courses offered to the U.S. Military Services by civilian distributors such as the National Technological University. Nearly all of these courses use video presentations. The data elements for this database include information on the distributors of each course, television network affiliation, course duration, military occupation code supported, and credit offered. Its primary users are training system developers and system program managers.

This list of databases is still being developed. It does not yet include databases that may be found in other countries and additional databases to which the DoD has access may be identified. Still, the list shows that there are many databases and data elements of various types available to address the issue of advanced technology to aid training design.

By virtue of its size, even this preliminary collection of databases presents a significant challenge to those who may wish to use it to answer specific questions. What it requires is a framework or structure to link the data elements in the various ways that NATO users might employ to address issues of interest to them. An initial framework based on the standard elements of systems was suggested in Section II of this paper. Others based on the six entry points (authoring systems, models, embedded training and simulation, advanced hardware technology, cognitive approaches, and expert systems) discussed by Seidel and Chatelier in the Introduction to this book or the categories of software technology, hardware technology, and operational platforms also discussed in the Introduction could be used. For that matter, all three of these frameworks could be used together to structure the many data elements that may be accessible to the users of a NATO database.

Other data elements are missing. Even in this large collection of data elements there is too little to be found on front-end analysis for training systems and on costs and effectiveness of various training approaches. If there is to be a science, technology, or engineering of training systems so that known approaches yield predictable, quantifiable, and roughly equivalent results in all hands, then the relationship between specific choices in the analysis, design, development, implementation, and assessment of training systems and their consequences in terms of specific training outcomes must be known. The role of a NATO database may also be to highlight what is needed to answer the questions that the training community should be asking. If we ask, then:

Which (if any) of the data elements in these databases should be included in the NATO database?

The answer appears to be all of them. And all of them could be included if permission to access them by a NATO-serving organization is given and if they can be properly linked in a hierarchical fashion to a framework that is sufficiently comprehensive to resolve most issues of training, sufficiently user-friendly to be used, and sufficiently efficient to be responsive.

IV. IMPLEMENTATION ALTERNATIVES FOR THE NATO DATABASE

How should the NATO database be accessed by its users?

There are a variety of ways that the NATO community might access the information in these databases. Among these are the following:

(1) Direct access. Independently designated individuals in each NATO country might access each database separately and directly using telephone dial-ups and modems. Users would have to know what data elements are available and how to access them on their own. As Table 1 suggests, this approach would provide access to all data elements in all accessible databases, the data elements would be always up to date, and the expenses for setup and maintenance would be minor. However, access would require multiple connections using multiple access protocols and multiple user interfaces, and there would be little access protection provided to the managers of the database thereby making it less attractive for them to allow NATO users access to their database.

(2) NATO gateway. A NATO gateway might be provided so that all the above databases and any others that might be made available to the NATO training community could be accessed through a single connection that would switch among the databases as needed and would control access to any data elements that database managers might wish to protect. As Table 1 suggests, this approach would allow all data elements to be accessed using a single connection and a single access protocol with a single user interface. The data accessed would always be up to date, and some protection would be provided for the database manager to control access. On the other hand, a NATO gateway of this sort would be expensive to set up, and it would require on-line access.

(3) NATO database. A separate NATO database might be created by extracting and periodically updating information from existing databases. It would contain some or all of the information in the above databases, as well as information from additional databases of interest to the NATO training community. Access to the NATO database would be provided directly using telephone dial-ups and modems. As Table 1 suggests, this approach would allow all data elements to be accessed using a single connection and a single access protocol with a single user interface. It would also allow a moderate level of protection to be exercised by the database manager. On the other hand, a NATO database would be expensive to set up, its data elements would be updated only periodically and not constantly, and it would be moderately expensive to maintain.

(4) NATO directory of databases. Rather than compile a separate database for the NATO community, it may be more reasonable to develop a database that points to data elements that can be found in relevant databases elsewhere -- along with information about how each database may be accessed. As discussed by Seidel and Chatelier in the Introduction to this book, the potential user would go through a process of on-line inquiry to focus his/her search on (1) the specific information required and (2) the identification of which, if any, existing databases hold the required information. In this way, users would be led through a structured series of questions to determine what data elements are available and how they might be accessed. Retrieval of the specific information needed would still be left up to the users as it is in the Direct Access approach discussed above. However, users would not access the data elements directly through dial-up connections. They would find what data elements were available to address their concerns and how these data elements

TABLE 1. CHARACTERISTICS OF DIFFERENT WAYS TO ACCESS THE NATO DATABASE

Type Access	Accessible Data Elements	Required Dial-Ups	Required Protocol(s)	Access Protection	Update Frequency	Feasbility	Cost of Set-up	Cost of Maintenance
Direct	All	Multiple	Multiple	Low	Constant	Low	Low	Low
NATO Gateway	Selected	Single	Single	Moderate	Constant	Low	High	Low
NATO Database	Selected	Single	Single	Moderate	Periodic	High	High	Moderate
NATO Directory of Databases	Selected	Single	Multiple	High	Constant	High	Low	Low

might be accessed indirectly through the database manager and not through direct dial-ups. As Table 1 suggests, this approach would provide access to all data elements in all accessible databases, the data elements would be always up to date, and the expenses for setup and maintenance would be minor. While this is the least efficient alternative for database access, it may be the most realistic because of its low cost and high feasibility.

V. CONCLUSION

Only the last alternative combines the features of low cost and high feasibility. The low cost is reasonably apparent. Feasibility is determined by the currently limited access to exiting databases through telephone dial-ups and the difficulty of providing protection of the data elements in them. It appears that a NATO directory of databases is the most practicable alternative given current technical functionality and available resources.

INDEX

Kolibri 360 GmbH
Ferdinand-Jühlke-Straße 7
99095 Erfurt, Deutschland
produktsicherheit@kolibri360.de